GAIAN SYSTEMS

CARY WOLFE, SERIES EDITOR

60 *Gaian Systems: Lynn Margulis, Neocybernetics, and the End of the Anthropocene*
BRUCE CLARKE

59 *The Probiotic Planet: Using Life to Manage Life*
JAMIE LORIMER

58 *Individuation in Light of Notions of Form and Information*
Volume II: *Supplemental Texts*
GILBERT SIMONDON

57 *Individuation in Light of Notions of Form and Information*
GILBERT SIMONDON

56 *Thinking Plant Animal Human: Encounters with Communities of Difference*
DAVID WOOD

55 *The Elements of Foucault*
GREGG LAMBERT

54 *Postcinematic Vision: The Coevolution of Moving-Image Media and the Spectator*
ROGER F. COOK

53 *Bleak Joys: Aesthetics of Ecology and Impossibility*
MATTHEW FULLER AND OLGA GORIUNOVA

52 *Variations on Media Thinking*
SIEGFRIED ZIELINSKI

51 *Aesthesis and Perceptronium: On the Entanglement of Sensation, Cognition, and Matter*
ALEXANDER WILSON

50 *Anthropocene Poetics: Deep Time, Sacrifice Zones, and Extinction*
DAVID FARRIER

49 *Metaphysical Experiments: Physics and the Invention of the Universe*
BJØRN EKEBERG

48 *Dialogues on the Human Ape*
LAURENT DUBREUIL AND SUE SAVAGE-RUMBAUGH

47 *Elements of a Philosophy of Technology: On the Evolutionary History of Culture*
ERNST KAPP

(continued on page 330)

Gaian Systems

Lynn Margulis, Neocybernetics,
and the End of the Anthropocene

- - - - - - -

BRUCE CLARKE

posthumanities 60

 University of Minnesota Press
Minneapolis
London

The University of Minnesota Press gratefully acknowledges the financial assistance of Texas Tech University for the publication of this book.

Portions of the Introduction and chapter 1 are adapted from "'Gaia Is Not an Organism': The Early Scientific Collaboration of Lynn Margulis and James Lovelock," in *Lynn Margulis: The Life and Legacy of a Scientific Rebel*; copyright 2012 by Dorion Sagan; reprinted by arrangement with Chelsea Green Publishing, White River Junction, Vt., www.chelseagreen.com. Portions of chapters 1 and 2 are adapted from "Rethinking Gaia: Stengers, Latour, Margulis," in *Theory, Culture, and Society* 34, no. 4 (2017): 3–26; copyright 2017 by SAGE Publications. Portions of chapters 1, 4, and 7 are adapted from "Mediating Gaia: Literature, Space, and Cybernetics in the Dissemination of Gaia Discourse," in *Imagining Earth: Concepts of Wholeness in Cultural Constructions of Our Home Planet*, ed. Solvejg Nitzke and Nicolas Pethes; copyright 2018 by Transcript Verlag. Portions of chapter 3 are adapted from "Autopoiesis and the Planet," in *Impasses of the Post-Global: Theory in the Era of Climate Change*, volume 2, ed. Henry Sussman, published 2012 by Open Humanities Press. Portions of chapters 3, 4, and 5 are adapted from "Steps to an Ecology of Systems: *Whole Earth* and Systemic Holism," in *Addressing Modernity: Social Systems Theory and U.S. Cultures*, ed. Hannes Bergthaller and Carsten Schinko, 259–88 (Rodopi, 2012), published by Brill. Portions of chapters 1, 3, 4, 5, and 6 are adapted from "Neocybernetics of Gaia: The Emergence of Second-Order Gaia Theory," in *Gaia in Turmoil*, ed. Eileen Crist and H. Bruce Rinker; foreword by Bill McKibben, 293–307; copyright 2009 by Massachusetts Institute of Technology, reprinted by permission of The MIT Press. Portions of chapter 7 are adapted from "The Planetary Imaginary: Gaian Ecologies from *Dune* to *Neuromancer*," in *Earth, Life, and System: Evolution and Ecology on a Gaian Planet*, ed. Bruce Clarke; copyright 2015 by Fordham University Press. Portions of chapters 4 and 8 are adapted from "Planetary Immunity: Biopolitics, Gaia Theory, the Holobiont, and the Systems Counterculture," in *General Ecology: The New Ecological Paradigm*, ed. Erich Hörl with James Burton, published 2017 by Bloomsbury Academic, an imprint of Bloomsbury Publishing Plc. Portions of chapter 9 are adapted from "'The Anthropocene,' or, Gaia Shrugs," *Journal of Contemporary Archaeology* 1, no. 1 (2014): 101–4, published by Equinox Publishing, Ltd.

Permission to excerpt materials from Lynn Margulis's papers granted by the Estate of Lynn Margulis.

Published by the University of Minnesota Press
111 Third Avenue South, Suite 290
Minneapolis, MN 55401-2520
http://www.upress.umn.edu

Printed in the United States of America on acid-free paper

The University of Minnesota is an equal-opportunity educator and employer.

Library of Congress Cataloging-in-Publication Data
Names: Clarke, Bruce, author.
Title: Gaian systems : Lynn Margulis, neocybernetics, and the end of the anthropocene / Bruce Clarke.
Description: Minneapolis : University of Minnesota Press, [2020] | Series: Posthumanities ; 60 | Includes bibliographical references and index. | Summary: "A groundbreaking look at Gaia theory's intersections with neocybernetic systems theory" —Provided by publisher.
Identifiers: LCCN 2020020431 (print) | ISBN 978-1-5179-0911-6 (hc) | ISBN 978-1-5179-0912-3 (pb)
Subjects: LCSH: Margulis, Lynn, 1938–2011. | Gaia hypothesis. | Systems theory. | Cybernetics.
Classification: LCC QH331 .C73 2020 (print) | DDC 570.1—dc23
LC record available at https://lccn.loc.gov/2020020431

UMP BmB 2020

For Donna, with cosmic love

CONTENTS

INTRODUCTION An Epistemological Transition   1

PART I. GAIA DISCOURSE

CHAPTER 1. A Paradigm Shift   23

CHAPTER 2. Thinkers of Gaia   47

CHAPTER 3. Neocybernetics of Gaia   83

PART II. THE SYSTEMS COUNTERCULTURE

CHAPTER 4. The Whole Earth Network   101

CHAPTER 5. The Lindisfarne Connection   139

CHAPTER 6. Margulis and Autopoiesis   157

PART III. GAIAN INQUIRIES

CHAPTER 7. The Planetary Imaginary   183

CHAPTER 8. Planetary Immunity   213

CHAPTER 9. Astrobiology and the Anthropocene   243

ACKNOWLEDGMENTS   275
NOTES   279
BIBLIOGRAPHY   297
INDEX   315

An Epistemological Transition

You could say that I found Gaia by way of chaos theory. In my part of academe, chaos theory arrived in 1987.[1] By the 1990s, inspired partly by the avid interdisciplinary reception of this more technically denominated *dynamical systems theory,* I began in earnest to cultivate a post-tenure specialization in literature and science. But as I set about to reschool myself in physics, chemistry, and biology, to come up to speed on chaos and complexity theory, thermodynamics and information theory, and then cybernetics and systems theories, where Gaia was concerned, not much came to hand. Even after it had crossed my threshold, for a while I was reluctant to take it seriously. I had formed the nebulous impression that what "Gaia" named in scientific context was not quite real science but some kind of New Age notion connected to god knows what exactly. I took it to be the sort of idea that I, a recent interloper into the discourse of the sciences, in order to establish or maintain some minimal credibility, should avoid.

Around 2000, I was searching for an accessible introduction to biology for my undergraduate literature and science classes, something in the vein of medical researcher Lewis Thomas's celebrated text of 1974, *The Lives of a Cell: Notes of a Biology Watcher,* but more recent.[2] *Lives of a Cell* did not mention Gaia by name, but in retrospect, it closely anticipated early Gaia discourse: "I have been trying to think of the earth as a kind of organism, but it is no go. I cannot think of it this way. It is too big, too complex. . . . I wondered about this. If not like an organism, what is it like, what is it *most* like? Then, satisfactorily for that moment, it came to me: it is *most* like a single cell."[3] The face and figure of the Gaia hypothesis, I would learn, also shifted about like this, appearing now as "a single cell," other times as "a kind of organism" or as a "complex" entity of some sort.

As it turned out, what I did find was Lynn Margulis's popular exposition of her evolutionary theories in the recently released paperback edition of *What Is Life?*, written with her first son, Dorion Sagan.[4] I recollected then that *Lives of a Cell* had copped many of its best riffs from Margulis's early work. Popularizing her first book, *Origin of Eukaryotic Cells,* Thomas's *Lives of a Cell* also vetted its arguments for the starring role of symbiosis in cell evolution.[5] By the time Thomas was writing his book, in fact, Margulis had already formed a decisive association with the independent British scientist and inventor James Lovelock, and they were at work on their original run of coauthored papers on the Gaia hypothesis. Here in *What Is Life?* was Margulis's own expansive updating of her evolutionary narrative, set forth in equally vigorous and elegant coauthored prose.

I began teaching *What Is Life?* flanked by various works of bioscience fiction. It also introduced the Gaia concept, and this was probably my first encounter with an authoritative account. However, it did not bring Gaia forward so emphatically that one had to confront it head-on. I taught this text for several years, concentrating on its main account of deep evolution while sweeping Gaia off to the side. Then the semester arrived when instead of assigning the relatively lengthy and intricate *What Is Life?* I went with Margulis's terse 1998 memoir, *Symbiotic Planet.* Its final chapter, simply titled "Gaia," retells the name-of-Gaia origin story, with a cautionary twist:

> The term *Gaia* was suggested to Lovelock by the novelist William Golding, author of *Lord of the Flies*. . . . Lovelock asked his neighbor whether he could replace the cumbersome phrase "a cybernetic system with homeostatic tendencies as detected by chemical anomalies in the Earth's atmosphere" with a term meaning "Earth." "I need a good four-letter word," he said. On walks around the countryside in that gorgeous part of southern England near the chalk downs, Golding suggested Gaia. . . . The name caught on all too well.[6]

Following her intensive collaborations with Lovelock in the 1970s and 1980s, by the later 1990s *Symbiotic Planet* intimated Margulis's longstanding concern that "Gaia" as a trademark had exposed the science it covered to severe misconstructions. In an interview with Canadian science broadcaster David Suzuki, Lovelock acknowledged Margulis's mixed feelings: "Nobody, not even Lynn, liked it. She tolerated it, and was very understanding of its origins and went along with it. But her first reaction was that it wasn't a very good idea at all. It kind of brought up the idea of pagan goddesses and all and didn't fit at all with the atheistic view of science. But of course it almost instantly appealed to the New Age. And I don't say the New Age in any pejorative sense. Because in those days it *was* a new age."[7] Lovelock's remarks capture my own experience prior to encountering Margulis's advocacy for the Gaia concept.

In *Symbiotic Planet,* however, one passage in particular finally made the idea of Gaia click for me. It began with this statement—possibly a distant echo of Thomas, whom she knew well—regarding the proper sense of Gaia theory: "As detailed in Jim's theory about the planetary system, Gaia is not an organism" (119). Nor was it a "single cell." Margulis's negative propositions regarding Gaia proper (not "the Earth" altogether) began to cut away my misunderstandings, what I had been vaguely worrying about regarding the fringe metaphysics or planetary vitalisms kept alive, so to speak, by the name of Gaia itself having "caught on all too well." In this passage, Margulis rehearsed the finer points of the developed presentation of the theory, tethering metaphors tightly to the science, and gave her own articulation of the concept. "Gaia itself is not an organism," she continued, "directly selected among many. It is an emergent property of interaction among organisms, the spherical planet on which they reside, and an energy source, the sun" (119). Thus it happened that my initiation to Gaia theory did not come directly from the work of Lovelock, Gaia's primary author, but from the science writing of Margulis. *Gaian Systems* goes deeply into Lovelock's own Gaia discourse, but it also retains this initial orientation in tracing more fully than previous studies have done the particular signature of Lynn Margulis on the evolution of Gaia theory.

Margulis's point at that moment was that, even if considered as a "living" entity of some sort, one still could not reasonably submit Gaia to standard evolutionary expectations of reproduction, random variation, survival in competition, and natural selection. Rather, she countered, Gaia is a *system* that incorporates living systems. Having placed the organic metaphor into this more abstract perspective, Margulis then figured Gaia's status, not as an "organism," precisely, but as a *body*: "Gaia, the system, emerges from ten million or more connected living species that form its incessantly active body" (119). Then it clicked. If Gaia is a system, then Gaia theory is systems theory. Not only that: in the fullness of her engagement with it, Margulis would go on to treat Gaia through autopoietic systems theory. She would also incorporate the concept of autopoiesis into a range of her popular expositions on living systems at all pertinent scales. *Gaian Systems* explores the concept of autopoiesis—the centerpiece of a systems discourse that arose in the 1970s as "second-order cybernetics"—in particular relation to Margulis's pronunciation of Gaia. By the 1990s, Margulis was coordinating the Gaia concept with a suite of autopoietic systems theories also making their paradigm-changing way against institutional and ideological headwinds. In chapter 6 we will examine a number of Margulis's explicit autopoietic descriptions of Gaia, in passages such as these:

> Cells and Gaia display a general property of autopoietic entities: as their surroundings change unpredictably, they maintain their structural integrity and internal organization, at the expense of solar energy, by remaking and interchanging their parts.[8]

> Whereas the smallest recognizable autopoietic entity in today's biota is a tiny bacterial cell, the largest is Gaia, the organismal-environmental regulatory system at the Earth's surface, comprised of more than thirty million extant species.[9]

Gaian Systems will follow Margulis's lead to see what the autopoietic turn has added to Gaia's conception and description.

Neocybernetic systems theory (NST) has developed by expanding the concept of autopoiesis beyond its origins in biological systems theory.[10] For its inventors, the Chilean biologists Humberto Maturana and Francisco Varela, the premier instance of an autopoietic system is the living cell. Regarding the original, biotic form of the concept, living cells are autopoietic in that they produce their own production. The fundamental processes of living systems are recursive. Their operations are primordially self-referring. Living systems continuously select and transform the elements they take from their environmental mediums to produce their own continuation and transformation out of their own continuing production of selective transformations. By such incrementally renovating means, they maintain both their operational form and their metabolic and reproductive processes. They maintain the possibility of consorting and coupling with other material, biotic, and metabiotic systems and their environments. Further, in the most rigorous extension of the concept of autopoiesis beyond this biotic application, Niklas Luhmann's social systems theory differentiated self-producing systems into living and nonliving, or biotic and metabiotic registers. While "nonliving," the autopoietic products and processes of psychic and social systems (events of consciousness and communication) can emerge only with the environmental participation of living systems.[11] Chapter 3 will look into the logic of these metabiotic amplifications of autopoietic functioning in greater detail. Whether focused on the material-energetic functioning of living systems or the virtual and formal functioning of psychic and social systems, NST boosts immunity to silicon insolence by maintaining the operational reach and remit of such natural systems, biotic and metabiotic, in their couplings to the innovations of the technosphere.

Thus, when Margulis and Sagan state in *What Is Life?* that "the biosphere as a whole is autopoietic in the sense that it maintains itself" (20), an autopoietic conception of Gaia as a system may denote not a living system, precisely, but rather another kind of metabiotic system. In this view, Gaia is a self-generating, self-maintaining planetary constellation emerging from the interactions of living and nonliving components— systems and structures, embodying their integrated intermodulations. You could say that I really found Gaia once these two seemingly separate strands of autopoietic systems theory came together. I could now construct Margulis and Sagan's evocations of autopoietic Gaia in a fully neocybernetic sense. Margulis's biotic strand, responding to Maturana and Varela's original conceptuality, joined NST's metabiotic strand to resolve one of Gaia theory's most persistent equivocations—the matter of whether Gaia is itself "alive." This indetermination has been a highly productive issue for the elaboration of Gaia discourse. However, as an autopoietic system in the metabiotic register, one need not identify Gaia with the form of life per se. Rather, *Gaia participates in an essential quality of individual living systems—the autopoietic form of organization,* an emergent, recursive form of self-production and self-maintenance, *within a metabiotic coupling of abiotic and biotic dynamics.* Autopoietic

Gaia arises as a "property of interactions" from the interpenetration of the biota with the seas, the skies, and the rocks, after eons of their own extrabiotic commerce with the dying generations of living forms.

Gaian Systems reviews and assesses the different dialects of systems theory brought to bear on the discourse of Gaia. Its narrative of Gaia's conceptual evolution will also map the development of cybernetic systems theories, particularly as these discourses first approach and then overcome the debut of the first cybernetics in the heuristic equation of mechanical contrivances and biological systems. This finer history teaches that despite its mainstreaming as all things computer-scientific, the concept of cybernetics goes beyond both technological systems as well as its nominal relations with later computational developments and their popularizations. Rather, it develops from its point of origin at the machine/organism interface into a transdisciplinary discourse comprehending differentiated operational interrelations among Earth, life, mind, and society.[12]

"Cybernetic systems employ a circular logic which may be unfamiliar and alien to those of us who have been accustomed to think in terms of the traditional linear logic of cause and effect."[13] Observed afresh in systems-theoretical hindsight, Lovelock's presentation of Gaia as employing a "circular logic" anticipates the second-order cybernetic turn toward constitutive recursion. Lovelock and Margulis's initial Gaia research was concurrent and conceptually parallel with the new discourse of self-referential systems that emerged within NST.[14] Heinz von Foerster's exposition of cybernetics sketches the conceptual unfolding of NST from first-order circularity to second-order self-referential recursion: "Should one name one central concept, a first principle, of cybernetics, it would be circularity. Circularity as it appears in the circular flow of signals in organizationally closed systems, or in 'circular causality,' that is, in processes in which ultimately a state reproduces itself or in systems with reflexive logic as in self-reference or self-organization, and so on. Today, 'recursiveness' may be substituted for 'circularity,' and the theory of recursive functions, calculi of self-reference, and the logic of autology, that is, concepts that can be applied to themselves, may be taken as the appropriate formalisms."[15] Whether in the description of organic cells, recursively producing themselves from moment to moment by maintaining metabolic operations that constitute and bind their living processes, or of social ensembles, looped together by serial communications, the concept of autopoiesis anchors NST's modes of observation.

Autopoietic operation yields cognitive capacity. Already in Maturana and Varela's theory of autopoiesis, cellular life's self-referential processes produce a kind of biotic cognition, a reception of molecular forms as distinct cellular events. In *What Is Life?* Margulis and Sagan name this ubiquitous self-feeling of living systems *sentience.* Other authors call it "sense-making," a mode of basic knowing placing any living system in enactive relation to its own communal and material environments. In addition, the nutritive, metabolic, and excretory processes of any living system are at all times, if even infinitesimally, remaking their own niche. These microdynamics parlay up to the worldly macrocosm. Bruno Latour has recently stated this Gaian recognition in his own idiom: "Each agency modifies its neighbors, however slightly, so as to make its own survival slightly less improbable. . . . [T]he concept of Gaia captures the distributed intentionality of all the agents, each of which modifies its surroundings for its own purposes."[16] Autopoietic Gaia taps its own modes of planetary cognition from the deep wells of these microcosmic points of biotic sensation.

In Earth's long planetary distillations of biotic autopoiesis, the matter coursing through the Gaian system is itself transformed and continuously redeposited and repurposed.[17] NST regards the conditioning of matter by autopoietic form to mark the threshold where elemental cognitive capacities turn upon operations of systemic self-production. The sum effects of Gaia's metabiotic couplings of Earth and life processes are sensitive at ever-larger scales. These premises suggest that from the merest living cell to the widest Gaian iterations, autopoietic systems observe so as to select those elements of their environment that both maintain their self-bounded self-productions and best open out to material and meaningful alliances and exchanges. Materially, moreover, these premises also suggest that *our* Gaia emerged from the unique peculiarities of its cosmic situation. They suggest that wherever life may happen to come into existence, it will be fully contingent upon local peculiarities as well as universal conditions. To each living planet its own Gaia. Although Gaian processes, "other Gaias," are entirely conceivable on other planets harboring some sort of life, our Gaia is a planetary one-off.[18] Perhaps this Gaia will put forth exfoliations taking shape as space-faring populations housed in closed environments launched from Earth or its vicinity. In the meantime, as a population of one, Gaia's systemic form may be the most universal property of its organization. With whatever chemistries life may come forth in worlds beyond our own, the formal blueprint of planetary autopoiesis may well be their common denominator.

When Lovelock singly or in concert with Margulis brought the Gaia hypothesis forward in the early 1970s, it was introduced at specialized locations in normal corners of geoscience. However, the incursion of Margulis's biological considerations concerning early life and its evolution, filling out Lovelock's original geological, chemical, and thermodynamic arguments for a contemporary planetary state of atmospheric homeostasis, carried the Gaia hypothesis into uncharted regions of disciplinary hybridity and conceptual heresy. The cybernetic framing implied some operational ordering of collective behavior that closed around the maintenance and circulation of Gaian effects. Even prior to Margulis's collaboration, no previous scientific argument had suggested the presence of a planetary "homeostat" composed of circular worldly mechanisms by which life modulated its own environment. Yet within a few years, this rangy and seemingly improbable biocybernetic concept began to infiltrate and provoke mainstream scientific discussion.

"The presence of a biological cybernetic system able to homeostat the planet for an optimum physical and chemical state appropriate to its current biosphere becomes a possibility."[19] This is the initial form of Lovelock's hypothesis regarding the self-regulation of the planetary atmosphere, at the moment that it started its scientific career under the name of Gaia. This sketch captures Gaia's introduction as an application of cybernetic systems theory. We could say that cybernetics is in Gaia's DNA. But the genome would be a misleading figure for the system itself, because Gaia has no genome devoted to its own reproduction as such. This alone makes it different in kind from any living "organism," all of which reproduce in one way or another according to genetic guidelines. As cybernetics has mutated and diverged over the decades, so have the descriptions of Gaia as a system. Lovelock and Margulis's seminal writings were also the first to bring multiple lines of systems discourse to bear on the Gaia concept. It turned out to be particularly significant for Gaia's discursive evolution that a primary outlet for its hypothesis was *CoEvolution Quarterly,* the periodical successor to the

Whole Earth Catalog. This venue ensured that early in their mutual developments, Gaia theory intersected with second-order cybernetics, in its own right a leading edge of systems theory's own epoch of counter-cultural transformations.

The Gaia hypothesis also underwent an unusually vigorous extra-scientific development into a spectrum of Gaia figures ranging from naive and outlandish to inspired and indispensable.[20] Two reasons at least stand out for the unusual and abiding mobility of this scientific turn on the name of an archaic female deity. The first, of course, is *Gaia,* the name itself, and its manifold resonances as a term with which to conjure a wide swath of cultural responses. The other reason is equally profound but more personal—the creative chemistry of the scientific collaboration fashioned by Lovelock and Margulis. Combining his expertise in chemistry, geology, physiology, and cybernetics with her profound grounding in microbial evolution and ecology, over time they established a fully geobiological Gaia concept whose worldview-shaking import was not lost on its progenitors. Nor should it be lost sight of now.

Contemporary scientific and scholarly attention and debate at large have precipitated a bona fide discourse of Gaia theory.[21] The academic mainstreaming of many of Lovelock and Margulis's formerly controversial ideas has coalesced in research fronts such as Earth system science and astrobiology. Informed evocations of Gaia theory now accompany a growing sense of emergency over an Earth system in peril of entering a new regime inconducive to many current life-forms, including, of course, our own. The rise of Gaia theory preceded and prepared for the current recognitions of a global climatic and environmental crisis. These trends have run together with the arrival of a discourse of the Anthropocene through which to acknowledge that the massive accumulation of humanity's activities has now altered the functioning of the Earth system.[22] Nevertheless, whatever its current state may be, we now effectively observe the Earth system of our present concern through the Gaia concept, a massively complex but newly concrete presence for our planetary imagination to grasp. The current phrase "Earth system" is just the normalized locution for and the legitimized offspring of the biological cybernetic system upon which Lovelock's thought of Gaia originally speculated over half a century ago. Gaia is systems thinking at and for the planetary level.

Concocted from atmospheric chemistry, exobiology, and microbiology, the Gaia hypothesis echoed ecosystem ecology despite its arrival outside of ecology proper. Historian of science Joel Hagen has noted that "the idea that the biosphere is a homeostatic or cybernetic system of living and nonliving components was a central feature of the ecosystem concept. . . . Although Lovelock rarely used the term ecosystem, many of his ideas meshed perfectly with [a] broad systems approach to ecology."[23] Joining environment to life, Lovelock and Margulis's collaborative descriptions factored Gaia's ecology into the microbes' planetary role in the constitution of the atmosphere. Unlike this or that eukaryotic species, prokaryotes are the fundamental form of life. Bacteria and archaea are everywhere all at once. They perfuse the planet from top to bottom. The microcosm is the most planetary and arguably the most consequential component of the biosphere.[24] In the phylogeny that follows from Margulis's serial endosymbiosis theory, permanent mergers among the microbes determine the basic framework of evolutionary relationships.[25] Indeed, as Margulis writes in one of the most poetic statements in *Symbiotic Planet:* "Symbiogenesis was the moon that pulled the tide of life from its oceanic depths to dry land and up into the air."[26]

By such Gaian interactions the domains of life have unfolded into five kingdoms interrelated by their origins in microbial mergers. When placed into planetary view, this phylogeny yields not a branching tree but a reticulated web, woven from strands composed by the prokaryotes, of interactive life coevolving with its abiotic and postbiotic environments: "Planetary physiology—Gaia . . . is symbiosis seen from space."[27] Gaia theory is thoroughly ecosystemic in its interpenetration of life with its planetary and cosmic environments. The planetary mesh of the microbes networks the coevolution of life altogether with the parts of the Earth reached by living processes and the solar radiation in our neighborhood of the cosmos. Symbiogenetic dynamics couple Gaia to the emergence and maintenance of the biogeochemical cycles arising out of and returning to the "critical zone" of a modulated planetary surface.[28]

The microbes have driven life's major evolutionary developments and continue as always to prop up the rest of the biosphere. However, neither life nor its evolutionary history tells the entire story. When Gaia fell into place, a bacterial biosphere had already been interacting for hundreds of millions of years with a previously nonbiotic yet increasingly postbiotic milieu. Three billion years later, the forms of living organisms continue to find constitutive relations with both the abiotic and the biogenic factors of their environments, what Tyler Volk has felicitously named the *gaian matrixes* of air, ocean, and soil—the geobiological media of Gaia's compounded processes.[29] Lovelock has recently restated this extended Gaian perspective: "When Darwin came upon the concept of evolution by natural selection he was almost wholly unaware that much of the environment, especially the atmosphere, was a direct product of living organisms. Had he been aware I think he would have realized that organisms and their environment form a coupled system . . . *what evolved was this system, the one that we call Gaia.* Organisms and their environment do not evolve separately."[30] This theoretical view supersedes strictly biocentric constructions of the Gaia hypothesis.

Thanks to the particular and diverse sensibilities of its progenitors, the development of Gaia discourse has enjoyed an unusual emancipation from the usual domains of normal scientific cultivation. By the 1980s the dissemination of the Gaia concept had already generated a wide range of philosophical reflections. This does not mean that it simply rode the waves of cultural free association. In particular, Margulis encountered the discourse of autopoietic cognition at Lindisfarne Association meetings attended by biocyberneticians Humberto Maturana, Henri Atlan, and Heinz von Foerster, the neuroscientist Francisco Varela, and the poet and essayist William Irwin Thompson.[31] Focused on the Gaia hypothesis and the concept of autopoiesis through a philosophy fostering planetary cultural dynamics, these intellectual events culminated an era of thought I treat as the time of the *systems counterculture.* Gaian thought at Lindisfarne in the 1980s sees perception as a co-construction or distributed operation, a "repeating pattern" immersing mutual observers.

Consider this passage from *Symbiotic Planet*:

Analogous to proprioception, Gaian patterns appear to be planned but occur in the absence of any central "head" or "brain." Proprioception, as self-awareness, evolved long before animals evolved, and long before their brains did. Sensitivity, awareness, and responses of plants, protoctists, fungi, bacteria and animals, each in its local environment, constitute the repeating pattern that

ultimately underlies global sensitivity and the response of Gaia "herself." (126)

With scare quotes around Gaia "herself," Margulis undercuts the gendered anthropomorphism in the name. The entity so named is not a
gendered being but a system for which the sexual dynamics of its biotic components occur within a cauldron of other contingencies. Notwithstanding, its theory posits the production of systemic responses
to its environment. Gaia's planetary responses join the nonrandom
productions of the particular sensitivities of which worldly systems
are capable. Gaian responses both define and transform environments
with changes that rebound upon the system. They are recursive, both
regulative and creative, thus unpredictable. Gaian thought is also the
thinking and tracing of the forms of closure binding the planetary system. Systemic closures allow the patterns to form. Looking through the
distinction of closure, lines become loops that become lines again, fractal and nonreturning, nondeterministic. By thinking the dynamism of
systemic closures, Gaian thought stays alert to the open evolution of
Gaia's operations. Thanks to the material or virtual closure that makes
them possible, every autopoietic being perceives some aspect of its environment, however intimate or local, crucial to its continuation, and
so alters its dwelling as well. By these modulations of the environment,
each also participates in producing what others perceive.

Margulis performs Gaian thought as the thinking of "global sensitivity," in other words, planetary cognition, Gaia's responsiveness to,
its proprioception of, the flux of its own cycles and their environmental
consequences. Now consider what some call the Anthropocene stratum
of humanity's residual works and effects. One can anticipate that Gaia
will be responding to humanity's geological force one way or another
with the emergence of a revised planetary system. What will happen to
evolve upon a post-Anthropocene Earth? What is the repeating pattern
here, if any? Where do the boundaries lie between the operations of the
biosphere and those of the technosphere? Does Gaia encompass these
phenomena as well with its geobiological rejoinders? Will Lovelock's
superintelligent informatic entities promulgated in *Novacene* make
Gaia over in their own electronic image? Or will organic determinations
remain the bottom line of planetary viability? Here I follow Margulis in
binding the operations of the technosphere, however far-flung they may

become, within Gaia, not without. Whatever upheavals we may bring about, Gaia will factor its current technosphere into its own continuation over future geological time.

In a prescient 1987 think piece, "Gaia and the Evolution of Machines," an article we will turn to several times throughout this book, Sagan and Margulis offer some seminal reflections on Gaian thought. "A whole Gaian style of thought is emerging in which perception is seen as a participatory phenomenon, and with which we become more aware of the sum of organisms within the biosphere. . . . Traditional human ideas are in contrast with Gaian perceptions that link people inextricably, and in subordinate fashion, to the biota, that is, to the sum of plant, animal, and microbial life" (16). Gaian thought is, so to speak, nondominational—there is no definitive or singularly dominant construction of knowledge. And yet, they go on, "All the weight of Western history and success attach to political groups that subscribe to the idea of man's domination of nature. The Gaian thought style, however, extends 'horizontally' to other organisms and 'vertically' beyond human history. In it, human beings and technology may be seen as environments in the biosphere" (16).[32] Perceptual participation *with* one's environment overcomes the idea, if not always the act, of standing over and apart from it. Stated otherwise, to think "man's domination of nature" demands the nonreception of the Gaian perception of participatory panbiotic couplings, including those that couple the bio- and technospheres together.

The Gaian thought in the passage just quoted also acknowledges (Sagan and Margulis wrote several decades before the onset of Anthropocene discourse) the interpenetration of the technosphere with the biosphere, while also observing the heterogeneity of their functions and effects, the plurality of their environments. Simultaneous but differential closures and linkages traverse social and technological systems and their environments. While political systems may enforce social enclosures, Sagan and Margulis continue, "Because a Gaian view increases public awareness of our dependence upon other life forms, it is extremely valuable in battling the prevailing ideologies of selfishness: that nature is either pristine and should be preserved or is simply a bunch of resources to be plundered. The truth is that we are deeply connected to all other organisms, cannot help altering them, yet must be conscious of and responsible for our actions" (16). This discourse of Gaian participation calls out the paradox of protection that undercuts even well-intended notions of stewardship.[33] In endowing nature with a self that must be protected, one is still complicit with its domination, albeit in an attitude of patronizing distress. Nevertheless, nothing about our immersion in Gaia absolves us from the consequences of our collective effects on the biosphere. Through the perception of Gaia as the horizon of planetary viability, we may hope to own those consequences and so improve those effects.

Guided in particular by these and other texts of Margulis and Sagan, *Gaian Systems* gravitates to Gaia discourses that are also vehicles of Gaian thought. It narrates this textual history with the conviction that this mode of self-orientation of mind and feeling participates to some small degree in the thing that it contemplates and so is worth the labor of its communication. Gaian thought maintains the sensitivity of that registration by not hardening into human prescriptions. Lovelock already saw this clearly in his first book: "There can be no prescription, no set of rules, for living within Gaia. For each of our different actions there are only consequences."[34] Except that, as Latour suggests, especially in the "new climatic regime" into which we have precipitated ourselves, there is no description of Gaia that is not also an implicit prescription. To describe the current state of Gaia's response is virtually to prescribe a commensurate human reply of some sort: "Such is in fact the paradox of the invocation of 'nature': a formidable prescriptive charge conveyed by what is not supposed to possess any prescriptive dimension."[35] I will be content to describe what I can discern and let others draw out what prescriptions they may. Gaia matters. The further communication of its discourse can make positive differences in the thoughts and actions of those who encounter it.

Finally, some especially candid remarks, again from "Gaia and the Evolution of Machines": "The reader may wonder whether we are advocating belief in an unproven assertion: Gaia, the modulated biosphere. We are, but only so far as it is necessary to replace outmoded thought styles. Since perception is impossible without assumptions (i.e., belief), and since all science is the result of perception, the objection that such a view is unscientific is vain" (17). Perception is impossible without assumptions.[36] Whether we care to see Gaia in operation as a cognitive system of planetary self-reference rests on our choice of worldview. Gaia's periodic ups and downs have borne the growing pains of shifting paradigms. And with the current upheavals in the Earth system as observed with reference to the place of the human within planetary dynamics, the new climatic regime has made the stakes of these throes of transformation even clearer. To render the technosphere fit for long-term cohabitation with its enabling planet, outmoded thought styles must give way to pervasive redistributions of planetary knowledge. A systems-theoretical observation of our geobiological situation upon and within the planet we inhabit is a good place to start.

Scheme of the Book

Gaian Systems explores anticipations of Gaia discourse in ecological theory, Gaia's incubation within NASA's projects for planetary exploration, the career of its early formulations, its significant countercultural encounters and mainstream scientific debates, and the transition of the Gaia hypothesis into current Gaia theory. Part I treats Gaia discourse as a literature in which ideas in the orbit of Gaia theory contribute to the gathering of Gaian thought. At its textual core are Lovelock's and Margulis's own technical and popular presentations of their scientific ideas. Chapter 1 sketches a series of conceptual phases in the arc of Gaia discourse. It reviews selections from the initial correspondence and first collaborative articles of Lovelock and Margulis on the Gaia hypothesis. It then asks, what's in the name of Gaia? Seeming to broker relations between scientific and mythological ideas, that fateful name has surely brought about the unusually visible public face of the Gaia concept. Next, if Gaia is a system, what kind of a system is it? Lovelock and Margulis are consistent in their positioning of Gaia theory as an application of either first- or second-order cybernetic systems theory. From these affirmations and exigencies I extend my own systems-theoretical synthesis under the phrase *metabiotic Gaia*.

Chapter 2 focuses on key issues of recent Gaia discourse in the critical humanities and social sciences. Donna Haraway's cyborg version of the autopoietic description of the Gaian system in her 1995 text "Cyborgs and Symbionts" testifies to the power as well as the problematics of the idea of autopoiesis. Gaia discourse in Haraway anticipates Isabelle Stengers's and Bruno Latour's timely philosophical engagements with Gaia theory. Discursive figures of Gaia appear where Latour's sociology of science practices intersects Stengers's discourse of cosmopolitics, her critique of the modern sciences' hegemonic hold on contemporary knowledge production.[37] Latour's long-standing engagement with Gaia has brought new prominence to this discussion. I engage Latour on the issue of its systems description and hope to have sharpened the terms of this debate. Chapter 3 then expands the exposition of neocybernetic systems theory within a short history of wider cybernetic thought. Margulis appears to have encountered NST in the later 1970s, during the first decade of developments in second-order cybernetics. Moreover, the neocybernetic theorization of Gaia eludes its insertion into holistic schemes, in favor of system differentiations that factor the system-environment distinction, the alterity of the inside relative to the outside of the system, into the conceptual equation. The autopoietic theory of Gaia moves beyond its original occasion in biological systems theory once we observe Gaia as a metabiotic autopoietic system.

Part II traces the theoretical and historical strands that led to Margulis's mode of Gaia discourse. Chapter 4 details the significant commerce of the Gaia hypothesis with the remarkable collegial network established by the *Whole Earth Catalog* and its periodical successor, *CoEvolution Quarterly.* Some of the earliest publications of the Gaia hypothesis and its computational avatar, Lovelock's Daisyworld program, appeared before this scientifically informed, ecologically astute public familiar with cybernetic ideas. *CoEvolution Quarterly* also published Gerard K. O'Neill's proposals for space colonies in high orbit. These images of environmental closure are significantly contemporaneous with the Gaia hypothesis. They translate its terrestrial implications into idealized technological vessels of dubious ecological merit but powerful emotional appeal.[38] Channeled through these venues and issues, the Gaia concept will take on the aspect of a monumental project of thought crossing the lines of ecosystem ecology and cybernetic philosophy. The systems counterculture presented here as constituted by and documented within the Whole Earth network includes *Whole Earth Catalog* mastermind Stewart Brand, the cybernetic anthropologist Gregory Bateson, second-order cyberneticist Heinz von Foerster, the British mathematician George Spencer-Brown, and the polymathic systems thinker Francisco Varela.

Chapter 5 treats a further permutation of the systems counterculture around Lovelock and Margulis in the Gaian connections of an intellectual gathering called the Lindisfarne Association. Its founder,

William Irwin Thompson, dedicated Lindisfarne to the pursuit of a planetary culture. Thompson received significant participation from a renowned group of thinkers that included Lewis Thomas and Stewart Brand. In the later 1970s, Bateson and Varela were consecutive residential fellows at Lindisfarne's Manhattan campus. In 1981, Thompson invited Lovelock and Margulis to join the conversation with Gaia as the focus of discussion. From this meeting and a second one held in 1988, Thompson edited two volumes approaching Gaia discourse from a largely neocybernetic angle and adding his own resonant cultural reflections on the gathered developments. *Gaia: A Way of Knowing* and *Gaia 2: Emergence* record Gaia's primary theoreticians, Lovelock in particular, in intensive dialogue with Varela's cutting-edge discourse of autonomous systems.

Chapter 6 examines Margulis's discourse of autopoietic Gaia in a series of writings dating from the mid-1980s to the early 1990s. The proximate cause of this discourse is Margulis's exposure to neocybernetics at Lindisfarne. During this period, Margulis also began her important writing collaboration with Dorion Sagan, guaranteeing the consistently high quality of their coauthored series of popular volumes, beginning with *Microcosmos*. At this moment, Margulis championed the concept of autopoiesis with a determination that matched her previous efforts on behalf of Lovelock's Gaia. The two strands wove together as *autopoietic Gaia*. The countercultural stamp on this synthesis is especially clear in two noteworthy single-authored essays from 1990. For Margulis, an expanded conception of autopoiesis was to be the philosophical antidote for what she diagnosed as "big trouble in biology."[39] Challenging the leading evolutionary narratives at that moment, this irruption of autopoietic Gaia theory incubated in the lab of the systems counterculture was not well understood and not always well received. I hope to have given it an adequate reception here.

Part III begins by inquiring into the new planetary imaginary crystallized by NASA imagery of the Earth viewed from space. Some materials selected in chapter 7—the novel *Dune* and Bateson's *Steps to an Ecology of Mind*—are items of Gaia discourse after the fact. These texts transmit the cultural moment of Gaia's evolution in the 1960s at the intersection of cybernetics and ecology. They made significant contributions to the broader planetary imaginary in which the early Gaia concept also participated. Other materials—the *CoEvolution Quarterly* articles on

O'Neill's space colonies in high orbit and the literary uptake of such materially closed artificial ecologies in the novel *Neuromancer*—are contemporaneous with the career of the Gaia hypothesis, which lies in the immediate background of their production. To bring this overview of the planetary imaginary in the Gaian era into the current moment, we then welcome the arrival of the new geocentrism as a significant contribution to Gaian thought.

In the rest of the book I relate metabiotic Gaia discourse to current issues in neighboring theoretical conversations, including biopolitics, immunity, symbiosis, astrobiology, the Anthropocene, and the geological turn. Chapter 8's first inquiry concerns Gaia's boundaries. From the standpoint of metabiotic Gaia as an autopoietic system, as part of its self-production it will also make and maintain boundaries that cut it out of its environment as a distinct systemic identity. In fact, the issue of Gaia's boundaries goes back to the first formulations of the Gaia hypothesis in Lovelock's earliest thermodynamic descriptions, and boundary issues recur as well in Latour's recent treatment of Gaia as the "critical zone" as a more focused approach to the "Earth system." Next, how does Gaia theory stand with the discourse of biopolitics? These themes merge on the matter of systems and their immunities. In a reprise of the Lindisfarne connection, Varela explicitly framed a neocybernetic approach to Gaia through his concurrent theorization of autonomous networks in the immune system. Finally, what does the newly prominent idea of *the holobiont* have to do with Gaia theory? For starters, both are deeply grounded in the scientific innovations of Margulis, who also coined the term. The holobiont functions biopolitically to bind the microbiota to the symbiotic planet. The operations of Gaia's boundaries would be one aspect of metabiotic Gaia as an agent of planetary immunity.

Chapter 9 concludes *Gaian Systems* by turning toward the cosmological side of Gaia. Its first inquiry asks what the cultural reading of the geological concept of the Anthropocene lacks that contemporary Gaia discourse might supply. Concurrent with the onset of Anthropocene discourse is a new profile for the science of astrobiology—the study of life in cosmological context. More so than the post-Gaian consortium of Earth system science, contemporary astrobiology owns its debts to Lovelock and Margulis for its conceptual roots in Gaia theory. What happens if we put Gaia and the Anthropocene into a shared astrobiological context? These planetary formations now unfold against larger cosmological developments including the origin of life and the destination of intelligence. As a domain of NASA science, astrobiology also studies materially closed artificial environments in the context of space exploration and habitation. Were we to leave our home planet for extended periods, would it be possible to take Gaia with us? Kim Stanley Robinson's 2015 novel *Aurora* meditates profoundly on these issues, as systems operations and dysfunctions on the journey of a generation ship toward another sun parallel current ecological imbalances produced by the Anthropocene technosphere. The contemporary technosphere comes forward in a selection of its recent discourses ranging from the geological to the geopolitical and the astrobiological. For some thinkers, Gaia seems to be ready for an Anthropocene makeover with a smart technosphere taking up the biosphere's controls. To me, that outcome seems unlikely—control theory for an unsteerable system. Even more problematic, in the end, is the vision of Gaia in Lovelock's most recent book, *Novacene: The Coming Age of Hyperintelligence.* Several decades earlier, Margulis also addressed the place of technology in relation to Gaia, and a comparison of their approaches offers a concluding contrast between the informatic and the autopoietic sensibilities.

PART I
GAIA DISCOURSE

CHAPTER 1

A Paradigm Shift

In "The Independent Practice of Science," James Lovelock describes his earlier professional milieu as a salaried researcher at the National Institute for Medical Research (NIMR) in London in 1961, prior to his emancipation as an independent scientist. It was then that NASA sent him "an invitation to be an experimenter on the first lunar Surveyor mission. It was well known at the NIMR that I regarded science as a way of life in which science fiction was reduced to practice."[1] In U.S. patent law, *reduction to practice* technically means to move an invention beyond the initial stage of conception to the testing and application of a prototype. Lovelock speaks at the end of the 1970s as the inventor who engineered the Gaia hypothesis. Single-handedly and in collaboration with the microbiologist and evolutionary theorist Lynn Margulis, Lovelock would bring the Gaia concept forward as applied systems science. His Gaia discourse is the speculative practice of a systems engineer steeped in the technological imaginary of cybernetics and information theory. In his most recent book, *Novacene: The Coming Age of Hyperintelligence,* Lovelock admits that "I have never really been a pure scientist, I have been an engineer."[2]

The philosophical sociologist Bruno Latour and Earth system scientist Timothy M. Lenton note in "Extending the Domain of Freedom, or Why Gaia Is So Hard to Understand," how "Gaia was discovered through a level of human technology and the self-awareness of the planetary consequences of that technology."[3] Their particular reference is to Lovelock's invention in 1958 of the electron-capture detector, a device exquisitely sensitive to vanishing bits of atmospheric molecules, from industrial emissions to pesticide residues. Reviewing philosopher of science Sébastien Dutreuil's groundbreaking research into Lovelock's pre-Gaian activities, Latour and Lenton draw out Gaia's debt to the electron-capture detector. Its ability to determine and distinguish natural and anthropogenic aerosols also coincided with early warning signs of the Anthropocene.

> His inventions were used to detect the global spread of anthropogenic pollutants. . . . And it was Lovelock's resulting reputation for instrument design that led NASA to employ him in the design of life detection experiments for what were to become the Viking missions to Mars. . . . In many ways, his insight to look at the Earth as if from Mars was to extend to all life forms the analogy that their disseminations of chemical byproducts were like those of modern factories. (662)

As a researcher into air pollution making fine measurements of emissions from humanity's fossil-fueled industrial civilization, Lovelock had already gained an understanding of life as a geological force sheerly through its measurable effects on the composition of the atmosphere. Apparently, he surmised, the biosphere with or without its current abundance of modern humanity is equally involved in supplying the gaseous part of the planet. This immanent account of Earth-bound activities, however, still wanted what Lynn Margulis would bring to the development of Gaia—the energy budgets and metabolisms of the microbes and the rest of the biota—to complete the view of a planetary entity in which these gaseous fluxes attained an effective degree of operational closure as the "biological cybernetic system" of Lovelock's initial formulations.

Lovelock's expertise in atmospheric monitoring adds an indigenous element to Gaia's primary derivation by exobiological contrast with Venus and Mars. Throughout the 1960s, technological developments associated with the U.S. space program also incubated the science from which the Gaia concept would spring. Lovelock periodically worked in Pasadena at the Jet Propulsion Laboratory (JPL) on contracts to develop scientific instrumentation for Mars landers. The life-detection schemes put forward at JPL by his biologist colleagues assumed for Mars Earth-style life in a watery medium, the detection of which demanded probes making contact with the surface of Mars. Lovelock recalled, "At this time scientists still seemed to think that life flourished on Mars. I recall Carl Sagan enthusing over the wave of darkness that crosses Mars when winter ends. He and many others saw this phenomenon as indicative of the growth of vegetation. . . . This image of Mars sustained their belief in biological life-detection techniques."[4] However, by 1964 Lovelock had devised a life-detection technique based on a different principle, the search for an entropy reduction—that is, for a signature of some ordering of matter and energy commensurate with the organization of forms of life. Entropy is fundamentally a thermodynamic and thus a physical concept, so the title of Lovelock's first recognizably Gaian paper, "A Physical Basis for Life-Detection Experiments," declares an alternative to *biological* life-detection schemes.[5] How could one find such a signature?

The crucial turn toward Gaia proper came in 1965 when Lovelock received newly detailed infrared spectrographs of the atmospheres of Venus and Mars. They showed atmospheric entropies off the charts. Both of these planetary atmospheres are dominated by CO_2 and are chemically inert, virtually at thermodynamic equilibrium. Whatever combustion or reduction of chemical potential among the components of their atmospheres had ever been possible there, it has long since burned out. According to Lovelock's physical life-detection scheme, the verdict concerning Mars was obvious: it harbors no life. When the Viking explorers landed on Mars a decade later, their probes found what Lovelock had predicted—no life. Born out of that prediction of the lifelessness of Mars was a theory regarding the self-regulating nature of a "living Earth." Lovelock conceived this idea by turning his interrogation of nearby planetary atmospheres back upon Earth and noting with new eyes that our atmosphere is in a cosmically improbable state of chemical disequilibrium. Rather than burning out, Earth's highly combustible mixture of reactive gases has remained in a far-from-equilibrium state for hundreds or thousands of millions of years. The idea of Gaia as a planetary system responsible for maintaining and regulating such an energizing chemical imbalance over geological time ignites in the vessel of this conceptual conundrum over Earth's atmospheric composition.

Lovelock's next proto-Gaian paper, coauthored with philosopher Dian Hitchcock, put this conundrum to work. "Life Detection by Atmospheric Analysis" makes the crucial move out of normal science at mid-twentieth century and into the Gaian cosmos.[6] Although there were some exceptions, normal science at that time generally assumed Earth's atmosphere to be a largely geological and hence primarily an abiotic phenomenon. Lovelock consolidated the countervailing idea that the atmosphere of a planet on which life is happening will be to a significant if not exclusive extent the *product* of those living processes—enough so for that atmosphere to present, in the current parlance, biosignatures for external inspection. His scheme is now normal astrobiology. The brilliant thing about it is its economy. One does not need to go to Mars or any other place to apply it. Along with the other planetary bodies of our immediate vicinity, eventually we will assess the atmospheres of the exoplanets detected in succeeding decades by similar means.[7]

Early Gaia

Lovelock first encountered Lynn Margulis, then a microbial evolutionist in her late twenties and the ex-wife of his JPL colleague Carl Sagan, in the spring of 1968 at the second in a series of exobiology conferences sponsored by NASA, "Origins of Life: Cosmic Evolution, Abundance, and Distribution of Biologically Important Elements."[8] Thanks to a conversation with William Golding the year before, he had already been contemplating "Gaia" as a name for his thinking regarding Earth's atmosphere as a self-regulating planetary system. Two years later, Margulis sent Lovelock a query about his work on planetary atmospheres along with information about her own research.[9] Shortly after beginning to collaborate with Margulis but prior to the appearance of their co-authored Gaia papers, Lovelock published "Gaia as Seen through the Atmosphere," a two-page letter in the specialist journal *Atmospheric Environment.* This was not the first time he had ventured to publish a version of his speculation regarding the possible existence of what he called there "a biological cybernetic system able to homeostat the planet for an optimum physical and chemical state appropriate to its current biosphere."[10] However, it would mark the first time that he publically presented that hypothesis under the name of "Gaia."

Lovelock's first letter to Margulis, his reply to her inquiry, is dated September 11, 1970. It does not mention "Gaia," only the scientific idea behind its initial formulation, delimited specifically to the atmospheric envelope. Lovelock thanked his correspondent for sending him materials on the early evolution of cells and affirmed his growing conviction that the primary components of Earth's atmosphere are maintained biologically—in other words, produced and regulated in their proportions by living processes.[11] By the spring of 1971, Margulis was reading Lovelock's essays and manuscripts related to the as-yet-unpublished

"Gaia" hypothesis. As we noted, these proto-Gaian papers addressed the open issue at that time, regarding the relation of planetary atmospheres and life, whether and to what extent Earth's atmosphere is biogenic. A letter to Lovelock on March 31 indicates how her microbial understanding at that time came up to the edge of Lovelock's geochemistry.

> Several of your charts are fascinating and very comprehensible (Table 1, Table 2 of planetary atmospheres) but where do these estimates come from? I'd really like to learn your methods for making these sorts of estimates as well as your sources of original data. Microbes strongly interact (i.e. take up, give off) hydrogen, nitrogen, ammonia, methane, carbon dioxide, oxygen, hydrogen sulfide at least and we are just beginning to know enough about the bugs in which these reactions occur to order them in an early-to-late evolutionary sequence. But how the gases themselves act in the environment.... I really don't know what I need to learn or where to begin.[12]

Nevertheless, in a letter dated September 17, 1971, Lovelock asked whether she would consider coauthoring a paper on the topic of the atmosphere. The proposal for a formal collaboration came from Lovelock at this precise moment. His letter dated September 27 of that year thanked her for considering while not yet agreeing to his offer of collaboration until after more conversation. They first met in person when Lovelock came through Boston around Christmas of that year. On January 3, 1972, Lovelock wrote to thank her for the warm welcome and lengthy discussions. He then mentioned in passing that with regard to the idea of a living planet, William Golding had suggested the name of "Gaia."[13]

The abiding contribution that Margulis brought to the Gaia hypothesis in the first years of her collaboration with Lovelock was the addition of deep time, evolutionary depth. Lovelock preferred to study systems available for current inspection.[14] The main emphases of Margulis and Lovelock always diverged to some degree. Nor are the details of the chronology of Margulis's early involvement with Lovelock's Gaia project particularly well known. For instance, an otherwise veracious obituary of Margulis posted on the website of the British newspaper *The Telegraph* committed several misstatements on this particular topic: "It was Lynn Margulis's expertise in microbes that led her, in the mid-Seventies, to the British atmospheric chemist James Lovelock, who had come to suspect that living organisms had a greater effect on the atmosphere than was commonly recognized. Together they proposed a theory that earth itself—its atmosphere, the geology and the organisms that inhabit it—is a self-regulating system in which living organisms help to regulate the terrestrial and atmospheric conditions that make the planet habitable."[15] This is a serviceable description of current Gaia theory, but several statements here warrant historical correction. In their first papers, Lovelock and Margulis presented the Gaia *hypothesis,* a different, preliminary form of the theory. Their collaboration began in 1971, not in the mid-1970s. Margulis initiated their correspondence "in the summer of 1970 . . . to ask about atmospheric oxygen."[16] However, Lovelock's first recognizably Gaian papers had appeared in the mid- to late 1960s.

Lovelock published under the name of Gaia before any coauthored Lovelock and Margulis papers came out. Lovelock's definition of Gaia at that moment as "a biological cybernetic system" indicates the proper form of the Gaia hypothesis at the outset of their collaboration. The hypothesis hinged on what was then a heretical notion that *life controls the environment*: Gaia is a *biological* cybernetic system. The atmospheric chemist Lovelock had independently arrived at this biocybernetic orientation. It may be that, just to be on the safe side, Lovelock did not utter the name of Gaia until after he had secured Margulis's agreement to be his coauthor. That decision appears to have been the immediate professional outcome of the social success of their first face-to-face meeting.

On January 13, 1972, Lovelock wrote to confirm his sense that they were now ready to move ahead with a well-developed scientific paper.[17] Margulis must have already been at work on it, as four days later he writes to thank her for sending a second draft.[18] His previous letter covered the copy of a just-submitted manuscript for "Oxygen in the Contemporary Atmosphere," a short essay coauthored with James P. Lodge Jr. appearing later that year in *Atmospheric Environment* immediately preceding Lovelock's letter on "Gaia as Seen through the Atmosphere." On January 24, Margulis wrote back to comment on it. Her manner manifested her own professional drive and stringent editorial bent. Her remarks also indicated that she had now grasped Lovelock's Gaian argument:

The mail will probably cross again. Anyway, I have read your oxygen article five times and finally not only do I dig it but I find it brilliant. Have you sent it in? Even so, I think you should strongly consider reducing the size, outlining the argument (I'll do this if you want my collaboration) and submitting it as a technical comment to Science in response to Lee [Leigh] Van Valens article. Van Valen raised this issue well but did not perceive the solution. I would also change the wording in several places to make it more transparent to the potential ecological and general biological audience. Please let me know soon what you think of this possibility.[19]

This letter may well be the original expression of Margulis's own Eureka moment at which the elements of Lovelock's biocybernetic Gaia concept fell into place for her: "I have read your oxygen article five times and finally not only do I dig it but I find it brilliant." To tell from this testimony, what Margulis has finally wrapped her head around is Lovelock's cybernetic scheme, his envisioning of self-regulation emerging from the circular functions drawn from the operational closure of one of Gaia's feedback circuits, here, a component of the Gaian system in the effective form of a negative feedback loop between the anaerobic and aerobic portions of the biosphere. She goes directly on to describe his scheme back to him:

Methane producers as far as I know are fermenting anaerobes. If the local environment gets too aerobic they turn off. Therefore they release less methane into the atmosphere. Therefore, according to you, less gets transported up to circuitously loose hydrogen (via water, according to you) and the mechanism for keeping aerobic shuts off. This provides more anaerobic niches and the methane bacteria go to work again. Your basic conceptual plan here must be correct.

By January 1972, Lovelock and Margulis were collaborating on a co-authored paper buttressing his theories of atmospheric self-regulation with her knowledge of microbial metabolisms. On February 1, Margulis indicated further progress on their mutual manuscript.[20] Then Lovelock decided to submit, as a supplement to the oxygen paper coauthored with Lodge, the letter to the editors of *Atmospheric Environment* that

would go public with "Gaia" for the name of a hypothetical entity "with
the powerful capacity to homeostat the planetary environment."[21] At
some point before February 16, Lovelock wrote Margulis regarding the
manuscript of "Gaia as Seen through the Atmosphere" to confirm that
her contributions to his thinking were already in evidence there. He expressed concern about getting out in front of their collaboration in a
way that did not credit her explicitly. He would have misgivings about
publishing it if Margulis felt that it would steal the thunder from their
coauthored article in progress. Moreover, he was aware that his discussions with Margulis had changed some of his current thinking.[22] He
encouraged her to coauthor and sign this professional communication
as well. Lovelock went on in this long and crucial letter to Margulis to
worry an issue that remains intrinsic to Gaian science, one of its key
issues: how to negotiate the distinction between living and nonliving
systems.[23] The letter in *Atmospheric Environment* states it this way:

> As yet there exists no formal statement of life from which an exclu-
> sive test could be designed to prove the presence of "Gaia" as a living
> entity. Fortunately such rigor is not usually expected in biology. . . .
> At present most biologists can be convinced that a creature is alive
> by arguments drawn from phenomenological evidence. The persis-
> tent ability to maintain a constant temperature and a compatible
> chemical composition in an environment which is changing or is per-
> turbed if shown by a biological system would usually be accepted as
> evidence that it was alive. Let us consider the evidence of this nature
> which would point to the existence of Gaia. (579)

Margulis addressed Lovelock's concern over the distribution of credit
by encouraging him to publish the "Gaia" letter on his own, on the consideration that it would help to prepare the reception for their more detailed coauthored presentation of the hypothesis. Her own unconcern
on this score may also be due to her having already grasped Gaia's big-
ger picture and their roles in it as scientific revolutionaries. As Margulis
pointed out in a letter to Lovelock on February 16, the Gaia hypothesis
was no small idea, nothing that any one paper could exhaust as a topic:

> As for Gaia I do not in any way think you are preempting our mutual
> paper. On the contrary the more already in print and justified the bet-

ter off we are. After all we are involved in attitudinal (scientific paradigm, Kuhn) change. Furthermore I really have not done the methane argument for myself in the detail I would like to before signing on. Go ahead and get it out on your own.[24]

Margulis referred here to the philosopher of science Thomas Kuhn's notion of a paradigm shift.[25] Kuhn's classic example of such an intellectual event was the epochal replacement of the Ptolemaic cosmos by the Copernican system, producing a massive reordering of fundamental ideas about the world.

The philosopher of aesthetics Peter Sloterdijk has given an even wider-angled view on the cultural repercussions induced by such a demolition and rebuilding of world pictures. Regarding the elaborate vision of the universe embodied in the scheme of Ptolemaic spheres, "In the encompassing orbs, the ancients discovered a geometry of security." However,

> these sublime imaginary constructs of wholeness were doomed to vanish with the beginning of the Modern Age, while the human location, the planet Terra, took on increasingly explicit contours. In a dawn that took centuries, the earth rose as the only and true orb, the basis of all contexts of life, while almost everything that had previously been considered the partnered, meaning-filled sky was emptied. This fatalization of the earth, brought about by human practices and taking place at the same time as the loss of reality among the once-vital numinous spheres, does not merely provide the background to these events; it is itself the drama of globalization. Its core lies in the observation that the conditions of human immunity fundamentally change on the discovered, interconnected and singularized earth.[26]

The cosmological planet of Gaia theory is itself a revolution in our understanding of the "conditions of human immunity." Gaia theory has also risen to the occasions Margulis sensed, overturning venerable habits of geological and biological practice and reconstituting large portions of normal science going forward. Gaia theory has also provoked a broader cultural rethinking of terrestrial security and viability, of the immunitary contingencies and biospheric services humans and other living organisms rely upon for the conditions of their being. Margulis's passing remark to Lovelock in 1972 indicated a prescient appreciation of the potential energy folded up inside their fledgling Gaia hypothesis. Should it flourish as an explanatory and evidentiary model, it could bring "attitudinal change" as a seminal rethinking of planetary science and of humanity's place in relation to its home world.

Lovelock sent off his single-authored Gaia letter to *Atmospheric Environment*, and around July Margulis was ready to submit their first coauthored Gaia paper: "When the enclosed has received both your additions and your blessings I think it will be ready for *Science* (and when rejected minor modifications may fix it up for *Nature*)."[27] On September 22, 1972, the editor of *Science*, Philip H. Abelson, sent Margulis a rejection letter along with three readers' reports on the submission of the manuscript "The Earth's Atmosphere: Circulatory System of the Biosphere?" The reports listed Lovelock as its lead author. None mentions "Gaia" one way or another. The reports—two short notes, one detailed write-up—were not hostile or closed-minded. Their gist was that to pursue its thesis adequately the paper needed more work. One of the two short readings dropped a negative judgment overall: "The support offered by the authors for the primary thesis is not at all convincing. However, the article is reasonably well written and offers food for thought. Some specific statements on the early earth history are clearly incorrect (see text notations) and will undoubtedly get 'shot down.'"[28] The other short report submitted a positive evaluation along with a pointed recognition of the daring nature of the article's argument: "The general intent of the article is of broad interest. Its most important feature is the suggestion that the atmosphere has been adapted by organisms to their needs, as well as organisms being adapted to it. This would of course be of immense interest if it could be substantiated." Its critical remarks were narrowly focused: "I think that some tightening of expression is desirable. I am not too happy with the use of 'homeostat' as a verb." This was followed by the reviewer's own suggestion: "I think a reference to J. B. A. Dumas and M. J. B. Boussingault's *Leçon sur la statique chimique des êtres organisés,* Paris 1841, which really started the idea, would be desirable." The third, detailed reading mixed praise for the authors' ambitions with criticism of their execution. It began with a summary that underscored the potentially seminal significance of their thesis: "The paper develops an idea that is of interest and potential significance to scientists of a number of fields (and the human future). With all respect to the question that is raised, I feel that the reasoning and writing of the paper is not now adequate to that question."

In 1974 the journal *Tellus,* edited by the Swedish meteorologist Bert Bolin, published "Atmospheric Homeostasis by and for the Biosphere: The Gaia Hypothesis," with Lovelock as lead author, and the journal *Icarus,* edited by Carl Sagan, published "Biological Modulation of the Earth's Atmosphere," with Margulis as lead author.[29] After its opening abstract, the *Icarus* article presents an epigraph drawn from Dumas and Boussingault (1844): "The atmosphere, therefore, is the mysterious link that connects the animal with the vegetable, the vegetable with the animal kingdom." The *Icarus* essay delves deeply into Margulis's stock-in-trade, "early earth history." Its tables show histories of atmospheric gases and temperatures charted on timelines beginning with the origin of Earth, let alone the origin of life, followed by her trademark treatment of prokaryotic evolution preceding the advent of "larger (eukaryotic) life forms" by two billion years. "We emphasize the microbial contribution for two reasons: their metabolic versatility leading to profound environmental effects and because the regulation of the planetary environment was apparently proceeding long before the evolution of the larger (eukaryotic) life forms" (476).

Like the cybernetic discourses on which it was established, the co-authored Gaia concept of Lovelock and Margulis overflowed the notional boundaries of standard scientific disciplines. Both geology and biology had previously supposed, in addition to classical accounts of linear causality, reasonably neat separations between the abiotic and the biotic realms. However, as natural processes have tinkered, so to speak, with their own evolving elements, breeding all manner of emergent formations and reality-testing their stability or viability— producing life out of nonlife, then more complex out of less complex life, then bootstrapping life and nonlife together into an operational consortium called Gaia—those processes have done so without concern for the nice human constructions of peer-reviewed disciplinary distinctions. And even though these disciplines have existed in modern form only since the nineteenth century, they were codified well before the rise of the systems sciences. Lovelock's conceptual breakthrough came straight from cybernetics, the scientific metadiscipline that began by coordinating artificial and natural "contrivances" for systemic operation and self-regulation. His early letters mentored Margulis on this discourse.[30] As we will explore in detail in Part II, Lovelock's cybernetics put Margulis on the winding road to the neocybernetic conception of autopoietic Gaia. This was a destination beyond both the neo-Darwinist orthodoxy of that moment and Lovelock's own orientation toward control systems. Their collaboration coupled two great scientific outliers, connecting and strengthening their individual and mutual challenges to received disciplinary ideas.

The Name of Gaia

Isabelle Stengers's *In Catastrophic Times* refers to Lovelock and Margulis when she writes concerning Gaia: "I want to maintain the memory that in the twentieth century this name was first linked with a proposition of scientific origin."[31] Perhaps one still needs to articulate this reminder, for broadly speaking, after five decades of cultural traffic, many may have forgotten that *this* Gaia first intruded *as* a cybernetic hypothesis. Out in cyberspace and elsewhere, there are any number of Gaia notions advanced by persons with limited exposure to Gaia discourse or theory who yet desire to attach some mention of Gaia to their matters of concern. In these instances, it may be that a kind of diffuse popular scientism mediates Gaia for an audience unlikely to be abreast of the scientific headwinds against which the Gaia hypothesis and its theoretical developments have had to negotiate their bona fides. Or it may also be that they have been moved by some genuine article of Gaian thought. Here are two such examples. The first is from the website of the magazine *Motorcyclist*: "I don't know if there's anything to the Gaia theory—that the world is one living organism with a conscience. But I do know this: The day I rode my Honda VFR home for the first time, the weeds in my garden were doing high-fives. I'm not saying I've neglected *everything* since getting my long-awaited bike, but I'm pretty sure my motorcycle has a lot more hours on it than my lawnmower."[32] A second is from the website of a South African business magazine: "James Lovelock's Gaia theory—that Earth and its entire species constitute one living organism—is applicable to South Africa. Though of different races and cultural origins, we are one big family. If one member of the family is not well, the whole family suffers."[33]

A motorcycling homeowner dude neglecting his lawn has a Gaian vision of his weeds as they celebrate their Earthly reprieve. He takes the idea of Gaia to mean that "the world is one living organism with a conscience." This is absolutely classic, a perfect articulation of a broadly popular Gaia notion. This manner of moralizing the idea of Gaia is rarely stated so explicitly, and it attaches itself to a localized instance of the Gaian thought, however seriously held, of humanity's nondomination of nature. A related conscientious notion recurs in the South African example. Riven by "different races and cultural origins," human beings cannot feel their family ties with each other, let alone with the rest of nature. Gaia stands in for the principle of a human unity that humanity has yet to achieve in its actual behavior. Here the Gaian thought of "Earth and its entire species" follows Margulis's coauthored prescription that "Gaian perceptions . . . link people . . . to the biota." What it may lack is the further specification that in its Gaian intuition, this connectedness occurs "inextricably" and situates humanity in relation to the biota "in subordinate fashion."[34] In both of these samples, however, the matter of Gaia's multiplicity gives way to a vision of oneness. For a considerable portion of the wider public, "Gaia" must be "a single organism" or "one living organism" that comprises "one big family." In these popular conceptions, Gaia is an amorphous planetary essence admonishing selfish human squabbling. These found usages underscore that the name "Gaia" has always been a magnet for the ambient bits of mythic response that circulate in modernity's secular atmosphere. Diffuse notions of holistic totality verging on divine agency have been prone to stick to it.

Spiritual concerns are unlikely to have been on Lovelock's mind in 1965 when he had his initial premonition of the entity he came to call Gaia. Nonetheless, according to his account, its image did come to him in the form of an "organism": "It dawned on me that somehow life was regulating climate as well as chemistry. Suddenly the image of the Earth as a living organism able to regulate its temperature and chemistry at a comfortable steady state emerged in my mind. At such moments, there is not time or place for such niceties as the qualification 'of course it is not alive—it merely behaves as if it were.'"[35] By the later 1960s, "a planet-sized entity, albeit hypothetical, had been born, with properties which could not be predicted from the sum of its parts. It needed a name."[36] Giving his planetary entity a proper name would accord due recognition to a vastly intricate, geologically persistent system.

Lovelock has told the ensuing story many times, with different emphases, but I think never so charmingly as during his interview with David Suzuki:

> William Golding said, "If you're going to have a big idea like that you'd better give it a proper name." So I said, "Good, what would you call it?" He said, "I'd call it Gaia." . . . And we went on walking for twenty minutes, talking at complete cross-purposes, because I didn't have a classical education. I didn't know anything about Gaia, the Greek goddess. But I did know about g-y-r-e, gyre, the great whirl in the ocean or in the atmosphere, and this made sense of course, this was a fed-back system, and this is what he's talking about. And he said, "No no no no no, I mean the Greek goddess of the Earth." And then it clicked, of course.[37]

Lovelock accepted Golding's gift horse of this archaic name and, as we have noted, debuted it publicly in his 1972 letter to *Atmospheric Environment*. "Gaia" went forth as a hypothesis about a homeostatic system holding "climate as well as chemistry" within viable limits. Two years later, with Lovelock as first author co-writing with Margulis, this basic description received a more felicitous but also more problematic phrasing:

> This paper examines the hypothesis that the total ensemble of living organisms which constitute the biosphere can act as a single entity to regulate chemical composition, surface pH and possibly also climate. The notion of the biosphere as an active adaptive control system able to maintain the Earth in homeostasis we are calling the "Gaia" hypothesis. . . . [T]he word Gaia will be used to describe the biosphere and all of those parts of the Earth with which it actively interacts to form the hypothetical new entity with properties, that could not be predicted from the sum of its parts.[38]

This early presentation manifests two conceptual tensions that run through Gaia discourse and theory. One of these is a wavering between a biotic and a metabiotic model. Is Gaia essentially biotic, "the total ensemble of living organisms which constitute the biosphere"—a phrasing often compressed to "the sum of the biota" or, even more holistically, "a living organism"? Or is Gaia essentially metabiotic, "the biosphere *and* all of those parts of the Earth with which it actively interacts"—a description to be underscored by indicating Gaia's status as a "coupled system"? Lovelock and Margulis will go back and forth on this issue, with Margulis understandably tending more toward the biotic side, but with both working out different but compatible routes to a metabiotic characterization. A second conceptual tension is the alternation between a cybernetic and a holistic description of Gaia, an oscillation of emphasis between the technical heterogeneity and the formal totality of the assembled system. The "oneness of Gaia" certainly has adequate warrant in the primary annals of the Gaia hypothesis. However, placing the stress on the singularity of this "hypothetical new entity" has tended to undervalue the complexity of Gaia's planetary aggregation and to blur the manifold of elemental cycles and ecological subsystems needed to buffer the operations of the "whole system."[39]

For a description of Gaia in operation, the preferable terms are *coherence* in relation to its emergent cybernetic *functions* as a systemic ensemble. Even as a coordinated ensemble, "the biosphere can *act as* a single entity . . . *as* an active adaptive control system." In the passage above, Lovelock and Margulis state the cybernetic contingency of Gaia's systemic self-constitution, but at the same time they open the door for a holistic reification of the mythic personification already on offer in the name of Gaia. It is thus helpful to recall Lovelock's initial ignorance of Gaia's classical provenance when Golding first pronounced that name in his presence. It just means that *the name of Gaia is a rhetorical mediation,* the inspired brainchild of a literary artist coining a catchy title. The story of Gaia is a definitive case of the mediation becoming the message, the signifier overtaking the signified. Even while the name of Gaia has been tremendously effective as a deliberate branding device and as a sliding signifier in cultural communication, it is of little use for understanding Lovelock's and Margulis's *concepts* of Gaia. For that, one can only study the details of their discourse. These start from the observation that the entity named Gaia is best conceived as a system—as a living system or, more precisely, as an autopoietic system.

Gaia and Systems Theory

Gaia theory gathers chemistry, biology, geology, and physics into a multidisciplinary consortium that contains while surpassing presystemic scientific programs dominant since the seventeenth century.[40] If systems theory seems counterintuitive at times, the problem lies largely with tired intuitions and conceptual biases in need of updating through closer contact with the heterogeneity of mature systems thinking. For instance, one wing of classical cybernetics is the aforementioned subject of control theory. The control concept was basic to the first cybernetics, founded by Norbert Wiener's seminal work of 1948, *Cybernetics: Or, Control and Communication in the Animal and the Machine*. Wiener and his collaborators showed how certain aspects of physiological systems ("in the animal") and technological systems ("in the machine") could be considered formally equivalent, insofar as both natural and designed systems could exhibit "control" in the form of self-regulation produced through circuits or closed loops of negative feedback. One could also submit social systems, steered (however erratically) by the feedback of communications, to cybernetic analyses of this sort. Still, when it leaves the arena of technological or automatic mechanisms, the very idea of "control" can seem to be an affront to values of freedom and self-determination. As self-possessed individuals we may hope to be "in control" or to have things "under control," and we also dislike the idea of being under another's control, or worse, subjected to control by "the system." However, control theory is a delimited area within systems discourse. Part II will explore a radically different mode of cybernetics—a neocybernetics of autonomy—as equally applicable to a description of Gaia.

Here is another example of the occasionally antithetical semantics of systems theory. Typically, "negative" stands to "positive" as bad to good, deleterious to desirable. In the operation of systems, however, negative functions can be desirable and positive ones deleterious. Take feedback. Negative feedback denotes regulatory processes in which operational consequences counterbalance or are subtracted from that same suite of operational effects. Here, what is called "circular causality" generally produces desired regularities—think of a thermostat turning the heat off when it's hot enough, back on when it's too cold. Cybernetics' name for this form of system regulation is *homeostasis*—correcting discrepancies so that operations stay the same on average or within range of a desired level. In contrast, positive feedback denotes a loop in which operational consequences compound or are added to that same sequence of operations. Here the circular flow generally produces runaway increases—as when a microphone is placed so close to an audio speaker that the "positive" amplification of the amplification drowns out the desired sound.

Closely related to the counterintuitive semantics of feedback is the distinction between systems that are "open" and systems that are "closed." Liberal political values laud that which is open and shun that which is closed. However, when it comes to the self-regulation of systems through negative feedback, only a closed loop will complete the circuit. The same with homeostasis: operational closure is necessary to maintain the regulatory function of the loop. Moreover, autopoietic systems in particular are sufficiently complex to surpass this very distinction: they are open and closed at the same time, albeit in a transversal fashion separating their outer and inner dimensions. Even while autopoietic systems are environmentally open to material-energetic fluxes or semiotic mediations, their operations are internally closed so that the system sequesters its integrity as a functional unity. The system concept rests on schemas of closure, simultaneity, and reciprocity. Systemic closure produces a finite interior, an actual or virtual space and time for tiers of differential operations. However, once operational boundaries form, there will be all manner of cross-boundary translations.

For one more example of semantic irony in systems-theory discourse, consider the opposition of the terms "top-down" and "bottom-up." "Top-down" typically connotes dictatorial, hierarchical, or undemocratic power structures, whereas "bottom-up" connotes participatory and egalitarian arrangements. However, as Lovelock has noted in the context of Gaia theory, "the top-down holistic view, which views a thing from outside and asks it questions while it works, is just as important as taking the thing to pieces and reconstituting it from the bottom up."[41] In an analysis of systems released from doctrinaire constructions, "top-down" can name not just the exercise of authority *de haut en bas* but also, through a quite different semantics, a whole-systems perspective attuned to the emergent behaviors of discrete ensembles and protective of the integrity of the ensemble under observation. Similarly, from a systems-theoretical perspective, the term *bottom-up* calls out the wolf of reductionism from under the sheep's clothing of egalitarianism. "Bottom-up" can denote not only the location of power within the polis but also, in mainstream scientific application, a reductionist perspective that takes things to pieces.

Addressing this terminological imbroglio—the unstable mix of scientific and political semantics in the standard opposition of "holism" and "reductionism"—Lovelock has construed its defects from his scientific perspective in noting how "We are also in an adversary contest between our allegiance to Gaia and to humanism. In this battle, politically minded humanists have made the word 'reductionist' pejorative, to discredit science and to bring contumely to the scientific method. But all scientists are reductionists to some extent."[42] NST can train our intuitions of systems to expect and negotiate such inversions of sense. For instance, just as autopoietic systems are at once environmentally open and operationally closed, so too is Gaia theory at once a top-down view placing worldly systems in their widest environmental contexts and a bottom-up perspective on emergent processes arising from the coupling of living systems to nonliving environments under constant Gaian reconstruction.

The Gaia concept first coalesced in the 1960s at the intersection of three streams of systems theory. These are the thermodynamics of mechanical and natural systems first developed in the mid-nineteenth century, the cybernetics of self-regulating control systems first developed in the mid-twentieth century, and its contemporaneous companion discourse of information theory derived by transposing statistical mechanics from matters of heat flow to matters of informatic transmission. "There is little doubt that living things are elaborate contrivances," Lovelock and Margulis write in one of their first coauthored papers. "Life as a phenomenon might therefore be considered in the context of those applied physical sciences which grew up to explain inventions and contrivances, namely, thermodynamics, cybernetics, and information theory."[43] Lovelock's original conception of Gaia as a cybernetic system indicated a *natural* contrivance produced in the coevolution of the biota with their abiotic environment.

Let us return to the thread of homeostasis. The engineering discourse of homeostatic feedback mechanisms informed Lovelock's earliest presentations of Gaia as a planetary control system. In this conception, Gaia was located in first-order cybernetic fashion at the interface of biological systems, mechanical contrivances, and computational devices. In Lovelock's first book, *Gaia: A New Look at Life on Earth,* published in 1979, the chapter titled "Cybernetics" addresses homeostasis as a form of cybernetic steering. "The primary function of many cybernetic systems is to steer an optimum course through changing conditions towards a predetermined goal."[44] Walter Cannon originally coined the term *homeostasis* to name the way heart rates, body temperatures, and other physiological processes are internally regulated to return when perturbed to reliable and predictable norms.[45] Through the classical cybernetic metaphor heuristically equating bodies and machines, one could extend the physiological concept of homeostasis to any system that exhibited self-regulation around a set point. In *The Wisdom of the Body*, Cannon himself already ventured sociological applications of homeostasis. A related physiological process is proprioception—the faculty of internal self-perception guiding locomotive balance.

In the same chapter of *Gaia: A New Look at Life on Earth*, Lovelock worked with this corporeal analogy to present Gaia as the proprioceptive or internal self-balancing feedback system of the biosphere as a planetary body. He then reverted to the mechanical realm to offer the homeliest of mechanical cybernetic analogies: Gaia in operation performs like the thermostat of a kitchen oven. Earth is the oven box, the sun is the heat coil, and Gaia is the regulator that keeps the temperature at an "optimum." A few years later, homeostasis was further instantiated in a "cybernetic proof" of the Gaia hypothesis that first came forward in the early 1980s with Lovelock's Daisyworld computer simulations.[46] Lovelock later noted how his toy model of Gaian homeostasis also exhibited the paradoxical description of its real-world counterpart. Daisyworld "is reductionist and holistic at the same time. The need for reduction arose because the relationships between all the living things on Earth in their countless trillions and the rocks, the air, and the oceans could never be described in full detail by a set of mathematical equations. A drastic simplification was needed. But the model with its closed loop cybernetic structure was also holistic."[47]

Summing up, the Gaia hypothesis began as a thought experiment drawing on a basic first-order cybernetic model of self-regulation using negative feedback to correct deviations from a desired state of operation. With analogies drawn from garden-variety engineering cybernetics, Lovelock's Gaia hypothesis suggested the operation of planetary feedback regimes that hold certain climatic conditions, such as global temperature over geological time, within viable limits. Lovelock reasoned thus from cybernetic conceptuality to systemic Gaian function while keeping open the bridge between mechanistic and organic functions: "The key to understanding cybernetic systems is that, like life itself, they are always more than the mere assembly of constituent parts. They can only be considered and understood as operating systems."[48] As "operating systems"—not as the digital innards of computers, of course, but simply, unlike nonfunctional structures or nonoperating objects, as systems that operate in real time to maintain a desired state or produce some ongoing outcome. If such were found to be the case, then Gaia would name a form of planetary agency with the capacity to constitute variably viable conditions for and thus prolong the evolutionary run of the biota generally. In this sense, as biophysicist Harold Morowitz later stated in ratifying Lovelock's idea of the Gaian system, "life is a property of planets rather than of individual organisms."[49]

Metabiotic Gaia

The neocybernetic discourse of autopoiesis supersedes earlier holistic models. Dirk Baecker has summarized this aspect of NST's second-order theorization, its postholistic conceptuality: "The unity of which systems theory speaks is not the unity of a whole but the unity of systems that are ecologically linked with each other, lacking any 'super-system' to ensure and organize that ecology, let alone direct it teleologically to a better future."[50] And as Anna Henkel has observed regarding the current conceptual resources of NST, "the concepts of meaning, interpenetration and communication can be further developed . . . to open up a new way to view the boundaries of the social and the observation of materiality and sociality."[51] In parallel with the interpenetration of psychic and social systems in Niklas Luhmann's metabiotic transposition of the autopoiesis of living systems, in metabiotic Gaia matter and meaning consort while producing and maintaining operational distinctions. Let us read NST's metabiotic conceptualization of autopoiesis back into the Gaian occasion.

In our corner of the cosmos at least, the emergence of life on Earth ends the era of the preautopoietic cosmos. The autopoietic processes of living systems partake of the nonautopoietic processes of the physical cosmos—atomic valences, chemical bonds, radiant electromagnetic fields, and dynamical and thermodynamical systems—while they also modulate Earth's particular geological formations, such as hydrological dynamics, atmospheric and meteorological systems. Moving to the metabiotic arena, in NST, psychic and social systems coordinate their respective operations through a common formal milieu, the medium of meaning. In the Gaian instance, however, the counterpart to the medium of meaning is the medium of matter. Akin to the curiously virtual, insubstantial form we call meaning, materiality is also infinitely transformative. In the metabiotic formula for Gaian operation, Gaia's own autopoiesis couples the open material-energetic processes of abiotic systems and the metabiotic extrusions of technological structures to the self-bounded operational processes of biotic systems within the physicochemical medium of matter.

Just as meaning in incessant motion holds minds and societies together, in metabiotic Gaia the flux of matter couples abiotic portions of the Earth to living processes. On the one hand, with regard to meaning systems, "The world itself, as co-occurring other side of all meaning forms, remains unobservable."[52] This formulation anticipates the paradox of autopoietic Gaia as an observing system that posits an ahuman order of planetary cognition. On the other hand, the corresponding implication here is that, when in the medium of *matter*—in which biotic and abiotic systems interpenetrate each other without totalizing merger—the world of *meaning* "remains unobservable." Gaia per se is not a *meaning* system; its planetary cognition does not make "sense." Rather, planetary cognitions maintain the conditions through which Gaia constructs its own continuation, and thus, the possible continuations of its systemic elements, such as living beings and meaning-making minds participating in communications. The autopoietic production of minds and societies, it then follows, has also nested itself, sequentially and differentially, within this cosmic matrix of planetary operations. Nevertheless, in our own time, Gaia's body has intruded corporeal meanings—of which global warming and its attendant ills is perhaps the most sore—sufficiently resonant to disrupt prior orders of insular human thought and sociality.

Figure 1 diagrams the metabiotic redescription of Gaia through a set of differentiations for a Gaian ecology of systems juxtaposed within an enabling matrix of material and virtual possibilities. *System differentiation* denotes a situation in which "the whole system uses itself as environment in forming its own subsystems and thereby achieves greater

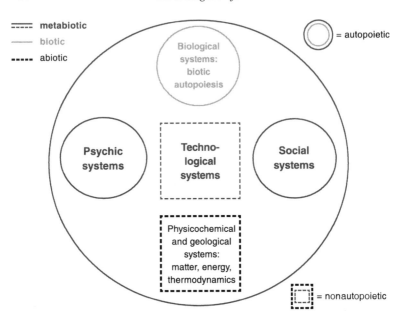

FIGURE 1. *Differentiations in the Gaian system.*

improbability on the level of those subsystems by more rigorously filtering an ultimately uncontrollable environment."[53] The evolution of Gaia over geobiological time yields a series of system differentiations and transformations. At first, as the accretion of a cosmic object coalescing to a semi-solid sphere 4.5 billion years ago, planet Earth is a scene of strictly abiotic physicochemical processes and nascent geological systems. At some point, living systems spring forth and proliferate. Earth transforms from a pre-living place with chemical potential into an environment coupled and looped into the evolutionary flux of microbial life. Biotic autopoiesis within the planetary microcosm proceeds to explore both its naturally recombinant possibilities and its coevolving niches while extending a microbial patina into the crust and mantle. Well before the appearance of eukaryotic organisms about 1.5 billion years ago, once the Archean microcosm had achieved planetary distribution, that biota looped itself all the way into geological processes it had already radically renovated. Likely lasting some millions of years, the moment comes when metabiotic Gaia emerges by closing the sum of the loops among the material and cognitive mediations linking these abiotic and biotic systems into nested ecosystems.

Lovelock has imagined the primal buildup to the origin of Gaia in this way:

> At some time early in the Earth's history before life existed, the solid Earth, the atmosphere, and the oceans were still evolving by the laws of physics and chemistry alone. . . . Briefly, in its headlong flight through the ranges of chemical and physical states, it entered a stage favorable for life. At some special time in that stage, the newly formed living cells grew until their presence so affected the Earth's environment as to halt the headlong dive towards [thermodynamic] equilibrium. At that instant the living things, the rocks, the air, and the oceans merged to form the new entity, Gaia. Just as when the sperm merges with the egg, new life was conceived.[54]

In this account the analogy of sexual reproduction figures Gaia's status as a biotic system, a superorganism, or living being of some sort, but without a definitive declaration to that effect. My account suspends that indeterminate simile. For NST, Gaia is shot through with biotic processes but is itself metabiotic. It arises from the structural coupling of living operations with the abiotic dynamics of their cosmic, solar, and Earthly elements. In due evolutionary time, however, the forms of distinction between nonautopoietic, physicochemical, and geological systems on the one hand and autopoietic, living systems on the other are reentered into Gaia to form the further reticulated metabiotic systems that now arise within Gaia's geobiological matrix. Viewed in Gaian context, psychic and social systems are metabiotic autopoietic forms in which events of distinction generate virtual boundaries in place of the material membranes of biotic systems. Language and meaning appear in their midst as metabiotic environmental mediations coupling minds and societies.[55] Such formal echoes and operational parallels among systems biotic and metabiotic, autopoietic and nonautopoietic, may account for the powerful ways that Gaia has shown its different faces within vastly diverse cultural codes. That circumstance would suggest that the Gaian system's overlapping forms of life and Earth deeply resonate with the forms and mediums of human psychic and social systems. It would suggest that these system-environment frequencies go all the way down and all the way out. It may be that Gaia's systemic resonance for consciousness, mediation, and communication has produced both its mythic and, in time, its scientific faces—its primal intuitions, its historical articulations, and its belated recognitions. Gaia's planetary cognition is thus an ur-medium of corporeal meaning awaiting adequate remediation. We seek to construct and communicate the sense of Gaian systems and forms. For Stengers, Gaia will name the relinquishing of one history, in which time flows blindly toward a horizon of economic growth detached from ecological sensitivity, and the constitution of another regime of human time: "Naming Gaia is therefore to abandon the link between emancipation and epic conquest, indeed even between emancipation and most of the significations that, since the nineteenth century, have been attached to what was baptized 'progress.' Struggle there must be, but it doesn't have, can no longer have, the advent of a humanity finally liberated from all transcendence as its aim. We will always have to reckon with Gaia, to learn, like peoples of old, not to offend her."[56] As we will contemplate later, Gaia can also bring Anthropocene humanity to recognize the limit of its capacity to order the environment at whim. Latour writes, echoing Sloterdijk, that the Crystal Palace of human immunity has now shattered from within: "No immunology—in Sloterdijk's expansive sense—is possible unless we learn to become sensitive in turn to these multiple, controversial, mutually entangled loops. Those who are not capable of 'detecting and responding rapidly to small changes' are doomed."[57] Here at the center of the universe constituted by our cosmic coupling to an animate but indifferent Gaia, we had better become biopolitical animals keenly sensitive to our systemic entanglements. The figures of Gaia discourse may cultivate a planetary imaginary adequate to this intuition of Gaian being.

CHAPTER 2

Thinkers of Gaia

Intellectually serious extrascientific engagements with Gaia theory have been gathering under the shadow of climate change and global warming. This chapter reviews the state of Gaia discourse as represented by three of its most significant recent commentators. Arriving ahead of the moment of the Anthropocene, the feminist theorist and historian of science Donna Haraway found a place for Gaia in her cyborg discourse of the 1990s.[1] Lovelock's own recent forecast of an Earth given over to postbiological cyborgs adds some piquancy to Haraway's earlier treatment.[2] Additionally, in the two last decades, figures of Gaia have animated the writings and scholarly conversations of the Belgian philosopher of science Isabelle Stengers and the French sociologist Bruno Latour.[3]

Donna Haraway: Gaia as Cyborg

In the 1990s, the Gaia hypothesis advanced to Gaia theory. Lovelock's Daisyworld computer models had gone into circulation to give Gaia some computational cred just as the nonlinear recursions of chaos theory were going mainstream alongside the explosion of personal computing, email, and the internet. Lovelock's second book, *The Ages of Gaia: A Biography of Our Living Earth* (1988), consolidated Gaia's scientific discussion. Robust scientific meetings had convened by then to take up Gaia's place in the life and Earth sciences. In the mid-1990s, Haraway stepped back from the discourse of Gaia theory and the popular conceits of Gaia notions then at hand to construct her own sense of the Gaia phenomenon.

In bringing the Gaia concept into the fold of cyborg discourse, Haraway was particularly attentive to Gaia's peculiar cybernetic pedigree. In her foreword to *The Cyborg Handbook*, published in 1995, she noted that "Gaia is the name that James Lovelock gave . . . his hypothesis that the third planet from the sun, our home, is a 'complex entity involving the Earth's biosphere, atmosphere, oceans, and soil; the totality constituting a feedback or cybernetic system which seeks an optimal physical and chemical environment for life on this planet.'"[4] Haraway isolated another one of Gaia's conceptual tensions—or better, definitive equivocations—regarding traditional ontological categories: Is Gaia a planetary mechanism—"cybernetic" in the restricted sense—or a living being or organism, or some amalgamation of the two—that is, a cyborg? For Lovelock, Haraway writes, "the whole earth was a dynamic, self-regulating, homeostatic system; the earth, with all its interwoven layers and articulated parts, from the planet's pulsating skin through its fulminating gaseous envelopes, was itself alive" (xiii). To conceive Gaia as a veritable cyborg resolves the conceptual tensions surrounding the boundaries of life through an imagery of operational merger. "Lovelock's earth—itself a cyborg, a complex auto-poietic system that terminally blurred the boundaries among the geological, the organic, and the technological—was the natural habitat, and the launching pad, of other cyborgs" (xiii).

Haraway also noted some refraction in Lovelock's lines of vision. The first was, so to speak, horizontal—the view informed by his professional and disciplinary location within technoscience: "Lovelock's perception was that of a systems engineer gestated in the space program and the multinational energy industry and fed on the heady brew of cybernetics in the 1950s and 1960s" (xiii). And it was certainly odd that the evocation of a planetary hypothesis called *Gaia*—"named after the Greek goddess who gave birth (incestuously) to the Titans" (xii)—was the brainchild of an atmospheric chemist-cum-cyberneticist and freelance NASA and Royal Dutch Shell contractor, and "not, say, the intuition of a vegetarian feminist mystic suspicious of the cold war's military-industrial complex and its patriarchal technology" (xiii). Perpendicular to the first line of vision was the view informed by NASA's pinnacle attainment of a view of Earth from space. "In Lovelock's prescient perspective," Haraway writes, "the whole earth, a cybernetic organism, a cyborg, was not some freakish contraption of welded flesh and metal, worthy of a bad television program with a short run. As Lovelock realized, the cybernetic Gaia is, rather, what the earth looks like from the only vantage point from which she could be seen—from the outside, from above" (xiv). In chapter 7 we will explore some recent realignments of Gaian vision that counter the prior hegemony of the view from space. The orbital view of Earth is in fact not the only vantage from which to observe Gaia. However, what Haraway nails here is that the form of its communication determines the form Gaia takes for us. Gaia is an artifact of the locations of its discourse. At the moment of Gaia's inception at the end of the 1960s and for several decades thereafter, NASA-generated whole-Earth imagery was completely in the ascendant with regard to Gaia's planetary imaginary. And as Haraway developed its implications, the classic view of Gaia "from the outside, from above" went well beyond anything Lovelock himself had suggested up to that point in terms of sublime planetary vistas. For Haraway, Gaia's arrival formed the latest episode in the epic of the human species itself arriving at the cosmic threshold heralded by the annunciation of the cyborg.

Haraway's cyborg Gaia is not really a theory of planetary function or a hypothesis submitted to protocols of scientific verification. Rather, Haraway sees Gaia's ostensible figure covering over its subtext as an ideological project bearing the stigmata of its cybernetic birth. Cyborg Gaia is a potent allegorical operator, a station on the way to a distinctly virile form of the astronautical sublime rising above the womb of the world:

> Gaia is not a figure of the whole earth's self-knowledge, but of her discovery, indeed, her literal constitution, in a great travel epic. . . . The people who built the semiotic and physical technology to see Gaia *became* the global species, in which they recognized themselves, through the concrete practices by which they built their knowledge. This species depends on an evolutionary narrative technology that builds dramatically from the first embryonic tool weapon wielded by the primal hunter to the transformation of himself into the potent tool-weapon that seeds other worlds. To see Gaia, Man learns to position himself *physically* as an extraterrestrial observer looking back at his earthly womb and matrix. (xiv)

Cyborg Gaia would be a macrocosmic instance of the human microcosm envisioned by Manfred Clynes and Nathan Kline, who coined the term *cyborg* in 1960 "to refer to the enhanced man who could survive in extra-terrestrial environments. . . . Enraptured with cybernetics, they thought of cyborgs as self-regulating man-machine systems. . . . Space-bound cyborgs were like miniaturized, self-contained Gaias" (xv).[5]

Haraway's Gaia discourse is salutary in its demolition of the popular caricature of Gaia as, in the satirical formulation given by Lynn Margulis around the same time, "an Earth goddess for a cuddly, furry human environment."[6] Nevertheless, while Haraway reads Gaia as an ideological figuration of cybernetic provenance, and her text provides historical instances and contexts of cybernetics, there is no gloss on the term itself. It seems that one should already know all one needs to know about the topic. Droll metaphors consistently heighten her own references to cybernetics. Gaia is a sort of inspired cyborg hallucination induced by the addiction that had "technical and popular culture . . . shooting up with all things cybernetic in the 1950s and 60s in the U.S." (xvi). Cybernetics is figured as a mental agent that will addle your brains. It is a "heady brew"; its enthusiasts are "enraptured," and presumably, not in a good way. It would seem that Cold War technoscience was positively infested with cybernetics addicts craving the stuff to stave off the fever and shakes of too precipitous withdrawal from their systems-theory habits.

Meanwhile, Haraway withholds the same rough treatment from another systems-theoretical term at large in this text. Marginally glossed but largely unexplicated, it is *autopoiesis*. We have already been at work recovering the provenance of this term, since the concepts attached to it play a central role throughout our discussions of systems theory in relation to Gaia discourse. Three of the four appearances of the term *autopoiesis* in "Cyborgs and Symbionts" brandish a nonstandard medial hyphen: "auto-poietic." We have already seen two of them:

> Gaia—the blue- and green-hued, whole, living, self-sustaining, adaptive, auto-poietic earth. (xi)

> Lovelock's earth—itself a cyborg, a complex auto-poietic system that terminally blurred the boundaries among the geological, the organic, and the technological. (xiii)

Apart from a single instance, Lovelock himself never applied the term *autopoietic* to his descriptions of Gaia.[7] It does not appear in the Lovelock texts cited in the references of "Cyborgs and Symbionts." As we know, the basic cybernetic qualifier Lovelock consistently uses for Gaia is *homeostatic*. Regarding the text at hand, then, and especially at the time of its publication, one could have wondered about the source of the term *auto-poietic* as applied to Gaia. Haraway does not state its proximate source, but we already know the answer. She has extracted it from the texts of Margulis, "one of the formulators of the Gaia hypothesis" (xvii). "Cyborgs and Symbionts" now segues to Margulis's new account of cellular evolution based on symbiogenesis, the evolutionary assembly of different life-forms through permanent couplings of preexisting beings. In an endnote, Haraway thanks Margulis for having provided her with the manuscript of her then-forthcoming coauthored volume *What Is Life?*, "a rich exposition of the travails of the auto-poietic earth" (xvii).

In this and previous instances, Haraway leaves the definition of *autopoiesis* unarticulated until a minimal bracketed gloss appears in a passage she quotes from an earlier Margulis and Sagan text, *Origins of Sex*. "From an evolutionary point of view," Margulis and Sagan write there, "the first eukaryotes were loose confederacies of bacteria that, with continuing integration, became recognizable as protists, unicellular eukaryotic cells. . . . The earliest protists were likely to have been most like bacterial communities. . . . At first each autopoietic [self-maintaining] community member replicated its DNA, divided, and remained in contact with other members in a fairly informal manner."[8] We will return to *Origins of Sex* in chapter 6. This passage from that text coordinates the concepts of autopoiesis and symbiosis within Margulis's signature theory of symbiogenesis, the multiple endosymbiotic events in the evolutionary formation of the eukaryotic cell out of prokaryotic, or bacterial, components.

Margulis drew her presentation of autopoiesis out of crucial developments in biological systems theory that arrived in the mid 1970s, contemporaneously with the first wave of her joint Gaia publications with Lovelock. Humberto Maturana and Francisco Varela originally presented that concept as a criterion by which to distinguish living systems—minimally, prokaryotic cells—as *autopoietic* or self-producing, as opposed to mechanical, technological, or designed systems as *allopoietic* or other-producing, that is, as having the events of their existence and maintenance in being outsourced to some external agency. In other words, *the concept of autopoiesis deconstructed the original cybernetic splice between animals and machines, between living and nonliving systems.* Its virtual effect is to turn the trope of the cyborg inside out, cutting its halves apart once more through a distinction of operation. Moreover, throughout Margulis and Sagan's *What Is Life?* the concept of autopoiesis is coordinated not just with the form of life of organisms but also with the particular form of the geobiological operations of the system called Gaia. And it was this Gaian application of the term *autopoiesis* that appears to have caught Haraway's eye in Margulis's text, which she then retailed without further comment. In "Cyborgs and Symbionts," autopoiesis seems to hide in plain sight.

As later writings from *When Species Meet* to *Staying with the Trouble* attest, Haraway has gone on to contest the concept of autopoiesis.[9] Regarding her own reliance on this abjected operator, however, her recent work has struck a sort of compromise under the symptomatic name of *sympoiesis.*[10] NST brings out the precise conceptual twist that accounts for why this notion challenges the cyborg paradigm. We can look again at her crucial statement from "Cyborgs and Symbionts" in this regard: Gaia was "itself a cyborg, a complex auto-poietic system that terminally blurred the boundaries among the geological, the organic, and the technological" (xiii). Such "terminal blurring" has always been the battle cry of cyborg discourse. "Boundary breakdowns" and "leaky distinctions" are the very stuff of the postmodern, provocative, and critically productive "Cyborg Manifesto."[11] However, the intellectual liberations induced by the breakdown of "boundaries" around the human, the animal, and the machine, around the material and the semiotic, the actual and the virtual, the physical and the informatic, have condoned a lot of terminological haziness alongside the terminal blurrings.

Gaian Systems hopes to adduce some clarity to these matters of systems-theoretical description and to related nodes of conceptual distinction. The problem that autopoiesis brings into a cyborg world is precisely that it is a theory *that posits boundary production for those systems that exhibit the autopoietic form of organization and operation.* Thus, to call Gaia "a complex auto-poietic system that terminally blurred the boundaries among the geological, the organic, and the technological" is to let the technical sense of autopoiesis go by as well as to leave hanging what Margulis may have intended to convey by insisting on the characterization of Gaia as an autopoietic system. Without question, under other names of her own construction, Haraway has been a profound thinker of Gaia and an early explorer of Margulis's autopoietic Gaia conception.[12] Nonetheless, the cyborg description of Gaia blurs its boundaries as a matter of conceptual principle. The autopoietic description of Gaia has a different aim. Cyborg Gaia gives way to autopoietic Gaia when a factitious unity of system operations calls forth its own deconstruction into distinct structural couplings maintaining heterogeneity of functions. This mode of description observes material and operational distinctions among "the geological, the organic, and the technological" in order to bring out finer orders of attention in the construction of their systemic couplings and compositions.

Isabelle Stengers: Gaia the Intruder

Both Isabelle Stengers and Bruno Latour draw their figures of Gaia from the salient sources of Gaia theory, citing the Gaia hypothesis and its subsequent theory as formulated by Lovelock and Margulis. However, whereas Haraway acknowledges Gaia's first- and second-order cybernetic lineage, both Stengers and Latour take pains to extract the Gaia concept from the cybernetic nexus of its scientific origin. The discourse we have been detailing under the formal name of neocybernetic systems theory, or NST, is absent from their writings. Rather, both thinkers share a line of anachronistic separation from the classical cybernetics of the 1950s and 1960s. In a recent reprise of her own Gaia discourse, Stengers dismisses Gaia's cybernetic origins, in that they would commit her turn on Gaia too strongly to a "realist" description.[13] Nevertheless, given the "ticklish" and "touchy" responsiveness manifested in her figure of Gaia the Intruder—which is to say, in my own idiom, the autonomic movements of its cognitive reflexes—Stengers's argument traces a systemic conceptuality under erasure. Moreover, Stengers's insistence on "calling it Gaia" is perfectly sound. The entity so named possesses sufficient integrity or systemic autonomy to "hold together"—indeed, to twitch—against the irritations of *its* environment, and so, as with all things that perdure in a world in flux, deserves a proper name:

> Calling it Gaia is signifying that it is, and will remain, what can be called a "being," existing in its own terms, not in the terms crafted to reliably characterize it. It is not a living being, and not a cybernetic one either; rather it is a being demanding that we complicate the divide between life and non-life, for Gaia is gifted with its own particular way of holding together and of answering to changes forced on it

(here the charge of greenhouse gas in the atmosphere), thus breaking the general linear relation between causes and effects. (137)

Stengers favors an ontological notion of being over functional concepts of system. Nonetheless, this passage reinscribes an autopoietic conceptuality once we extend that concept to the formation of metabiotic systems.

Stengers and Latour mobilize the figure of Gaia as a mirror to humanity in its hour of need before climate change gets entirely too far gone. Both factor the Gaia concept into a redefinition of the stakes of political ecology. However, I would argue, it is not necessary to maintain that Gaia is *not* a system, or that "Gaia, the outlaw, is the anti-system," or that Lovelock's "version of the Earth System is anti-systematic," in order to stabilize a characterization of Gaia rightly taken as something other than a "unified whole."[14] Rather, what is needed is reference to adequate accounts and appropriate forms of contemporary systems theory. In any case, these authors' circumspect or emphatic circumventions of Gaia's cybernetics serve to reconstitute variant if unmarked forms of systemic descriptions. Allow the worldly agencies attributed to their Gaia figures a modicum of systemic specificity, and we may positively align the Gaia discourses of Stengers and Latour with both Lovelock's best cybernetic descriptions and, most importantly for my argument, the systems discourse that supports Margulis's autopoietic Gaia concept.

The reality of Gaia is not in question here. What is in question is the kind of entity Gaia may be said to be, and what is at stake in a given description. This book aims to trace the history of systems descriptions of Gaia while developing my own contribution to this form of theorization. Carrying forward Margulis's development of Gaia theory through Maturana and Varela, I have taken the further step of proposing an extension of autopoietic Gaia theory through NST's discourses of operational differentiation and interpenetration and Anna Henkel's suggestion regarding the "meaningful operating thing."[15] All systems, however autonomous in self-operation, are nonetheless finite and must be open to assemblage within environments that entirely exceed them, environments themselves rendered partial by the very distinction of the systems they embed. For these reasons, it important to place the biotic and metabiotic components of Gaia—its dedicated autopoiesis as bootstrapped from the sum of the biota—in relation to the abiotic thermodynamics of its cosmic environment. What counts in this metabiotic conceptuality is the emergence of operational coherence from differentiated couplings. In this regard, Gaia's bounded autonomy grounds the possibility of its yielding a coherent response to environmental provocations, a response by which Gaia has directed us toward the moral of our own responsibility.

Stengers's major statement in a Gaian vein is *In Catastrophic Times: Resisting the Coming Barbarism.* Here Stengers draws attention to Gaia's planetary autonomy, its transcendence of humanity. Industrial and extractive activities in the service of capital have provoked Gaia and produced her current "intrusion." A preliminary observation of Gaia also appeared in her earlier work *Thinking with Whitehead,* in the context of this remark: "The fact that endurance is a factual success without higher guarantee may be expressed as follows: may those who are no longer afraid that the sky might fall on their heads be all the more attentive to the eventual impatience of what they depend on."[16] This impatience of the world, its indifference regarding the duration of our human constructions, runs alongside what systems theory calls elsewhere the "autopoietic imperative," the impatience that drives living self-production altogether, as in this passage from Margulis and Sagan on "incessant self-organizing metabolism": "Autopoiesis, the chemical basis for the impatience of living beings, is never optional."[17] Such "endurance" is much more than "sustainability." It names the impatience of nonhuman worldly systems to maintain their self-production against either an entropic cosmos or a runaway rationalization of "resources" on the part of humans. *Thinking with Whitehead* then thinks Gaia in the terms of political ecology: "The new figure of Gaia indicates that it is becoming urgent to create a contrast between the earth valorized as a set of resources and the earth taken into account as a set of interdependent processes, capable of assemblages that are very different from the ones on which we depend" (163). For Stengers's Gaia, the contemporary Earth system itself, submitted to the accumulation of modern human activities, manifests its duress and resists: "Gaia is ticklish," an unidentified voice declares. "We depend on her patience, let us beware her impatience" (164). As a cosmopolitical trope or planetary figuration, this Gaia is a conceptual resource for cultural resistance to globalism's ecological violence.

In Catastrophic Times responds to capitalist globalization in its culpability for the Anthropocenic state of affairs—in her words, "the new grand narrative in which Man becomes conscious of the fact that his activities transform the earth at the global scale of geology, and that he must therefore take responsibility for the future of the planet" (9). Of course, "Man here"—with a capital M—"is a troubling abstraction. The moment when this Man will be called on to mobilize in order to 'save the planet,' with all the technoscientific resources that will be 'unhappily necessary,' is not far off" (9–10). The "coming barbarism" of Stengers's subtitle captures her concern to resist the violence and destruction that Earth and its inhabitants have in store if the "Man" of the Anthropocene turns his technoscientific prowess to "saving the planet" simply for the sake of the continuation of capitalist regimes of extraction and private accumulation. But that is the future predictable from the course of globalization under capitalist expansion, the version of history that "has economic growth for its arrow of time" (17).

Facing this unsustainable but nonetheless seemingly inevitable course of further global development, Stengers contemplates a second, alternative history. She takes a snapshot of the Gaia of Lovelock and Margulis showing the coherent planetary face of the climate crisis and calls it "the Intruder," an unpredicted, unpredictable, and rapidly evolving counterphenomenon that all contemporary living beings must abide as the horizon of their possibility. "The intrusion of Gaia" manifests itself precisely as the current global environmental crisis provoked by human activity. Yet Gaia's intrusion upon the first history founds a second history and transforms the narrative of resistance to the former, increasingly presumptuous and delusory narrative of the triumphal capitalist conquest of Nature for the benefit of Man. Stengers details her discursive gambit:

> Naming Gaia and characterizing the looming disasters as an intrusion arises from a pragmatic operation. *To name is not to say what is true but to confer on what is named the power to make us feel and think in the mode that the name calls for.* . . . Naming Gaia as "the one who intrudes" is also to characterize her as blind to the damage she causes, in the manner of everything that intrudes. . . . Gaia is neither Earth "in the concrete" nor is it she who is named and invoked when it is a matter of affirming and of making our connection to this Earth

felt, of provoking a sense of belonging where separation has been predominant, and of drawing resources for living, struggling, feeling, and thinking from this belonging. It is a matter here of thinking *intrusion, not belonging.* (43–44)

Counter to more typical wishful appropriations of the figure in the mode of maternal return, Stengers's intrusive Gaia does not beckon back to the lost garden, but neither is it, as Lovelock sets forth on occasion, a vengeful and punitive Mother Nature bent on dispensing chastisements. Figuring Gaia as the Intruder enforces the thought of humanity's *distinction from* Gaia, even as we necessarily participate in the effects of a phenomenon to which we both do and do not belong.

The conceptual innovation in Stengers's discourse is to develop the figure of Gaia not in the mode of immanence—as, for instance, one finds depicted in Eywa, the Gaia figure in the planetary imaginary of James Cameron's *Avatar*—but in the mode of a kind of mundane transcendence. Stengers's Gaia

> makes the epic versions of human history, in which Man, standing up on his hind legs and learning to decipher the laws of nature, understands that he is the master of his own fate, free of any transcendence, look rather old. Gaia is the name of an unprecedented or forgotten form of transcendence: a transcendence deprived of the noble qualities that would allow it to be invoked as an arbiter, guarantor, or resource; a ticklish assemblage of forces that are indifferent to our reasons and our projects. (47)

Stengers vets the intrusion upon human presumption of a transcendent yet desublimated Gaia in order to demarcate its opponent, capitalism, as a different and opposed form of globalized transcendence, one that is all too human and yet, for the moment if not for all time, treated as a veritable force of nature beyond human controls: "Capitalism does, in effect, have something transcendent about it, but not in the sense of the laws of nature. Nor in the sense that I have associated with Gaia either, which is most certainly implacable, but in a mode that I would call properly materialist. . . . Capitalism's mode of transcendence *is not implacable, just radically irresponsible,* incapable of answering for anything" (52–53).

Gaia appears to Stengers as the most appropriate planetary figure to oppose to capitalism under globalization, because Gaia is both commensurate in scale and comparable in its autonomous relation to the minutiae of its environment, such as individual living beings taken one by one, including us humans. Gaia transcends humanity in its planetary response to the environmental provocations produced by as-yet-unchecked human activities of industrial and extractive processes in the service of the amplification of capital:

> That I have been led to characterize both the assemblage of coupled material processes that I named Gaia and the regime of economic functioning that Marx named capitalism by a mode of transcendence highlights the particularity of our epoch, that is to say, the global character of the questions to which they oblige us in both cases. The contemporaneity of these two modes of transcendence is evidently no accident: the brutality of the intrusion of Gaia corresponds to the brutality of what has provoked her, that of a development that is blind to its consequences, or which, more precisely, only takes its consequences into account from the point of view of the new sources of profit they can bring about. (53)

Stengers's Gaia presents a deliberate cosmopolitical trope in which the resistance of the Earth system itself and the manifestation of its duress when submitted to the modern accumulation of human activities provide the deep ground for a figuration of possible and necessary cultural resistance to capitalism's irresponsibility.

Bruno Latour's Gaia Theory

Latour's explicit treatments of the Gaia concept go back at least to *Politics of Nature.* His first mention of Gaia in this earlier text occurs in a riff on the vexed cultural politics of a prior ecological discourse built on the shaky opposition of "nature" and "society." He notes "how much difficulty ecology movements have always had finding a place on the political chessboard. On the right? The left? The far right? The far left? Neither right nor left? Elsewhere, in government? Nowhere, in utopia? Above, in technocracy? Below, in a return to the sources of wisdom? Beyond, in full self-realization? Everywhere, as the lovely Gaia hypothesis suggests, positing an Earth that would bring all ecosystems together in a single integrated organism?"[18] This "lovely Gaia hypothesis" would seem to be disparaged for sharing in the general vexation of the impasses of ecological praxis based on holistic models. In fact, what Latour's locution identifies is the caricature those wider vexations work on the specificity of Lovelock's Gaia. When this Gaia reappears toward the end of this text, Latour gives Lovelock his due. Older ideas of Mother Earth fall away along with the traditional hierarchies mobilized by the Western discourse of Nature: "On the contrary, we can benefit from the fundamental discovery of the ecology movement: no one knows what an environment can do; no one can define in advance what a human being is, detached from what makes him be" (197). To this statement Latour places an endnote that posits Lovelock's model of Gaia's ecological becoming as a way past prior politics of nature:

> Lovelock, the inventor of the Gaia *hypothesis,* is quite careful, moreover, not to make this an already constituted totality. His books lay out the progressive composition of the links between scientific disciplines, each charged with a sector of the planet and gradually discovering with surprise that they can define one another mutually.... By forcing the issue, we can say that Lovelock's Gaia is the complete opposite of nature, and that it bears closer resemblance to a Parliament of disciplines.[19]

Latour's evocations of Gaia in *Politics of Nature* already cite the problem of totality. *Gaian Systems* shares this crucial concern with Latour's critique—the propensity of the Gaia concept to fall under simplistic holistic descriptions, such as being "in essence" a "single integrated organism." This is the general form of the metaphor for Gaia that most often short-circuits more careful and differentiated approaches to its systemic complexity. But in this 2004 work Latour also indicated Lovelock's care in his writings to undercut the construction of Gaia as an "already constituted totality." In Latour's description, Lovelock's text holds off conceptual dynamics that could otherwise undo Gaia's utility for political ecology, specifically the problem of "premature unification" of totalities at the expense of their networked heterogeneities.[20] Latour's more recent face-to-face encounters with the Gaia concept continue to air these precise issues. The political as well as ontological problem remains the one named by his recent Gaia paper, "Why Gaia Is Not a God of Totality."

Latour's most extensive treatment of Gaia to date is *Facing Gaia: Eight Lectures on the New Climatic Regime.* John Tresch offers a neat summation of *Facing Gaia*'s virtues and trials: it "marks significant new turns in Latour's bold, generous, and tirelessly creative philosophical project. The book's challenges mimic those of the situation it seeks to describe: just as climate change unleashes unpredictable, tangled, contradictory forces that we struggle even to observe, readers may stumble as Latour races to catch Gaia—or at least to tag it."[21] *Facing Gaia* was first developed as the Gifford Lectures delivered in Edinburgh in 2013 for a venerable lecture series endowed with the hoary Victorian theme of "natural religion." Latour has made Gaia the occasion for a new contemplation and critique of the relations between science, religion, and politics. To that end, he declares his own religious sympathies, describing himself as one who respects "people who sing the glory of God as much as people who celebrate the objectivity of the sciences" (171). The irony in this self-description is simply that for Latour, the modes of being of the separate objects of both chants, God's glory and the objective world, depend in the final instance on the efficacy of the chants. That is, for Latour a belief in "the objectivity of the sciences" is essentially a form of religious conviction: as he put it in the Gifford Lectures, "Even though you claim that your entity isn't a god, it doesn't mean you don't belong to a religion."[22]

Latour's larger point and philosophical refrain is that scientific practices are as roundly political as those of any religious institution. Moreover, Western modernity's persistent confusion of God with Nature, as in the phrase "natural religion," has always undermined modern science's self-profession of its strictly secular status. In the case at hand, this very confusion has always also complicated the reception of the Gaia hypothesis as a scientific idea. And yet, regarding Gaia, a "finally worldly, secular" figure for nature's materiality, "the paradox of the figure that we are attempting to confront is that the name of a proteiform, monstrous, shameless, primitive goddess has been given to what is probably the *least religious entity* produced by Western science. If the adjective 'secular' signifies 'implying no external cause and no spiritual foundation,' and thus 'belonging wholly to this world,' then Lovelock's intuition may be called *wholly secular*."[23] Latour's discourse explains Gaia's cultural propensities for mythic digressions in order to set Lovelock's Gaia apart from them. But to do this effectively, to finally demarcate a worldly, secular Gaia, it takes the sort of discourse of political theology Latour brings to the occasion to provide the conceptual counterpoint needed to hold the distinction in place. Latour's own nonmodern audacity in declaring the form of his own religious commitment lends authority to his project for a secular Gaia.

Latour's secular Gaia enjoins participatory rather than alienated stances of observation that acknowledge Earth's cosmic centrality as, for all we yet know, the sole unmistakably living planet:

> Lovelock brought his reader down to what should be viewed once again as a *sublunary world*. Not that the Earth lacked perfection, quite the contrary; not that it hid the somber site of Hell in its entrails; but because it held—alone?—the privilege of being in disequilibrium, which also meant that it possessed a certain way of being *corruptible*—or, to use the terms of the previous lecture, of being, in one form or another, *animated*.
>
> In any case, it seems capable of actively maintaining a difference between its inside and its outside. It has something like a skin, an envelope. (*Facing Gaia: Six Lectures*, 78)

Although Latour does not put it this way, his phrasing here reprises the definition of an autopoietic system.[24] In an autopoietic conception, Gaia's planetary envelope results from the active production of an operational boundary separating "its inside and its outside," separating Gaia's sheer immunitary sphere of habitability and material viability from the abiotic complexities of its terrestrial and cosmic environments. There is already general agreement that at the profound base of Gaia's biotic operations are the original riders on the planetary storm, the minutely differentiated profusion and globally dispersed metabolic residues of our evolutionary ancestors, the microbes. Latour underlines our task of recognizing the biotic animation of the Gaian microcosm by analogy with the late nineteenth century's astonishment at Pasteur's revelation of the existence and consequentiality of humble bacteria and recondite fungi.

For Lovelock, everything that is located between the top of the upper atmosphere and the bottom of the sedimentary rock formations— what biochemists aptly call the critical zone—turns out to be caught up in the same seething broth. The Earth's behavior is inexplicable without the addition of the work accomplished by living organisms, just as fermentation, for Pasteur, cannot be started without yeast. Just as the action of micro-organisms, in the nineteenth century, agitated beer, wine, vinegar, milk, and epidemics, from now on the incessant action of organisms succeeds in setting in motion air, water, soil, and, proceeding from one thing to another, the entire climate. (*Facing Gaia: Eight Lectures,* 93)

Secular Gaia anchors what we can call Latour's Gaian biopolitics. In *Politics of Nature,* as we saw, Latour stated: "No one can define in advance what a human being is, detached from what makes him be" (197). With that ground of possibility now identified as Gaia, in the Gifford Lectures Latour asked flat-out: "What sort of *political animals* do humans become when their bodies are to be coupled with an animated Gaia?"[25] Here would be gathered a posthumanist cohort of cognizant Gaian beings. Latour asks: of what political ecology would they be capable? NST would ask first about cognitive capacity: how would they experience and communicate such Gaian couplings? What Latour consistently calls Gaia's "animation," as in "being, in one form or another, *animated,*" I prefer to render through a systems-theoretical vocabulary as *cognition,* insofar as "actively maintaining a difference between its inside and its outside" means that the system at hand possesses sufficient operational closure to maintain that very distinction and conduct itself accordingly. Perhaps due to his larger polemics contra "natural religion," Latour prefers an idiom that reinstates a notion of *anima* banished by demystified scientific modernity. I would hold that the notion of *system* is the preferable term for a fully secular description of Gaia.

Within a Gaian biopolitics, human beings find themselves returned to a horizontal relation with and within an animate or cognitive geobiosphere:

What is *moving* in Lovelock's prose (and even more in that of his sidekick Lynn Margulis [1938–2011]) is that every element that we ignorant readers would have seen as part of the *background* of the majestic cycles of nature, against which human history had always stood out, becomes active and mobile thanks to the introduction of new invisible characters capable of reversing the order and the hierarchy of the agents. We knew that a substantial part of any mountain formation consists in the debris of living beings, but perhaps the same thing holds true for the cloud layer, manipulated by marine micro-organisms. Even the slow movement of tectonic plates might have been triggered by the weight of sedimentary rocks. (*Facing Gaia: Eight Lectures,* 92–93)

However, Margulis was more than Sancho Panza to Lovelock's Don Quixote. It was one of Margulis's particular contributions to underscore how microbial mobility is Gaian "animation" at the primeval level of sheer living motions. The "new invisible characters" name the composite agency of the numberless infinitesimal microbes perfused through Gaia's planetary envelope. These legions literally gather the clouds in the sky by the metabolic production of chemical products that, once airborne, seed cloud formation. Moreover, by the reproduction of their living generations yielding eons of accumulating remains, they literally move the mountains upon the tectonic plates.[26]

Later in *Facing Gaia: Eight Lectures,* Latour coordinates his presentation of Gaian animation with Stengers's description of Gaia as "a ticklish assemblage of forces." Both thinkers underscore an extension of worldly agency, one that manifests itself not only by sheer material dynamism but also by self-referential sentience, what I will call planetary cognition. Latour describes "what it means to live in the Anthropocene" as the recognition of a

"sensitivity" . . . applied to all the actors capable of spreading their sensors a little farther and making others feel that the consequences of their actions are going to fall back on them, come to haunt them. When the dictionary defines "sensitive" as "something that detects or reacts rapidly to small changes, signals, or influences," the adjective applies to Gaia as well as to the Anthropos—but only if it is equipped with enough sensors to feel the retroactions. Isabelle Stengers often says of Gaia that it is a power that has become "touchy." (*Facing Gaia: Eight Lectures,* 141)

A touchy Gaia may be prone to seizures it will be ill to suffer. If Gaia in the epoch of the Anthropocene "seems to be excessively sensitive to our actions," then we should "learn to become sensitive in turn to these multiple, controversial, mutually entangled loops" (141).

GAIA'S PERILOUS PATH

The diverse disciplinary and cultural uptake of the Gaia hypothesis is a textbook case of science in action, Latour's stock-in-trade. More importantly, in the full accounting of its theory in process, the complex of Gaian ideas developed by Lovelock and Margulis aligns with Latour's own philosophy of nonmodernity and its redistribution of natural and social agencies, its worldly sociology of quasi-objects and quasi-subjects. His success in moving informed Gaia discussion into new precincts of scholarly conversation has certainly been a welcome development for Gaian thought. Latour's making Gaia a matter of intellectual concern has been perhaps the most robust response to its concept among theoretical observers of his echelon. Over the last two decades, Latour has welcomed the wandering, at times outcast hypothesis and body of theory regarding Gaia into his intellectual polity, where it has enjoyed more hospitable entertainment and a more expansive writ of understanding than in many other realms of disciplinary thought.

How does the Gaia concept takes shape when Latour narrates its modes of existence through the analytic of actor-network theory (ANT)? How does that description align with Gaia theory in Lovelock's and Margulis's own presentations, the fullness of which makes manifest a wide range of systems-theory discourse? Are the separate conceptual goods of ANT, first-order cybernetics, and NST mutually exclusive, positively complementary, or what, precisely? As we have seen, Haraway, Stengers, and Latour submit the conceptual themes of cybernetics and systems theories to variously stringent rhetorics of dismissal. For instance, in the first section we noted how Haraway's "Cyborgs and Symbionts" admonished the same cybernetic discourses from which her own cyborg trope descended. But thanks to its recessive Margulis subtext, in that particular article Haraway took the term *autopoiesis* off her polemical hook. In *Facing Gaia: Eight Lectures,* that term is never mentioned. Niklas Luhmann's social systems theory does makes brief appearances in the footnotes of Latour's *Reassembling the Social*: "Luhmann's masterly attempt at respecting [social] differences through the notion of autonomous spheres was unfortunately wasted because he insisted on describing all the spheres through the common meta-language borrowed from a simplified version of biology" (241n338). Whereas Latour may write off the concept of autopoiesis as a "simplified version of biology," it would be more generous to consider it as a crucial *supplement* to standard descriptions of living systems that do not sufficiently observe the profound implications of operational closure as the basal precondition for life's self-maintenance from the cell onward.[27]

If Haraway's version of Gaia in 1995 is an ideological allegory of Earth's cyborgization, Latour's reading of Gaia is very much a semiotic and narrative entity in symmetrical relation to its nondiscursive or worldly modes of being. In this vein, Latour declares that the very name of Gaia transfers to Lovelock and Margulis's scientific hypothesis the archaic curse derived from its mythological namesake. In the previous chapter I suggested that the name of Gaia is a consequential rhetorical mediation rather than a concrete denotation. Perhaps the myth of Gaia "cursed" the *reception* of Gaia theory, with effects of daemonic enthusiasm that have at times been not unproductive. But that set of connections is matter for another history.[28] My point here is that the mythic lineaments of the archaic goddess Gaia have transferred almost nothing of *conceptual* substance to the *theory* of Gaia. But in this regard Latour's attitude is anything but nominal. *Facing Gaia: Eight Lectures* affirms that "existence and signification are synonyms" (70). This semiotic or virtual realism, in which "every possibility of discourse is due to the presence of agents in quest of their existence" (70), would also seem to inform the discursive strategy by which he shapes the figure of Gaia out of a selection from Lovelock's texts and then specifies the particular modes of Gaia's existence through his reading of a tightly edited set of significations.

Both Haraway and Latour approach Gaia discourse with certain epical inflections. In Haraway's version the questing protagonist is a collective persona, "the members of a voraciously energy-consuming, space-faring hominid culture that called itself Mankind." Latour's hero is a singular personage: James Lovelock.[29] Latour situates the Gaia hypothesis and its primary advocates in the midst of a Homeric trial of strength developed primarily with Lovelock as Odysseus and himself as Athena. For instance, Latour bids to defend Lovelock's Gaia theory from the contemporary curse of what he deems to be, unlike the letter of Hesiod's *Theogony*, a *false* myth. Now, he himself "could easily escape the curse" attached to the name of Gaia, Latour announces, "by claiming that the name of a theory is of no importance, and that, after all, serious scientists avoid the name Gaia as much as possible, preferring the euphemism 'sciences of the Earth System.' But this would be cheating; it would amount to passing from one ambiguous character to another that is even harder to define. 'System'? What weird animal is that? A Titan? A Cyclops? Some twisted divinity? By avoiding the real myth, we would land on a false one" (*Facing Gaia: Eight Lectures*, 85).

This passage is remarkable in several ways. Instead of the standard Latourian device of distributing agencies, it puts the references to "Earth" and "the sciences" off to one side in order to pick "System" out of the lineup as the perpetrator of false mythology. However, there is some anachronism in Latour's witticism. He draws this "System" signifier specifically from Gaia theory's later rebranding as Earth system science. This phase in the evolution of Gaia arose nearly three decades after Lovelock and Margulis initiated their own discourses precisely under the sign of systems theory and in the idiom of cybernetic developments. In chapter 9 we will measure Earth system science as a serious contender for conceptual preeminence, but nonetheless, it is a legitimate if at time delinquent offspring of the Gaian inspiration. In any event, in the text at hand Latour's playful evocation of the system monster as a "twisted divinity" is a foretaste of similar treatment in store for Lovelock's own classical cybernetic vocabulary.

Like the wily Odysseus passing unscathed by Scylla and Charybdis, Latour's Gaia hero must occasionally negotiate perilous pathways and sail safely through conceptual reefs on which his theory would otherwise be shipwrecked. For instance, Latour asks, "How did Lovelock manage to retrace the path between the twin pitfalls of reductionism and vitalism?" (*Facing Gaia: Eight Lectures*, 94). Elsewhere in *Facing Gaia* these opposed perils are termed *deanimation*—the rationalist reduction of the world to inert and indifferent material bits and pieces pushed around by physical forces—and *overanimation*—the ensouling of the world with one or another universal spirit holding sway over the Whole. For Latour, the greater danger for Lovelock is the latter peril,

> since the most common definition of the Gaia theory is that Gaia acts as *a single, unique coordinating agent*. Gaia would be the planet

Earth considered as a living organism. This is often the way Lovelock presented his discovery:

> Gaia is the planetary life *system* that *includes* everything influenced by and influencing the biota. The Gaia system shares with all living organisms the capacity for *homeostasis*—the regulation of the physical and chemical environment at a *level* that is favorable for life.

"System," "homeostasis," "regulation," "favorable levels," these are all quite treacherous terms.[30]

"Treacherous"—how so? For readers familiar with Lovelock's and Margulis's several texts, and as Latour also acknowledges, Lovelock's vocabulary in this passage retails standard terms from the technical discussion of the Gaia hypothesis, consistently evoked since the 1970s. But granting Latour's idiom for the moment, one could say that Lovelock would also remain fond on occasion of deploying the equally if not more treacherous metaphor of "organism" despite Margulis's long-standing apprehension that it detracted from the proper *systemic* understanding of the Gaia concept!

However, Latour's rhetorical rationale is clear enough: this invective inscribes these terms on the list of perils for the Gaia hero to overcome. Even though Lovelock used this cybernetic vocabulary—despite its "treacherous" unreliability, or risk of collapsing, or overall hazardousness—he nevertheless had resources at hand, gained over a hard course of trials and errors, perhaps, but in any event ready to hold off its danger. Lovelock's greatness is in part to have threaded the narrow way to *animation without holism,* to "effects of *connection . . .* without taking the Totality route": "The whole originality—and it's true, I recognize it—the whole difficulty—of Lovelock's enterprise is that he plunges head first into an impossible question: how to obtain effects of *connection* among agencies without relying on an untenable conception *of the whole. . . .* If he contradicts himself, it is because he is fighting with all his might to avoid the two pitfalls while trying to trace the connections without taking the Totality route" (*Facing Gaia: Eight Lectures,* 97). Here is the quest and accomplishment of Lovelock's Gaia odyssey in Latour's critical fabulation: to have tasted systems theory without succumbing to its duplicity. Lovelock listened to the Sirens' song of cybernetics and yet sailed on, fit for further battles with the dervishes of deep ecology and the gnomes of neo-Darwinism.

In the immediate continuation of the passage quoted above, Latour's serious charge regarding the danger in Lovelock's cybernetic idiom is indicated at first through a tremendously clarifying rhetorical question that cuts to the theoretical chase. If "Gaia is the planetary life system that includes everything influenced by and influencing the biota," as Lovelock states, "is there then," Latour asks, "a superior order in addition to living organisms?" (94). In other words, shall Gaia be described as an emergent system over and above the elements that it gathers together, or, on the contrary, as adhering to a relational ontology in which all "influences" in any direction are propagated on the same level? This issue animates Latour's entire intervention into Gaia discourse. As he reminds his reader in a nearby footnote, the "refusal to conceptualize organization on two levels is the fundamental tenet of the actor-network theory" (95n64). And what he reads in Lovelock's text is its author's struggle to extricate himself and to liberate Gaia from the holistic constructions that Latour considers to be the sole possible outcome of the cybernetic mode of its conceptuality. Latour's text constrains cybernetic discourse to one or another dialect of the holistic juggling of parts and wholes. It reads systemic unity as false totality. But there is no conceptual inevitability to these outcomes. On the contrary, the strongest cybernetic formulations in both first- and second-order systems theories place multiplicities, differences, and distinctions before or at least alongside oneness, wholeness, and totality. In its best discourses, contemporary Gaia theory has put its earlier organicist tendencies and biotic biases in proper proportion by following a neocybernetic course distinguishing the universalizing indistinctions of organicism and holism into discrete realms of system differentiation.

Facing Gaia documents that Latour's treatment of systems theory is to a considerable degree displaced from the rich history of the *sociological* abuse of organicist models of social organization and of untenably totalizing models of technological objects. That antipathy would then be transferred to the way that the foundational biotic components of systems thinking—the explicitly politicized organicism of earlier holism, perhaps, or the cybernetic transfer of homeostasis from physiology to control engineering, or the cellular grounding of autopoietic systems theory—may lend themselves to comparable kinds of intellectual and critical overreach in an organicist key. In the fullness of Latour's career, that habitual anti-organismal polemic in relation to the sociotechnical milieu has now been transferred to and put to the test of the Gaian instance. However, unlike the sociotechnical milieu of humans and their machines, at least half of Gaia's intricate yet planetary instance is an organismal affair of living activities! In other words, Latour's approach to Gaia flips the script of classical sociology. Instead of subjecting the analysis of society to a patently inappropriate organicist schema of organization, Latour views the worldly phenomenon of Gaia through the sociotechnical methodology of networked actors. So why is his gambit any less inappropriate than the holistic methodology it rejects? A lot of the argumentation in *Facing Gaia* is implicitly devoted to the following construction of this issue: once we deconstruct the notion of polar opposition in the modern binary of nature and society, we make available a level of symmetry in a world constituted by a vast legion of actors. The actor-network theorization of Gaia addresses itself to this heterogeneity of agencies.

My own analytical scheme roughly parallels Latour's in that, by applying concepts drawn from NST to Gaia theory, I am also loosening the distinction between social and natural theorizations. The crucial difference, of course, is that NST avails itself of a metalanguage designed to negotiate those very ontological divisions. The prime example here is the concept of autopoiesis. NST works out key modifications allowing that concept metabiotic range to cross the distinction between psychic and social systems. Meanwhile, also first coming to a head in the 1980s, a separate biotic discourse of autopoietic Gaia emerges in the work of Margulis. My own move, then, is to leverage a metabiotic construction of autopoiesis back to its own biotic base in the autopoiesis of living systems. Let us go on now to compare this reverse engineering of autopoietic conceptuality to the Gaian system to Latour's own theoretical intervention. Staging a confrontation between ANT and the first-order cybernetic side of systems theory, *Facing Gaia* primes Latour's Gaia discourse for this very interaction with NST—the specific conceptual moves from homeostatic to autopoietic to metabiotic instantiations that have also absolved Gaia from theorization as a self-absorbing whole overriding its parts. Rather, in the metabiotic description drawn from this line of theory, the Gaian system forestalls the holistic collapse of the system-environment distinction by multiplying indefinitely the subsystems with which it produces and maintains the planetary environment.[31]

A FINE MUDDLE

Picking up *Facing Gaia* where we left that text, Latour had just articulated the issue of whether Lovelock meant, by cybernetic terms as *system, homeostasis,* and *regulation,* to imply that there is in fact "a superior order in addition to living organisms" (94). As much as Gaia could appear to transcend its elements as a system that exerts some regulative control over that ensemble, Lovelock "fights to keep anyone from entrusting all the agencies he has detected to a new, higher level, that of the totality" (95). A historical gloss explains this phase of narrative complication in the unfolding of his Gaia plot:

> To understand why [Lovelock] has so much trouble expressing himself, we have to remember that sociology and biology have continually exchanged their metaphors, and that it is therefore extremely difficult to invent a new solution to the problem of organization. All the sciences, natural or social, are haunted by the specter of the "organism," which always becomes, more or less surreptitiously, a *"superorganism"*—that is, a dispatcher to whom the task—or rather the holy mystery—of successfully coordinating the various parts is attributed. (95)

With this latter figurative flourish regarding "holy mystery," Latour implies that lurking behind the "specter of the 'organism'" is the "God of Totality" that is written everywhere between the lines of the modern constitution that ostensibly separates sociological from biological matters while promoting all manner of illicit metaphorical traffic between them. Latour's religious vehicle indicates that the notion of a "superior order" derives not only from its ostensibly systemic occasion but also from the semantic field of theological distinctions. Vertical metaphors of ontological hierarchy drawn from the schemas of classical Christian cosmology seem to overflow into the discourse of emergent levels in systems theory! While notions of a "superior order" and "higher level" may occur in the discourse of systems, whether they are complicit in smuggling totalistic ordinations into the matters at hand or simply pertain to mundane distinctions in organizational structure would have to be

determined case by case. My own concept of metabiotic Gaia is a refined form of Lovelock's "coupled system" processed through Margulis's
autopoietic Gaia concept and positing not a "superior" but a more intricate and composite systemic assemblage, adjacent and "in addition to
living organisms." And on the scene of supposedly secular discourses,
such vertical figures are far more commonly abused in vulgar notions
of *evolutionary* "superiority," as in the declaration that this or that species, race, or ethnicity is more "highly" evolved than another, along with
the concomitant and historically lethal consequences of the transfer of
such defective Darwinisms to social thought and political programs.

"Now the problem Lovelock saw very well," Latour immediately continues, "is that, in the literal sense, in the objects that he studied, *there
are neither parts nor a whole*" (95). Latour's proposition follows after and
parallels other postholistic constructions of new solutions to "the problem of organization" that have occurred in complexity theory, in assemblage theory, as well as in the discourse of NST and the postholistic
theory of system differentiation. These considerations temper Latour's
suggestions that Lovelock's alleged quest for a nonholistic form of the
Gaian system was somehow unique to his own project or that the various research fronts in systems theory concurrent with Lovelock and
Margulis's Gaia theorizing were just going down the same old "Totality
route" in a fashion that offered Gaia's theorists nothing of use.[32] Margulis
in particular would take the concept of autopoiesis into her own Gaia
theorizing in a significant way, and Lovelock himself would never jet-
tison a robustly cybernetic vocabulary, however nuanced it may have
become or however it may have been supplemented by other developments in the larger domain of systems discourse.

Latour's Gaia discourse, then, must constantly fend off the inveter-
ate systems semantics of Lovelock's own Gaian formulations by granting his protagonist resources of rhetorical indirection and a sensitivity "to the tropisms of prose" (132) that are surely more Latour's than
Lovelock's. For instance: "Lovelock does talk about a control system, but
he goes on to be immediately suspicious of the perilous connotations
that the technological metaphor would bring with it" (132). Lovelock
certainly notes the technological purchase of the cybernetic concepts
he brings to bear on Gaia as a "living system," but he generally does so
to exploit them for his Gaian point rather than hold them suspiciously
at arm's length. Or again, we can note Latour's own rhetorical resources

when he draws a contrast with an Earth system scientist whom he describes as terminally obtuse to Lovelock's actual quest,

> seeking to capture, in the shifting of his convoluted prose, something that is seeking its path, like life on earth itself: something that produces order downstream yet that does not depend on a pre-established order upstream. The Gaia theory comes from an *inventor* talking about an *invention* that is difficult to describe.

>> The nearest I can reach is to say that Gaia is an evolving system, a system made up from all living things and their surface environment, the oceans, the atmosphere, and crustal rocks, the two parts tightly coupled and indivisible. It is an "emergent domain"—a system that has emerged from the reciprocal evolution of organisms and their environment over the eons of life on Earth. In this system, the self-regulation of climate and chemical composition are entirely automatic. Self-regulation emerges as the system evolves. No foresight, planning or teleology . . . are involved.

> It would be hard to be clearer about the absence of Providence.[33]

Lovelock is glimpsed traversing conceptual and metaphorical dangers and complications in quest of a vision "seeking its path, like life on earth itself." What is striking indeed about this moment of Gaian rhapsody is the way its motif frees its author's own romanticism to wax vitalistic, just for a light moment with seemingly no lasting conceptual weight, and as referred to Lovelock in motion across Gaia's body in search of the source and form of all living sustenance. But for all that, one can still read Lovelock's text in this passage as not convoluted but limpid as the day is long and meaning precisely what it says. Here is one of Lovelock's mature, post- or metabiotic statements of Gaia theory in which its living and nonliving components constitute what Luhmann calls "a differentiated system . . . composed of a relatively large number of operationally employable system/environment differences."[34]

Moreover, as an "evolving system," Lovelock's Gaia in this passage is also an "emergent domain," out of which "self-regulation emerges," and yet, Latour duly notes, with no providential dispatcher or holy mystery necessary to account for what can also be taken to be the effective outcome of the Gaian system's operational closure. In chapter 5 we will learn more about how this later formulation of Gaia theory was modulated by Lovelock's attendance at the 1988 Lindisfarne Fellows meeting, where, in keeping with the symposium theme of emergence, he delivered a paper titled "Gaia: A Planetary Emergent Phenomenon."[35] Latour himself gets with Lovelock's Gaia program by providing his own formulation for systemic emergence. As the production of "order downstream yet that does not depend on a pre-established order upstream," however, it comes with discernible echoes of Michel Serres's uptake of Henri Atlan's biocybernetic theorization of living systems.[36] Latour accepts Lovelock's affirmation of Gaian emergence, but a systemic accounting for this phenomenon is disbarred by ANT's flat scenography and the preordained immanence of the actors. Still, what does "a superior order in addition to living organisms" come down to in the Gaian instance if not the metabiotic domain of "a system that has emerged from the reciprocal evolution of organisms and their environment"?

Latour's own depiction of Gaia is also notable. Parts of the passage immediately below read like the opening phase of a systems description that then stops without pressing its elements toward systematicity. For instance, it is "not that Gaia possesses some sort of 'great sensitive soul,' but that the concept of Gaia captures the distributed intentionality of all the agents, each of which modifies its surroundings for its own purposes" (98). Yes, but we could now consider how the aggregation of these modifications produces a transformed environment ready to modulate the agents of its own transformation. Or again, "animation is the essential phenomenon," and "animation is immediately propagated at all points" (69, 99)—full stop. The concept of ubiquitous (totalized?) animation preserves the granular distribution of "all the agents," which are conceived to "propagate" all on their own, in their own way, and with no higher-order meddling in their affairs. As we will note in a moment, by "propagation" Latour must mean a spreading-out of influence rather than a mere reproduction of entities. The distribution of all the animated actors composes Gaia as the final horizon of their network:

> With Gaia, Lovelock is asking us to believe not in a single Providence, but in as many Providences as there are organisms on Earth. By generalizing Providence to each agent, he insures that the interests and profits of each actor will be *countered* by numerous other programs. The very idea of Providence is blurred, pixelated, and finally fades away. The simple result of such a distribution of final causes is not

the emergence of a supreme Final Cause, but a fine muddle. This *muddle* is Gaia. (100)

No systemic emergence here. No presumption of "superior orders" or "higher-level" regulative processes. As sociopolitical allegory of "interests and profits," this is a deregulatory regime. With regard to variant formulations of *Gaia politique,* however, some other Gaian figures lean more toward regimes of regulated organization with distributed benefits. In the annals of Gaia discourse, systems aggregations often take on forms of solidarity. In Lovelock's trade unionism of the biota, for instance, we humans have an invitation to join with the biotic proletariat to work out socialized accommodations for the communal distribution of Gaian labor and wealth: "We are not managers or masters of the Earth," Lovelock writes regarding humans in the concluding remarks of *Gaia: The Practical Science of Planetary Medicine,*

> we are just shop stewards, workers chosen, because of our intelligence, as representatives of the others, the rest of life of our planet. Our union represents the bacteria, the fungi, and the slime moulds as well as the nouveau riche fish, birds, and animals and the landed establishment of noble trees and their lesser plants. Indeed, all living things are members of our union and they are angry at the diabolical liberties taken with their planet and their lives. (186)

Or again, for a Gaian biopolitics we could ally ourselves with Margulis's internationalist symbiotic communism of the microbes. Wherever they coalesce, these numberless living entities form mutual-aid colonies, have almost no conception of individual prerogatives, and rise to a crucial worldwide microcosm that owns the primary means of trophic production and maintenance of planetary viability. We had previously assumed for humanity global impunity for our exploitation of other living things as worldly resources. Nonetheless, "I hear our nonhuman brethren snickering," Margulis writes in the last paragraph of *Symbiotic Planet*:

> "Got along without you before I met you, gonna get along without you now," they sing about us in harmony. Most of them, the microbes, the whales, the insects, the seed plants, and the birds, are still singing. The tropical forest trees are humming to themselves, waiting for us

to finish our arrogant logging so they can get back to their business of growth as usual. And they will continue their cacophonies and harmonies long after we are gone. (128)

INSIDES AND OUTSIDES

Following his articulation of Gaia as a nonsystemic muddle, Latour draws in Margulis alongside Lovelock as a fellow traveler with his Gaia constructions. He reads Margulis's discourse of symbiogenesis as confirming a state of nondifferentiation between organisms and their environment—specifically, hosts and their symbionts: "Properly speaking, for Lovelock, and even more clearly for Lynn Margulis, there is no longer any environment to which one might adapt. Since all living agents follow their own intentions all along, modifying their neighbors as much as possible, there is no way to distinguish between the environment to which the organism is adapting and the point at which its own action begins" (100).[37] Such a state of indistinction between organism and environment is the generalized application of Latour's earlier statement regarding Lovelock's putative recognition that "in the literal sense, in the objects that he studied, *there are neither parts nor a whole*" (95).

In fact, Margulis's mature Gaia writings are not a suitable reference for the point Latour wants to make. Rather, she takes an autopoietic conceptuality to a quite different place. Here is an opportunity to tease out a crucial line of divergence between Gaia's primary theorists. At heart, Lovelock is a control systems engineer: "Whether we are considering a simple electric oven, a chain of retail shops monitored by a computer, a sleeping cat, an ecosystem, or Gaia herself, so long as we are considering something which is adaptive, capable of harvesting information and of storing experience and knowledge, then its study is a matter of cybernetics and what is studied can be called a 'system.' There is a very special attraction about the smooth running of a properly functioning control system."[38] Translated into physiological terms, such an engineer becomes a physician tweaking organic systems and the homeostatic goal becomes "health." Nonetheless, Lovelock the engineer is also content to let the distinction lapse between bodies and machines:

The control system of the engineer is one of those forms of proto-life mentioned earlier in this book which exist whenever there is a sufficient abundance of free energy. The only difference between non-living and living systems is in the scale of their intricacy, a distinction which fades all the time as the complexity and capacity of automated systems continue to evolve. Whether we have artificial intelligence now or must wait a little longer is open to debate. Meantime we must not forget that, like life itself, cybernetic systems can emerge and evolve by the chance association of events. All that is needed is a sufficient flux of free energy to power the system and an abundance of component parts for its assembly.[39]

This testimony documents Lovelock's own cyborg formulation of Gaia as memorably registered later on by Haraway in "Cyborgs and Symbionts." It also indicates that Lovelock's cyborg discourse in *Novacene* is not an entirely new phase of thought but a final postbiotic flourish of its first-order amalgam of control theory and information theory.

In contrast, Margulis is a systems biologist. One can see why she would come to insist on the criterion carried by the concept of autopoiesis. Without it, the difference between nonliving and living systems is simply one of scale and not of kind, one of random material-energetic instance and not of bounded operational self-production—in other words, the distinction collapses. Once she finds the concept of autopoiesis, Margulis will hold it fast. Her conviction appears to be that, for all the transformative interchange, uptake, and outflow between organisms and their environments, the difference between life and nonlife rests on the fundamental distinction between them articulated by presence or absence of biotic autopoiesis. Writing with Dorion Sagan in the mid-1980s, Margulis responded in the article "Gaia and the Evolution of Machines" to Lovelock's residually mechanistic claims. This obscure but crucial testament indicates precisely how to bring NST to bear on the machine issue: "Although there is an ineffable continuum between the living and the nonliving, we are beginning to understand the functions and organizations that are common to living entities. Living systems, from their smallest limits as bacterial cells to their largest extent as Gaia, are autopoietic: they self-maintain. As autopoietic systems they are bounded—they retain their recognizable features even while undergoing a dynamic interchange of parts."[40] Margulis became even more radically committed than Lovelock to biocybernetic systems theory, and as that developed, her commitment fastened upon a mode of description founded on the self-produced, membrane-bounded operational closure of living systems in the midst of their ultimate capacity in the fullness of evolutionary time to arrive at higher-order consortia of pre-evolved components. Margulis's way beyond holism, then, was never to suppress the observation of systematicity but rather to apply its properly operative form to the matter at hand. For metabiotic Gaia, the effective way to withdraw the part-whole distinction is to replace it with the system-environment distinction in the description of subsystemic differentiations.

For instance, consider the role of the mitochondrion, the organelle or cellular subsystem that handles oxygen respiration for the entire eukaryotic cell.[41] When Lewis Thomas began popularizing Margulis's *Origin of Eukaryotic Cells* in the 1970s, his surprise focused on the unexpected independence of the mitochondrion within its eukaryotic abode: "The RNA of mitochondria matches the organelles' DNA, but not that of the nucleus. . . . The mitochondria do not arise *de novo* in cells; they are always there, replicating on their own, independently of the replication of the cell."[42] Although its origin as an endosymbiont had been suggested before Margulis, a monophyletic orthodoxy dismissed such a construction. By 1970 Margulis had marshaled convincing if not definitive evidence to explain that the mitochondrion's relative autonomy derives from its vestigial alterity. Molecular genetic confirmation of her thesis arrived a decade later, lending authority to her subsequent summary of mitochondrial origins: "Mitochondrial ancestors came, probably polyphyletically, from the proteobacteria (or 'purple bacteria'), a large group of eubacteria with many respiring and phototrophic members."[43] Despite their assimilation into host cells, they have also maintained their own complex membranes, the semi-autonomy produced by distinct operational boundaries.

Symbiogenetic mergers never completely undo a measure of autonomy in the incorporated entity. In other words, even when the outside takes up residence on the inside, traces of that original alterity persist.[44] However, the mitochondrion's subsystemic boundaries are highly permeable. Mitochondria are now operationally integrated with their eukaryotic hosts; they are not viable in detachment from them. In their evolution into eukaryotic organelles, the proteobacterial precursors of the mitochondria lost autopoiesis while still maintaining formal and reproductive autonomy. A large part of the genome of the mitochondrial ancestor has migrated to the host nucleus. Margulis explains: "Obligate symbionts relegate redundant or dispensable metabolic functions to the host. . . . For all organelles that began as free-living organisms . . . as natural selection reduced inherent redundancy, partners became progressively more interdependent" (307, 313). Yet even with their reduced genetic complement, as eukaryotic subsystems mitochondria retain sufficient autonomy to produce their own membranes and maintain their differentiated functions, including reproducing on their own schedule.

In Margulis's account of the formation of the eukaryotic cell, then, the process of symbiogenesis coupled different free-living autopoietic systems, different kinds of prokaryotes, one of which became the host, the others endosymbionts. Once that association became permanent and obligatory, the endosymbionts relinquished autopoiesis to their now-nucleated host and coevolved into organelles, subsystems for which the cell is now the living environment. However, eukaryotic cells as complex autopoietic unities cannot maintain *their* self-production without mitochondria. Moreover: "Fully developed mitochondria are products of the interaction of at least two genomes, mitochondrial and nuclear. . . . The peptides of several large proteins . . . of respiratory chains are made jointly by nucleocytoplasmic and mitochondrial protein synthetic systems" (314, 315). The point to note here is the *conjoined interactions* of different genomes—"mitochondrial and nuclear"—and separate protein synthesis systems—"nucleocytoplasmic and mitochondrial"—within the same systemically differentiated but functionally unified living cell. These operations are not just connections or concatenations but mutually orchestrated autopoietically viable and eukaryotically commonplace interactions. It would seem perverse to subtract from these distinct agencies or actors—genome, organelle, cell—a notation of "level" corresponding to their location within the imbricated unity of the autopoietic apparatus. Rather, *the processes carried on by autopoietic systems cross their own inner and outer boundaries.* That these processes inexorably modify their inner and outer environments is not contested here. That general proposition is bedrock Gaia theory.

As I have suggested, these particular theoretical formulations point to a significant divergence of emphasis between Margulis and Lovelock. Latour represents these authors as a composite subject rescuing Gaia from holistic organizational fallacies of part-whole aggregations. "Here we can see the particular charm of Lovelock's prose, and Margulis's," because for them, "The inside and outside of all borders are subverted" (*Facing Gaia: Eight Lectures,* 101). Yet, elsewhere in this same book, Latour preserves an inside-outside distinction for Gaia altogether. We noted Latour's characterization of Lovelock's Gaia, that "it seems capable of actively maintaining a difference between its inside and its outside. It has something like a skin, an envelope" (78), with particular interest in its shadow formation of an autopoietic description pertaining to Gaia as a systemic unity. However, in the statement that, with regard to the "living agents" in Gaia's composition, "The inside and outside of all borders are subverted," the term "border" wavers between a structural description of material limit and a systemic description of operational boundary. The technical meaning of this statement is ambiguous, but its polemical intention would be to reduce operational boundaries to sub-verted borders. The problem is that this construction empties the life out of living systems in order to render them as bits of stuff suspended in a molecular wash. Gaia so described lacks the locally organized autono-mies that the systems description deems capable of scaling up in emer-gent fashion and kicking into operational effects. The resulting portrait of Gaia is a kind of open and passive yet "sensitive" medium sufficiently formless to bear the "propagation" of Gaian events on its surface:

> The interaction between a neighbor who is actively manipulating his neighbors and all the others who are manipulating the first one de-fines what could be called *waves of action,* which respect no borders and, even more importantly, never respect any fixed scale. These over-lapping waves are the true actors that one ought to follow *all along,* wherever they lead, without being limited by the internal border of an isolated agent considered as an individual "within" an environment "to which" it would adapt. The term is awkward, it is not Lovelock's, and yet these waves of action are the real brush strokes with which he seeks to depict Gaia's face. (*Facing Gaia: Eight Lectures,* 101)

This exercise of the Gaian imaginary renders a time-lapse portrait of the dynamics of inter-animated material agencies as trains of Gaian con-nection. Still, in this portrait, Gaia's effective coherence frays once more. The actors here have been specified as organisms, but their interactions seem inorganic. In this model of Gaia—to use the term for intimate biotic mutuality to which Margulis dedicated both the microbial-evolutionary and the Gaian-planetary paths of her career—the effect of connections is never "symbiotic." Here Gaia's connectedness never transpires as a mutuality of interests; it is just "mutual interference." Gaia is just "the climate . . . the historical result of reciprocal connections, which interfere with one another" (106). Indeed, Gaia "may have an order, but it has no hierarchy; it is not ordered by levels; it is not disordered, either. All the effects of scale result from the expansion of some particularly opportunistic agent grabbing opportunities to develop as they arise" (106).

Physics-laden imagery counters Gaia's organismal overanimation in an imaginary that seems to be drawn from luminiferous ("light-carrying") ether theory, in which the movement of radiation through space was taken to be the propagation of wave motions set up in the ether medium.[45] While such field models do provide a conceptual matrix for "*connectivity without holism*,"[46] these models are strictly abiotic. In contrast, the autopoietic Gaia conception may be carried to a roundly metabiotic description. In a portrait of Gaia *with* operational closure, self-reference induces a bias in the "waves of action"—or whatever concatenations of effects one may want to trace here—toward taking form as self-sustaining operational cycles. If we allow the waves to curve toward circularities of interrelation, the ensemble would eventually define a boundary cutting an emergent system out of the circumambient ether. The self-action of the waves would then fall into a round of recursions and take on systemic form and operation. Such systemic connectedness or cognitive closure would then ground the possibility of Gaia's actual capacity for responsiveness to its internal and external environments.

At one point, however, in responding to the fertile imagery of Sloterdijk's spherology, Latour will add to his Gaia description a certain quasi-circularity in the form of *loops*. These he opposes to the prior default mode of the planetary *sphere,* which geometric form clearly suffers from the imputation of totality. These loops are open-ended entities but seem largely to be singular "tracings" of the paths taken by Gaia's waves of action:

> How can one manage to trace the connections of the Earth without depicting a sphere? By a movement that turns back on itself, in the

form of a *loop*. This is the only way to draw a path between agents without resorting to the notions of parts and a Whole. . . . But let's not hurry to identify this movement, which in the previous lecture I called *waves of action,* with feedback loops in the cybernetic sense: we would revert at once to the model with a rudder, a helmsman, and a world government! (137)

Latour's countercybernetic points recur in these passages that otherwise, with the evocation of *feedback,* mark one of his closest approaches to veritable cybernetic territory. "Feedback loops in the cybernetic sense" must be set aside, however, or they would close, allowing the components of the cycle to arrive at some manner of regulatory coordination. Let us turn back once more to the "Cybernetics" chapter of Lovelock's *Gaia: A New Look at Life on Earth*: "Even though we may find evidence for a Gaian system of temperature regulation, the disentangling of its constituent loops is unlikely to be easy if they are entwined as deeply as in the bodily regulation of temperature" (57). Indeed, "the greater part of our search for Gaia is concerned with discovering whether a property of the Earth such as its surface temperature is determined by chance in the open loop fashion, or whether Gaia exists to apply negative and positive feedback with a controlling hand" (61).

In Lovelock, the point of positing feedback loops precisely in the cybernetic sense is to set the stage for experimental protocols that might determine whether or not they actually close to produce regulatory effects as the Gaia hypothesis would predict. Gaian science continues to explore the extreme niceties of this issue. Meanwhile, one can certainly take Latour's point that "Gaia is not a cybernetic machine controlled by feedback loops but a series of historical events, each of which extends itself a little further—or not" (140–41) and still prefer to describe Gaia as a metabiotic system in which biotic and abiotic feedback processes come and go as the system evolves. Nor does this description need to involve a leap "up to a higher 'global' level to see them act like a single whole" (141). The unity of Gaia's numberless differentiations can be observed as the paradox of a system that both is and is not its own environment. It is well, then, as Latour advises, to stay with what can be discerned and measured, despite its countless loops, and to pursue the investigation of making "their potential paths cross with as many instruments as possible in order to have a chance to detect the ways in which these agencies are connected among themselves" (141). Gaia is not an apex toward which life on Earth aspires or conspires. It is rather, just as Latour describes, "a small membrane, hardly more than a few kilometers thick, the delicate envelope of the critical zones" (140), upon which our very viability as a life-form depends and is currently precariously balanced. Nevertheless, the texture and being of that envelope is not the happenstance product of a planetary miscellany. If many of Gaia's regulatory processes remain mysterious, they may well be explained in due time. While it demystifies, the systems description of Gaia also reanimates the object to be explained, toward which we must continue to train our instruments as we sort out with all due rapidity the best way to understand our own planetary integration.

CHAPTER 3

Neocybernetics of Gaia

Neocybernetic systems theory refers to a range of *second-order* cybernetic concepts—recursion, reentry, closure, autonomy, cognition, self-reference, and self-referential systems. We have noted how the earliest phases of cybernetic thinking around feedback control, homeostasis, and related mechanisms for systemic self-correction foregrounded the concept of circularity.[1] As far back as one cares to trace it, cybernetic discourse has attended to entities, such as Gaia in Isabelle Stengers's description, "breaking the general linear relation between causes and effects." James Lovelock's first-order description of cybernetic non-linearity applies to both cybernetic orders: "The over-long delay in the understanding of cybernetics is perhaps another unhappy consequence of our inheritance of classical thought processes. In cybernetics, cause and effect no longer apply; it is impossible to tell which comes first, and indeed the question has no relevance."[2]

Instrumental approaches to feedback did not observe the sort of operational closures that inform self-referential systems such as organic cells or self-conscious minds: "Early cybernetics is essentially concerned with feedback circuits, and the early cyberneticists fell short of recognizing the importance of circularity in the constitution of an identity. Their loops are still inside an input/output box."[3] Francisco Varela gave a more technical version of the same point, in relation to the concept of self-reference, in a 1976 interview we will return to in the next chapter: "The notion of feedback is a self-referential one," he explained there, "but it was seized by the engineers who made it appear hierarchical. They apply a reference signal, identify input and output, and the output affects the input with a little delay. So the self-reference becomes hidden underneath, because of the trick of dealing with it in time." Concerning the more complex temporality of self-referential systems: "Unless you confront the mutualness, the closure, of a system, you just lose the system. It is the simultaneity of interactions that gives whole systems the flavor of being what they are."[4] Constrained by the instrumental and engineering emphases of the first cybernetics, systems theorists like Varela "felt the need to clearly distinguish themselves from these more mechanistic approaches, by emphasizing autonomy, self-organization, cognition, and the role of the observer in modeling a system. In the early 1970s, this movement became known as *second-order cybernetics*."[5]

The designation "first-order" is a back-formation denoting the original cybernetic logic centered on operational circularity in natural and technological systems, in which, for instance, output effects are fed back as causal inputs, thus superseding a strictly linear description. In the earlier cybernetics, while circular causality is instrumental for the self-regulation of systems, the system in question may not be altogether recursive. For example, in the operation of the governor of a steam engine or the thermostat of a furnace, the feedback mechanisms proper are coupled but subordinated to linear (input-output) mechanisms—in the classical instance, heat engines. In these examples, the assembly at hand is not entirely self-regulating: control is determined outside the system, by another, external system, or by a closed environment to be regulated. Neocybernetics arose when Heinz von Foerster forged a "cybernetics of cybernetics" by turning the logic of operational circularity upon itself.[6] Following his account, the development of second-order cybernetics generalized circularity in the concept of recursion. Recursion was now explored in its own right as formal self-reference in those systems capable of rising to cognitive operations, to wit, "observing systems."[7] This description pressed living systems to the forefront: biological systems' self-referential maintenance of self-produced organizations and cognitive boundaries between internal operations and external environments received a formal blueprint in the theory of autopoiesis.

Von Foerster referred to first-order cybernetics as "the cybernetics of observed systems," that is, the cybernetics of things, such as natural or technological systems, while second-order cybernetics is "the cybernetics of observing systems," that is, the cybernetics of *cognitive* systems, those capable of producing observations in the first place.[8] In *Autopoiesis and Cognition* (1980), Maturana and Varela published their definitive case for considering autopoietic systems, such as living cells, as cognitive, or as restated in second-order cybernetic parlance, not merely as observed but more fundamentally as *observing* systems producing life-maintaining, self-making cognitions of their environments.[9] Maturana and Varela coined the term *autopoiesis* in 1971 to denote this group of interrelated concepts—circular organization, operational closure, and self-referring processes. The discourse of autopoiesis named the self-referential or recursive form of the "organization of the living" as coupled to a self-referential description of the cognitive processes that produce the discourse. Biotic autopoiesis is recursive self-constitution applied to the observation of cells and organisms. Observed both *as* and *by* an autopoietic operation, the minimal organization of life, the cell, takes the form of a closed circular process of self-production (autopoiesis) within a system open to selective environmental interaction (cognition). In other words, while the environment feeds and otherwise stimulates such a system, and can bring about the system's compensatory responses to perturbations, neither the environment nor the observers it contains can *operate* (or control) that system. All mythical tales and literary fantasies to the contrary, life cannot be endowed from without. The same can be said for consciousness and sociality.

Maturana and Varela published a paper on autopoiesis in Spanish in 1973. Varela, Maturana, and Ricardo Uribe coauthored the first English appearance in a 1974 number of *BioSystems,* in the same year that the first coauthored Gaia papers of Lovelock and Margulis appeared in *Icarus* and *Tellus.*[10] "Autopoiesis: The Organization of Living Systems, Its Characterization and a Model" gave a technical definition of the "autopoietic organization" emphasizing circular recursion in the production of "the cell as a material unity," that is, as a distinct concrete entity standing out from its environmental matrix as long as its autopoiesis continues:

> The autopoietic organization is defined as a unity by a network of productions of components which (i) participate recursively in the same network of productions of components which produced these components, and (ii) realize the network of productions as a unity in the space in which the components exist. Consider for example the case of a cell: it is a network of chemical reactions which produce molecules such that (i) through their interactions generate and participate recursively in the same network of reactions which produced them, and (ii) realize the cell as a material unity. (188)

In the living cell, autopoietic recursion, systemic organization in circular operation, produces the self-binding of the system.[11] Self-production produces for itself (by recursive operation) the conditions (here, a semipermeable membrane) that create and maintain the operational closure that ensures the viable while never perfect autonomy of the process (here, cellular metabolism) allowing self-production to continue. Autopoiesis is a paradigmatic case of the second-order cybernetic observation of closed recursions that constitute systemic identities.

In addition to recursive operation, the biotic concept of autopoiesis situates the operational continuation of the living system as both logically and biologically prior to its reproduction. "Autopoiesis: The Organization of Living Systems" begins here, with the primacy of self-production over reproduction:

> The great developments of molecular, genetic and evolutionary notions in contemporary biology have led to the overemphasis of isolated components, e.g. to consider reproduction as a necessary feature of the living organization and, hence, not to ask about the organization which makes a living system a whole, autonomous unity that is alive regardless of whether it reproduces or not. As a result, processes that are history dependent (evolution, ontogenesis) and history independent (individual organization) have been confused in the attempt to provide a single mechanistic explanation for phenomena which, although related, are fundamentally distinct.
>
> We assert that reproduction and evolution are not constitutive features of the living organization and that the properties of a unity cannot be accounted for only through accounting for the properties of its components. In contrast, we claim that the living organization can only be characterized unambiguously by specifying the network of interactions of components which constitute a living system as a whole, that is, as a "unity." We also claim that all biological phenomenology, including reproduction and evolution, is secondary to the establishment of this unitary organization. (187)

This corollary aspect of autopoietic logic would be especially significant for Margulis's appropriation of the concept as an explicit component of her own Gaia discourse, to the extent of presenting Gaia as the autopoietic planet: "Autopoiesis of the planet is the aggregate, emergent property of the many gas-trading, gene-exchanging, growing, and evolving organisms in it."[12] Margulis also drew on the discourse of autopoiesis in countering the neo-Darwinist worldview. Her strategy was to ally autopoiesis with symbiosis as a pervasive form of living organization and—in instances of "symbiogenesis"—as a source of evolutionary variation. In concert with the discourse of autopoiesis, symbiosis per se is *not* about reproduction or the supposed vagaries of genetic mutation. It is precisely about "the organization of the living" as that also takes the form of communal assemblies over and above the recursive formalisms of individual living systems. What her mantra "Gaia is symbiosis seen from space" foregrounds for any particular case of symbiosis, from the microcosm on up, is the *living* part of the living-together of extant organisms. Ongoing life co-maintains itself in ever-varied ensembles established by hazard and stabilized by mutual viability. As more recent studies of the "metagenome" and the holobiont formed by complex symbiotic communities have shown, the viable unit of reproduction can be the symbiotic consortium and not merely its symbionts taken individually. We will resume a Gaian appreciation of the holobiont in chapter 8.

Lovelock's initial Gaia hypothesis took shape before Maturana and Varela brought the theory of autopoiesis forward. As he was writing *Gaia: A New Look at Life on Earth,* the concept of autopoiesis was still relatively unknown. He would lament what appeared to be the absence of a scientific definition of life, cybernetic or otherwise. "Even the new science of cybernetics has not tackled the problem, although it is concerned with the mode of operation of all manner of systems from the simplicity of a valve-operated water tank to the complex visual control process which enables your eyes to scan this page. Much, indeed, has already been said and written about the cybernetics of artificial intelligence, but the question of defining real life in cybernetic terms remains unanswered and is seldom discussed."[13] In fact, until the introduction of autopoiesis as a neocybernetic definition of the form of the living, Lovelock's own Gaia discourse was among the most notable prior efforts to "define real life in cybernetic terms." It would be Margulis who eventually annexed autopoiesis to the cybernetics of Gaia.

Simply put, first-order cybernetics is about control, second-order cybernetics is about autonomy. NST takes recursive processes beyond mechanical and computational control processes toward the formal autonomy of natural systems. Neocybernetics aims in particular at natural systems in which *circular recursion constitutes the system*. In Varela's terms, circular causality constitutes a systemic identity. This finer appreciation of recursive self-constitution refines the choices of systemic observation. Lovelock's Gaia began as a first-order cybernetic model viewing the biota as a thermostat controlling the viability of the abiotic environment on behalf of life. Margulis also subscribed to this description in Gaia's first decade. As its critics were quick to point out, this scheme had several limitations. For one, by placing life altogether over and in control of its environment, it overcompensated for traditional evolutionary accounts in which life has always played the passive partner put upon to adapt itself to a capricious environment. For another, the early biocybernetic version of the Gaia hypothesis, treating the cybernetics of Gaia as one would the engineering of a control mechanism, prompted Lovelock in particular, during Gaia's earlier stages, to venture the first-order vocabulary of optimization, as in Gaia's production of an "optimum physical and chemical state appropriate to its current biosphere." Such claims brought accusations that the Gaia hypothesis imposed a teleological scheme on the evolution of the biosphere, as if the biota moved in concert with a collective purpose. Lovelock will invent his Daisyworld computer simulation of coupled feedbacks linking the biota and its environment to quell such complaints with a model of how Gaian homeostasis could arise as an automatic process.

However, by the later 1980s, in the development of the hypothesis into a theory, Lovelock and Margulis both subdued the rhetoric of optimization and brought life and Earth back into realignment as a coupled metasystem distinct from non-Gaian environments above and below Gaia's proper sphere. Gaia theory now integrated life with its terrestrial matrix into a geobiological system whose coevolution has been a composite phenomenon of co-emergence: "Through Gaia theory, I see the Earth and the life it bears as a system, a system that has the capacity to regulate the temperature and the composition of the Earth's surface and to keep it comfortable for living organisms. The self-regulation of the system is an active process driven by the free energy available from sunlight."[14] Nevertheless, Lovelock remained committed to "strong Gaia"—the conviction that Gaia is in some sense *alive* in its own right, even if only in virtue of its being a system functioning on the cybernetic model of homeostatic self-regulation: "To me it was obvious that the Earth was alive in the sense that it is a self-organizing and self-regulating system."[15] In chapter 5 we will examine a key moment when Varela queried Lovelock on the seeming animism of such rhetoric. And for her part, although she would remain adamant that Gaia is "not an organism," Margulis never consistently broke with her earlier habit of describing the Gaian system as some version of "the sum of the biota," a biocentric formulation that remained amenable to a strictly biotic deployment of autopoietic theory. In this regard, neocybernetic Gaia theory can move forward by combining aspects of both seminal Gaia theorists. This is why *metabiotic Gaia* takes Margulis's conception of autopoietic Gaia beyond the biotic occasion toward Lovelock's mature geobiological description. The systems thinker who comes the closest to limning this redescription of the Gaia concept turns out, once again, to be Varela, one of the primary links between the Whole Earth network and the Lindisfarne milieu.

The Evolution of Autopoiesis

The concept of autopoiesis is intriguing for its multifarious cultural history, itinerant discursive career, and contrarian stance, its persistent Continental and countercultural vogue and outsider status. From its inception as a cybernetic theory of biological form to its current presence on research fronts extending from immunology to sociology to architecture, from geobiology, artificial life, and cognitive science to a range of literary and cultural theories, the autopoiesis concept has developed on the margins, not in the strongholds, of mainstream Anglo-American science. A certain globe-trotting streak runs through the pedigree of autopoiesis. Leonardo Bich and Arantza Etxeberria note that the French philosopher of science Georges Canguilhem in his work *La connaissance de la vie* had already made a distinction between "heteropoetic" and "autopoetic" approaches to living systems.[16] Indeed, Canguilhem writes there in an essay on method that "man first experiences and experiments with biological activity in his relations of technical adaptation to the milieu. Such technique is heteropoetic, adjusted to the outside. . . . Only after a long series of obstacles surmounted and errors acknowledged did man come to suspect and recognize the autopoetic character of organic activity."[17] It is plausible that Varela knew this Canguilhem text.[18] However that may be, the template for Maturana and Varela's presentation of autopoiesis is Maturana's single-authored 1970 paper

"Neurophysiology of Cognition." In that essay, the physiology of cognitive processing in the nervous system is referred to basal operations of living systems in general: "Although the nervous system expands the domain of interactions of the organism by bringing into this domain interactions with *pure relations,* the function of the nervous system is subservient to the necessary circularity of the living organization."[19] Additionally, as the Italian theoretical biologist Pier Luigi Luisi has pointed out, "autopoiesis originated in a time-window (the early 1970s) when the world of biology was completely dominated by a vision of DNA and RNA as the holy grail of life. Alternative views about the mechanism of life didn't have much chance of appearing in mainstream journals."[20]

As the discourse of autopoiesis developed beyond its inventors' own discourses, both Maturana's and Varela's inclinations were to tamp down others' efforts to extend autopoiesis beyond biological systems, by insisting on its delimitation to the realm of molecular dynamics, on its material specificity as a membrane-bounded process of biological production. For his part, Varela wrote in 1981, "Autopoiesis is a particular case of a larger class of organizations that can be called *organizationally closed,* that is, defined through indefinite recursion of component relations"; however, it "is tempting to confuse autopoiesis with organizational closure and living autonomy with autonomy in general."[21] Varela's colleague Evan Thompson has rehearsed the complex coupling of openness and closure at the basal level of the biological autopoiesis of the living cell: "Metabolism is none other than the biochemical instantiation of the autopoietic organization. That organization must remain invariant, otherwise the organism dies, but the only way autopoiesis can stay in place is through the incessant material flux of metabolism. In other words, the operational *closure* of autopoiesis demands that the organism be an *open system.*"[22] With regard to social systems, Varela was definitive: "Unless a careful distinction is made between the particular (autopoiesis and productions)"—meaning that in autopoiesis proper, *the system produces itself by producing the very elements that compose it as a system*—"and the general (organizational closure and general computations), the notion of autopoiesis becomes a metaphor and loses its power. This is what has happened, in my view, with attempts to apply autopoiesis directly to social systems."[23] Varela's particular concern is certainly warranted for theories that posit *persons* as the elements "produced" by social systems. Placing the autopoiesis of social systems on a nonmetaphorical basis would have to locate an alternative rationale. Social systems theory arguably established a properly nonmetaphorical approach to the autopoiesis of "meaning systems," that is, those self-referential systems that operate in the "medium of meaning" as opposed to the milieu of matter and energy. "If we abstract from life and define autopoiesis as a general form of system building using self-referential closure," writes Niklas Luhmann, "we would have to admit that there are nonliving autopoietic systems, different modes of autopoietic reproduction, and that there are general principles of autopoietic organization that materialize as life, but also in other modes of circularity and self-reproduction."[24] So far this could appear to conform to Varela's notion of "the general (organizational closure and general computations)," that is, to no more than a metaphorical extension of closure in the absence of *self*-production.

What maintains Luhmann's social systemic appropriation as autopoiesis proper is that *communication* itself supplies "the particular (autopoiesis and productions)":

> Social systems use communication as their particular mode of autopoietic reproduction. Their elements are communications that are recursively produced and reproduced by a network of communications and that cannot exist outside of such a network. Communications are not "living" units, they are not "conscious" units, they are not "actions." Their unity requires a synthesis of three selections, namely information, utterance, and understanding (including misunderstanding). This synthesis is produced by the network of communication, not by some kind of inherent power of consciousness, nor by the inherent quality of the information.[25]

In Luhmann's own description, psychic and social systems are "nonliving autopoietic systems." For NST I call them *metabiotic* autopoietic systems: while nonliving, they emerge from and only from, and are environmentally contingent upon, living systems. Moreover, the co-operations of psychic and social systems produce a mutually contingent coevolution. They occupy a shared medium of meaning introduced by linguistic or other semiotic functions.[26] Due to their common imbrication with technical and linguistic infrastructures, the phrase "meaning systems" applies to both psychic and social systems. Social and psychic systems "interpenetrate" in the midst of their operational differentiation. Both are autopoietic, thus operationally autonomous, yet each co-emerges with the other and needs the other in order to operate. Each presents the immediate environment of the operation of the other. For their part, living systems' basal elements are the internally modulated molecular dynamics of atoms and molecules and their energy states. However, in meaning systems, the elements that carry out their own self-production are not material but virtual—they are *events,* more specifically, events of distinction and selection.

These meaning-systemic events take different forms: psychic systems produce and process events of consciousness; social systems produce and process events of communication. The nonautopoietic analogue in informatic systems is discrete events of transmission. This "event-character" of both the elements and the operations of meaning systems marks a crucial distinction from living systems with regard to their relative speeds of operation. Luhmann notes that "the formal definition of autopoiesis gives no indication about the span of time during which components exist. . . . Conscious systems and social systems have to produce their own decay. They produce their basic elements, i.e., thoughts and communications, not as short-term states but as events that vanish as soon as they appear. Events too occupy a minimal span of time, a specious present, but their duration is a matter of definition and has to be regulated by the autopoietic system itself: events cannot be accumulated."[27] In other words, metabiotic autopoiesis in meaning systems also produces what counts for them as time and operates on different temporal scales relative to the medium and duration of the pertinent elements.

Now, the Gaian system arises from the co-operations of biotic *and* abiotic, living *and* nonliving, elements. In the Gaian instance, the systemic elements and processes standing apart from the living instance are *not* virtual or technical—that is, are not structurally or linguistically mediated (psychic or social) distinctions of (conscious or communicative) forms. Rather, they are the sheer (nonliving) material and energetic bases out of which biotic operations arise.[28] Some three or so billion years ago, when a critical mass of abiotic, biotic, and biogenic elements fell into a closed loop locking in an emergent level of metabiotic autopoiesis, self-producing life coupled together with its increasingly modulated environment to induce a primal regime of planetary cognition. When placed within a Gaian temporal frame, over geobiological time, the metabiotic emergence of meaning systems—those systems specific to minds and societies for which cognition operates in semiotic mediums of meaning—can then be seen in their deep Gaian contingency as more recent epiphenomena. Our forms of consciousness and communication are ramified recursions of that Archean event. As formulated in these terms, however, Gaian cognition does not operate in the medium of meaning. In this construction, on the prosaic plane of systems theory, Gaia is a self-referential system of planetary cognition operating to produce globally regulative processes binding geological and biological processes and developments into a network of interdependent systems whose differential evolutions are mutually contingent in the final instance but not necessarily in the individual one. Gaia yields many degrees of freedom within ultimate limits to viability. At the end of their coauthored volume *Microcosmos,* Margulis and Sagan expressed this Gaian feedback scheme of biotic/abiotic reciprocation in a well-turned chiasmus: "On Earth the environment has been made and monitored by life as much as life has been made and influenced by the environment."[29]

Conceiving Gaia as autopoietic excludes it from a holistic or superorganic description. Metabiotic Gaia holds together the distinctions among its abiotic and biotic subsystems. With meaning systems, no higher-level super-system arises from the metabiotic interpenetration of psychic and social systems. This was the point Dirk Baecker made in our glance at his text in chapter 1: no higher-order autopoietic unity can subsume different *kinds* of autopoietic systems. Rather, as Luhmann affirms, "the difference of the systems is reproduced in the process of interpenetration."[30] The separate autopoieses of living, psychic, and social systems are *operationally* incommensurable. However, in the basal Gaian instance there is only one *kind* of autopoietic system on hand: biotic systems alone run the metabiotic autopoiesis of Gaia. Thus this theorization adheres to the interdiction of the merger of different modes of autopoietic operation. This also means that, despite various noospherical speculations, there can be no merger of Gaia's own operations with either its psychic or its social observers. Gaia's planetary cognition is neither psychic nor social—it is impenetrably and dedicatedly geobiological.[31]

With ecological symbiosis as a bridge, then, autopoiesis and Gaia enjoy a conceptual symbiosis as interlocking micro and macro modes of NST. Biotic autopoiesis defines the minimal formal requirements for the persistence of life in living systems, beginning with the cell, while Gaia captures "the network of interactions of components which constitute" the geobiosphere itself as a self-referential system for which the atmosphere is a kind of autopoietic membrane. The finitude of any possible observation is the formal correlate and the burden of operational closure in autopoietic systems: a system capable of cognition necessarily translates what it can know of its environment into the medium of its own elements.[32] Sagan and Margulis note that "What is remarkable is the tendency of autopoietic entities to interact with other recognizable autopoietic entities."[33] The Gaian system is the epitome of the autopoietic paradox of openness and closure already nested in living cells. For NST, the twin systems concepts of autopoiesis and Gaia epitomize a shift in the aims of scientific rationality, from instrumental control without due regard for environmental ramifications, to the observation and coordination of system–environment relations. They entail more reflective ethical stances toward such contingencies of interrelation. The metabiotic reading of Gaia theory scales up the recursive turn in neocybernetics from cellular dynamics to psychic, social, and planetary systematics. Threading the world with a common mode of operation-in-context, Gaia's autopoiesis links Earth, life, mind, and society, in the midst of their systemic differentiations. This fabric gathers up a conceptual framework large enough to contain and sufficiently complex to guide a Gaian thinking of differentiated interconnectedness.

Partial Earth

Luhmann's introduction to *Social Systems*, "Paradigm Change in Systems Theory," notes "a tradition stemming from antiquity" in terms of which, to address the concept of systems, one spoke "of wholes that are composed of parts."[34] Before and after the advent of systems theory proper, versions of *holism* populate the common idiom for the description of systems. However, "The problem with this tradition is that the whole had to be understood in a double sense: as the unity and as the totality of its parts. One could then say that the whole *is* the totality of its parts or *is more* than the mere sum of its parts. But this does not explain how the whole, if it be composed of its parts, plus something else, can count as a unity on the level of parts" (5). Holism names a systems discourse that is still framed to some extent in this traditional part/whole manner, and which, as a consequence, encounters these problematics of totalization. As we will see, with some important exceptions, holism in this sense is still at work in many of the American-based cybernetics and systems discourses that run through the Whole Earth network. Visions of globality are accorded ultimate value. *Parts* of the biosphere—organisms, species, societies, and their technologies—may coevolve, but it is the *whole* Earth that gathers them into an ecological union rendered as a singular totality. This counterreductionist orientation was both out in front of mainstream scientific and social thinking—*countercultural* in a good way—and at the same time prone to theoretical equivocations and impasses that NST both illuminates and goes beyond.

With regard to fitting systems theory with an improved set of conceptual terms, Luhmann writes, "The first move in this direction was to replace the traditional difference between *whole and part* with that between *system and environment*" (6). Let us pause to absorb the radicality of this shift. In the part/whole dichotomy, no account is taken of the *milieu* of the "whole" in question. The system-environment distinction brings the outside into constitutive relation with the inside of the "whole," where the "parts" are. Recursions of level need a logic adequate to the peculiarities of circularity. "This transformation, of which Ludwig von Bertalanffy is the leading author, enabled one to interrelate the theory of the organism, thermodynamics, and evolutionary theory. A difference between open and closed systems thereupon appeared in theoretical descriptions. Closed systems were defined as a limit case: as systems for which the environment has no significance or is significant only through specified channels. The theory concerned itself with open systems" (6–7). Of the two crucial "paradigm changes" in systems theory, the first concerns these environmental supplements to the observation of natural systems, in particular, Bertalanffy's application of the second law of thermodynamics to *open* systems. In the classical formulation of the second law, *entropy* measures the disorder of energies within closed physical systems. Maximum entropy denotes *equilibrium,* the perfectly random or completely disordered distribution of energic potentials.[35] The observation that living systems maintain and even increase the order of their organization led at first to speculations of a vitalistic nature that life somehow defies the second law. What this counterentropic outcome actually meant was that, thermodynamically considered, living systems are *open* systems using the uptake of free energy to decrease their entropy.

The arrival of the system-environment distinction heralds a second paradigm change, the theory of system differentiation: "What had been conceived as the difference between whole and part was reformulated as the theory of system differentiation and thereby built into the new paradigm. System differentiation is nothing more than the repetition within systems of the difference between system and environment. Through it, the whole system uses itself as environment in forming its own subsystems" (7). The system-environment distinction marks the difference between a bounded organization—the system "cut out" by its own operational finitude—and an unbounded matrix—the total environment as everything (unorganized matter, random energies, other systems) beyond the boundaries of any particular system. Moreover, the theory of system differentiation sets the stage for "a surpassingly radical further step. . . . It concerns contributions to a *theory of self-referential systems. . . .* The (subsequently classical) distinction between 'closed' and 'open' systems is replaced by the question of how self-referential closure can create openness" (8–9).[36] Luhmann gives this summary of the paradigm changes that inform his systems theory: "In the paradigm of the whole and its parts one had to accommodate inexplicable properties somewhere—whether as properties of the whole (which is more than the sum of its parts) or as the properties of a hierarchized apex that represents the whole. By contrast, in the theory of self-referential systems everything that belongs to the system (including any possible apex, boundaries, or surpluses) is included in self-production and thereby demystified for the observer" (10). For NST, the holistic short circuit of worldly complexities is defused by shifting to a more precisely located or appropriately complex system-environment framework.

NST favors distinction over totalization. Let us apply this rule to a Gaian example. As codified in *Ages of Gaia,* the refined observation of Gaia as a *coupled* system, a composed consortium of Earth and life, pivots on a prescient passage from Alfred Lotka's 1925 text *Elements of Physical Biology.* Lotka's statements anticipate an ecosystemic appreciation of the Earth system. On a photocopied page of Lotka's text that he sent to Margulis for her edification, Lovelock circled two passages.[37]

The first declared a recognizably "whole-system" conception comprehending the biosphere and the geosphere in their mutual submission to solar radiation:

> Application to Biology. The several organisms that make up the earth's living population, together with their environment, constitute one system, which receives a daily supply of available energy from the sun.

The second circled passage was the content of the footnote that Lotka appended to the phrase "one system" in the passage immediately above:

> This fact deserves emphasis. It is customary to discuss the "evolution of a species of organisms."... We should constantly take in view the evolution, as a whole, of the system [organism plus environment]. It may appear at first sight as if this should prove a more complicated problem than the consideration of the evolution of a part only of the system. But ... the physical laws governing evolution in all probability take on a simpler form when referred to the system as a whole than to any portion thereof.
>
> It is not so much the organism or the species that evolves, but the entire system, species and environment. The two are inseparable.[38]

This passage presages the conceptual tensions between unification and differentiation that run throughout Gaia discourse. Lovelock recognized Lotka's advocacy for what he now expounds as the Gaian consortium gathering "organisms and their environment" to "form a coupled system."[39] Both Lotka and Lovelock express their speculations within a holistic idiom of parts and wholes. Moreover, Lotka's vision of a holistic unification of "organism plus environment" places ecology under a physicalist ideal that would reduce biological processes to the laws of matter and energy. Applied to systems of due complexity, however, such descriptions obscure the imperative to maintain an operational boundary placing system processes apart and across from an environment that exceeds them. Only then can "self-referential closure ... create openness." Systems with operational closure maintain themselves apart from that of which they are also a part, precisely in order to correspond with it. Any worldly system is an operational unity that holds itself distinct from the environment out of which it emerges—not absolutely, of course, but sufficiently to produce and maintain its distinction, and, as such, its cognitive capacity. That environment is transformed in its own right by the system's presence in its midst and by the flux of its elements through the system's processes—this is classic Gaia theory, but that dynamic holds only as long as that distinction remains in operation. This is why NST goes in fear of unqualified intimations of totality and takes care not to de-differentiate systemic components. Such care might also be imputed to Lovelock's choice to draw from Lotka's passage the statement of a "coupled system" that maintains operational distinctions.

Finally, without the maintenance of the system in operational distinction from its enabling environmental matrix, even as there is any manner of open commerce across that boundary and interdependence among systems themselves, the phenomena that emerge from their couplings fall away. Whatever consortia come to be can find their emergent states and processes only within an environment that exceeds them and which they cannot fully know or master. In sum, while *Gaian Systems* honors the spirit of recent times that expressed itself in the "Whole Earth" as an ontological totality or image of unity, its aim is to promote the history of systemic developments that have led us to acknowledge tighter epistemological limits and sharper vital boundaries. Gaia does not operate according to a principle of completion and merger but as a finite holarchy of finitudes.

PART II
THE SYSTEMS COUNTERCULTURE

CHAPTER 4

The Whole Earth Network

The systems counterculture nurtured by the Whole Earth network cultivated the first drafts of neocybernetic systems theory. In turn, NST provided the conceptual space for the merger of the Gaia hypothesis with the theory of autopoiesis in Lynn Margulis's discourse of autopoietic Gaia. In autopoietic theory, recursive causality constitutes systemic identities. Autopoietic Gaia takes this mode of systemic observation to its planetary conclusions. Recovering neocybernetic Gaia theory can refresh our recollection of the considerable accomplishments of the ecological milieu and coevolutionary imperatives of the 1970s. Part II opens up some of the ways that the systems counterculture in the milieu of the Whole Earth network cultivated the parallel, sometimes intersecting developments of NST and the Gaia concept.

The systems counterculture was a loosely collegial group of seminal scientific thinkers whose developments of cybernetics and systems theories led them beyond mainstream doctrines and institutions. The systems counterculture constellated in this study includes Buckminster Fuller, Gregory Bateson, Heinz von Foerster, George Spencer-Brown, Humberto Maturana, Francisco Varela, James Lovelock, and Lynn Margulis. The abiding cultural effect of their work has been to detoxify the system concept of its military, industrial, and corporate connotations of command and control and to redeploy it in the pursuit of holistic ideals and ecological values. In the United States, this disparate cybernetic reformation came to a head in the later 1960s and remained well defined throughout the 1970s and into the 1980s. In these thinkers a broadly shared body of NST shuttled between mathematics and the natural and engineering sciences and migrated from there to new residences in the social sciences, humanities, and arts, challenging prior epistemological assumptions and infiltrating both high academic theory and popular culture. The systems counterculture entered alternative locations and venues where maverick collaborations became possible, where it could assemble and test idiosyncratic appropriations. Its coalescence was publically registered with the initial four-year run of the *Whole Earth Catalog* from 1968 to 1971. It arrived in full with its periodical continuation, *CoEvolution Quarterly,* from 1974 to 1984, followed by a gradual dispersal in the *Whole Earth Review.*

Fred Turner's canonical *From Counterculture to Cyberculture* has guided research in this area. But we may now be past our initial enthusiasms for digital utopia and ready to redirect our attention to an alternative series of nondigital but major intellectual developments that the *Catalog*'s creator and main editor, Stewart Brand, speaking about his discovery of the work of Gregory Bateson, termed *organic cybernetics.* "As a Bateson enthusiast and a publisher," Brand wrote in 1974, "I'll be printing sundry papers, speculation, gossip, tidbits, letters, etc. on cybernetics (well, organic cybernetics), in the periodic supplement to the revived *Whole Earth Catalog* . . . 'The CoEvolution Quarterly.'"[1] Moreover, looking over the larger growth of the Whole Earth network during these decades, one can track the emerging bifurcation noted in Brand's statement between first- and second-order cybernetics in the Whole Earth milieu of the 1970s, followed in the 1980s with the eventual eclipse of *CoEvolution Quarterly*'s bio-ecological orientation with the explosion of digital cyberculture.

Whole Earth Catalog

CYBERNETICS OF EARTHRISE

In their moment, the *Whole Earth* publications were the virtual house organ on the world stage for the informed general discussion of cybernetic ideas. These detailed systemic and ecological perspectives on local and global practices—in light of what was perceived as the planetary emergencies of *that* moment, such as explicit premonitions of global warming, the imminence of environmental devastation by nuclear war, rampant monoculture, and seemingly unsustainable human population explosion—are not just relics of the 1960s. They already had the planetary situation more right than not. The problems probed there, the sciences explored, the technological and political solutions debated, and the cultural and spiritual practices recommended have aged fairly well: many are as relevant as ever to our current abysmal quandaries.

The Gaia discourse within this venue is a case in point. Even prior to that, the *Catalog* cultivated cybernetics and systems theory in a prominent opening section titled "Understanding Whole Systems."

Between 1968 and 1971 the *Whole Earth Catalog* documented the NASA-inspired technophile wing of the American counterculture and motivated its perception of Earth as an ecological unity. As Lovelock was incubating the Gaia hypothesis at JPL in the decade before its formal introduction in 1972, this seminal countercultural publication— part magazine, part product and lifestyle guide—began its initial four-year run.[2] Seizing on a series of unprecedented NASA images as icons for a transformed ecological and environmental consciousness, the *Whole Earth Catalog*'s presentation of these celestial portraits premediated Lovelock's idea of Gaia as a "biological cybernetic system."[3] Apollo 8 brought back a strip of Earthrise images. What we now call the Earthrise photograph is the best of a series of shots taken on color film with a professional-grade camera. It was developed and transmitted to media outlets once the Apollo 8 mission returned home and then given a 90-degree rotation to the right to make the Earth rise over the Moon. No finer image of the Earth from space had ever been captured, and it was the first such image to enjoy universal distribution.[4] The *Whole Earth Catalog* smacked Earthrise on the front and back covers of its spring 1969 issue (Figure 2).

Aligning the Earthrise photograph with cybernetic themes, the *Whole Earth Catalog*'s visual rhetoric made a range of new observations on our planet's cosmic station ready for the taking. The *Catalog* capitalized at once on the iconography of this world seemingly seen whole—the living Earth observed from space as a systemic unity. Earthrise has assisted the way we have come to think about our planet astrobiologically, not as detached from but as bound up with the rest of the universe. Its gorgeous tableau inverted earlier perspectives by framing a distant Earth in relation to the near surface of its lunar neighbor. Other Earth-from-space images of that moment could seem to suggest that our planet just floats in space free of any attachments. By showing the Earth and Moon as gravitationally tethered and mutually constituting, Earthrise also evokes the solar system and the wider cosmos around us. At the same time, as has often been noted, it makes the difference starkly clear between a lifeless and a living world. In this spirit, the covers of nearly every iteration of the *Whole Earth Catalog* presented some NASA image

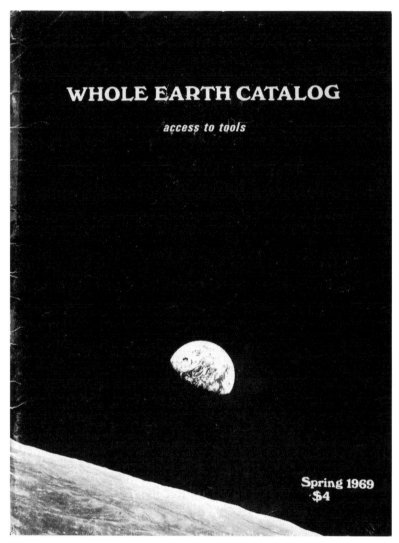

FIGURE 2. *The* Whole Earth Catalog, *front cover, spring 1969.*

of Earth taken from space. Here was a home planet newly visible in its own right and newly imaginable as a global ecosystem. Moreover, in parallel with nascent Gaian science, the *Whole Earth Catalog* pursued a whole-systems style of cybernetic thinking. Every iteration contained a section on "Whole Systems" that reviewed developments and retailed information about systems theory. This particularly heady bastion of the American counterculture framed its uptake of NASA's Earthrise photograph with systems vibrations and cybernetic fascinations.[5]

On the back cover of its spring 1969 issue, the Earth rising over the horizon of the Moon now has a caption. It reads: "The flow of energy through a system acts to organize that system." The statement is from Harold Morowitz's *Energy Flow in Biology*.[6] Two years later, on the inside of the front cover of the *Last Whole Earth Catalog*, Earthrise appears again, with a different legend: "The famous Apollo 8 picture of Earthrise over the Moon that established our planetary facthood and beauty and rareness (dry moon, barren space) and began to bend human consciousness." The Earthrise photograph perfectly fuses a cybernetic vision of the Earth as a planetary system with the whole-systems concept that informs the *Whole Earth Catalog*. Moreover, the systems concept orients one to a synoptic view of things, presses one to a conceptual elevation from which the boundaries of complex entities show forth against their environments. It is in this spirit that one might construe the psychedelic afflatus that leads off the purpose statement prefacing every edition of the *Catalog*—"We are as gods and might as well get good at it." In emulation and material fulfillment of classical visions of the divine, we have now created mind-blowing technological systems that lift us to a transcendent view of ourselves and the world around us, a view previously reserved to whatever deities might have been imagined to be looking down on mundane affairs. The *Whole Earth Catalog* received the gifts of the space-age engineering gods at NASA responsible for the rebirth of Apollo and his new adventures trafficking with the heavens.

It was only a few years after the arrival of the Earthrise photograph that Gaia joined this new cybernetic pantheon. In 1986 the PBS science series *Nova* aired the documentary *Gaia: Goddess of the Earth*. Over a series of images from the *Whole Earth Catalog*, the narrator intoned: "It was the pro-environmental movement who found the most reason to appreciate Gaia. . . . Words like *ecosystem* entered the vernacular. . . . These ideas were reflected in counterculture publications, which also popularized the science of automatic control systems—cybernetics—in both living things and machines. Gaia came in on this rising tide of interest in whole systems." Just as the *Catalog*'s NASA graphics resonate with the spirit of "Understanding Whole Systems," Lovelock's original thesis regarding Gaia as an automatic or self-regulating system for environmental homeostasis resonates profoundly with the *Catalog*'s cybernetic milieu.

Second-order cybernetics was still in embryo when the *Whole Earth Catalog* began operation. Yet the systems thinking the *Catalog* already purveyed—in reviews of works by Norbert Wiener, H. Ross Ashby, Warren McCulloch, Gordon Pask, and Ludwig von Bertalanffy, in addition to Fuller—gave voice to the most liberal and visionary wings of the first cybernetic thinking. Margulis and Lovelock's initial formulations of the Gaia concept would soon join this chorus. But first, the fall 1969 issue registered something of a tremor in the ether of such advanced thought with a curt and uncharacteristically confounded review of a work titled *Laws of Form* by a British mathematician named George Spencer-Brown.[7] Brand's entire commentary on it reads: "Jesus Christ. I'm not ready to review this book. Who the hell is? It merely starts over, remakes logic and mathematics from a different beginning, from the Tao's beginning of the prime distinction. It's too simple to grasp. All I can make is the notes at the end of the book, and they keep raising the hair on my head. The book is pure revolution."[8]

At the end of that same year, Heinz von Foerster, the founder and director of the Biological Computer Laboratory at the University of Illinois from 1957 to 1975, sent Brand, with whom he had no previous relation, a gift copy of the *Whole University Catalog*, a seminar publication project produced in his classroom in large-format emulation of the *Whole Earth Catalog*.[9] Two months later, Brand sent von Foerster a request: "This book, *Laws of Form*, has everybody spinning. Like John Lilly has bought and given away 6 copies and keeps getting knocked into trance by the material in the book. Our problem is that nobody will review it. Will you?"[10] Brand had found his man. In setting forth von Foerster's avid review, a virtual introductory lecture on *Laws of Form*, the *Whole Earth Catalog* for spring 1970 documents a seminal moment in the gestation of NST.[11]

Marking that something long awaited had now come into the world, von Foerster began with this exclamation: "The laws of form have finally been written!" In an unusually lengthy entry, he expounded some of the rudiments of this "radical further step" toward the discourse of self-reference:

Laws are not descriptions, they are commands, injunctions: "Do!" Thus, the first constructive proposition in this book is the injunction: "Draw a distinction!" an exhortation to perform the primordial creative act. After this, practically everything else follows smoothly: a rigorous foundation of arithmetic, of algebra, of logic, of a calculus of indications, intentions, and desires; a rigorous development of laws of form, may they be of logical relations, of descriptions of the universe by physicists and cosmologists, or of functions of the nervous system which generates descriptions of the universe of which it is a part.[12]

"The nervous system . . . generates descriptions of the universe of which it is a part" is to say that the operation of an observing system is self-referential in the first instance. Its possibility rests with a form of inner delimitation or closure that opens up a cognitive relation to its outside. As von Foerster's occasional colleague at the Biological Computer Laboratory, Gotthard Günther, had noted in a 1962 article on the logic of observation, "systems of self-reflection . . . could not behave as they do unless they are capable of 'drawing a line' between themselves and their environment. . . . This is something the Universe as a totality cannot do. It leads to the surprising conclusion that *parts of the Universe have a higher reflective power than the whole of it.*"[13] Concepts of observation and self-reference were already implicit throughout the *Whole Earth Catalog*'s considerations of whole systems by means of new observational technologies that allow us to look back at or down on our own world and see it and ourselves in a newly recursive way. Von Foerster then cited an excerpt from Spencer-Brown's notes to that volume that describes this self-observational dynamic at a powerful level of epistemological generalization. Here, too, as in Günther, what becomes clear is the inexorably partial outcome of any effort to achieve an encompassing view: "We cannot escape the fact," Spencer-Brown writes, "that the world we know is constructed in order (and thus in such a way as to be able) to see itself":

This is indeed amazing.

Not so much in view of what it sees, although this may appear fantastic enough, but in respect of the fact that it *can* see *at all.*

But *in order* to do so, evidently it must first cut itself up into at least one state which sees, and at least one other state which is seen.

In this severed and mutilated condition, whatever it sees is *only par-tially itself.* We may take it that the world undoubtedly is itself (i.e., is indistinct from itself), but, in any attempt to see itself as an object, it must, equally undoubtedly, act so as to make itself distinct from, and therefore false to, itself. In this condition it will always partially elude itself.[14]

In Spencer-Brown's technical terms, the production of observations—that is, the ability to know in the first place—depends upon acts of distinction that sever what is observed into marked and unmarked states. Acts of distinction produce a virtual division into the marked state of that specific indication and the unmarked state of everything else. To know what a distinction indicates demands in that moment the non-observation of what that distinction has left aside. Yet, as we noted in the previous chapter about all meaning-events, such states of cognitive division are evanescent: when the observer moves on to the next distinction, both possibilities remain ready for selection, or not. As a result, there will be a blind spot in any picture we can have of ourselves, at which point we cannot see the whole of which we are a part. This is a way of restating the self-referential basis of any heteroreferential observation. Any attempt at an objective view retains a subjective component that must be added into any full account of what cognition may perceive, even though the addition of that supplement undoes, or in-completes, the ostensible totality of what it supplements. And any attempt at an objective view of *ourselves* is an oxymoron or a paradox, necessarily throwing the part of ourselves we are viewing *with* into momentary eclipse. In brief, only self-referential meaning systems can reflect on the way that their inability to grasp the whole of what they are grounds their ability to know at all. With the inclusion of von Foerster's review of *Laws of Form,* the text of the *Whole Earth Catalog* virtually deconstructed its own holistic idealism. This was the "pure revolution" Brand intuited but could not yet articulate.

Many Cybernetic Frontiers

After bringing out the *Last Whole Earth Catalog* in 1971, Brand put *Catalog*-style publishing on a three-year hiatus to pursue other projects. One of these appeared in 1974 as the slim volume *II Cybernetic Frontiers,* compiled from two separately published articles plus appendices. Brand's introductory remarks placed him squarely within the systems counterculture he had already done so much to define and promote:

> I came into cybernetics from preoccupation with biology, worldsaving, and mysticism. What I found missing was any clear conceptual bonding of cybernetic whole-systems thinking with religious whole-systems thinking. Three years of scanning innumerable books for the *Whole Earth Catalog* didn't turn it up. Neither did considerable perusing of the two literatures and taking thought. All I did was increase my conviction that systemic intellectual clarity and moral clarity must reconvene, mingle some notion of what the hell consciousness is and is for, and evoke a shareable self-enhancing ethic of what is sacred, what is right for life.[15]

With a few strokes, Brand nailed the schizophrenia of modern American society—split then as now between military and corporate technocracies in the ascendant and ecosystems, communities, and psyches in splinters—and supplied the spiritual rationale for a countercultural supplement to the powerful but restricted "intellectual clarity" of mainstream control theory.[16] On the evidence of this presentation, the point of the whole-systems thinking Brand purveyed in the *Catalog* and beyond was to work out the possible forms of systemic integration between two cybernetic rationales, one intellectual, the other moral. "Tall order," Brand continued. Then he found Gregory Bateson:

> In the summer of '72 a book began to fill it for me: *Steps to an Ecology of Mind. . . .* Here in one single-minded book was highly original application of cybernetics, biology, linguistics, psychology, and formal logic to field work with New Guinea and Balinese natives, porpoises, alcoholics, schizophrenics, beetles, and national histories . . . a rigorous scientific refutation of the notion that rational science is adequate to save us. (9–10)

The first run of the *Whole Earth Catalog* foregrounded the writings of Buckminster Fuller, another whole-systems thinker of substantial proportions but quite different qualities. By 1973, Brand's primary cybernetic frontier was the work of Gregory Bateson. The chapter "Both Sides of the Necessary Paradox" in *II Cybernetic Frontiers* begins: "Cybernetics is the science of communication and control. It has little to do with machines unless you want to pursue that special case. It has mostly to do with life, with maintaining circuit" (9). Brand's declaration at this moment may now sound odd, since the last fifty years have only reinforced the notion that cybernetics has little to do with anything *other* than machines. However, it accurately registered the particular valences of Bateson's cybernetic worldview. In that world, the notion of "circuit" is a prominent mental operator. Bateson's "circuit" puns on the term's electronic sense and resonates with the informatic sense of a communications system in which messages circulate. However, his cybernetics of "circuit" expand to concern the overall circularity of closed loops or cycles as these convey meaningful differences in systemic ensembles of any sort. This milieu of communicative loops is more or less what Bateson meant by his "ecology of mind." Perhaps the best passage in Brand's Bateson chapter wrestles precisely with Bateson's idea of "circuit":

> I'm still getting used to the way Gregory uses the term "circuit." It's appealing to me because it is at once more general than "feedback loops," more accurate somehow, and more open-system . . . as if it can include cycles of interactive learning (student teaches the teacher to teach the student better), of material (flesh to ashes to flesh), of slow recurrence (every so often an ice age stresses the system), of standard homeostatic feedback (the chilled body shivers until warm), and of observer interference (the watched porpoise bedevils his observer). Without circuit, without continual self-corrective adjustment, is no life. (29)

II Cybernetic Frontiers then segues to "Fanatic Life and Symbolic Death among the Computer Bums," republished from a 1972 *Rolling Stone* article for a treatment of *Spacewar*—the ur-computer game developed at Stanford University in the predawn of the personal computer. His approach to this second topic still sounds slightly odd, when Brand recurs to the same framing device he used for the Bateson chapter: "And now, to pursue *the 'special case' of machine cybernetics*—computers and computer science—the sovereign domain of rational purpose, of explicit goal-directed behavior" (38; my italics). And yet, a larger sample of

Brand's statements on this matter include his headnote a decade later to James Lovelock's 1983 article "Daisyworld: A Cybernetic Proof of the Gaia Hypothesis": "Since cybernetics was kidnapped by computer science a couple decades back, there have been few working applied cyberneticians loose in the world. Lovelock . . . is one."[17] At the end of this chapter we will come back to Daisyworld, a computer model for Gaian self-regulation. However, with due respect to Lovelock's Daisyworld, was there ever really a time when organic cybernetics, a cybernetics "mostly to do with life," was *not* the special case relative to the general case of "machine cybernetics" and its vast array of computational creations? We can see that by 1983, Brand's Batesonian *CoEvolution Quarterly* of the 1970s—the *CoEvolution Quarterly* of von Foerster, Lovelock, Margulis, and Varela—was already morphing into Kevin Kelly's digitopian *Whole Earth Review* of the later 1980s and beyond.[18] Instead, let us consolidate the recognition that there was an extended moment when organic cybernetics, or a cybernetics of natural as opposed to designed systems, teemed with fresh creations—most notably, the concept of autopoiesis and the Gaia hypothesis—and scattered these hardy spores across some receptive regions of the intellectual landscape. Particularly in the Batesonian milieu of the systems counterculture of the 1970s, a cybernetics "mostly to do with life" helped these neocybernetic productions to unfold.

CoEvolution Quarterly

Brand was well primed to receive with some enthusiasm an authentic cybernetic theory approaching life in relation to Earth's atmosphere as a "whole system." The Gaia hypothesis would make numerous appearances during the eleven-year span of *CoEvolution Quarterly.*[19] Underscoring our focus on the development of systems discourses as they informed Gaia concepts, Brand's evocation of organic cybernetics precisely names the spirit of the periodical he would spin off from the *Whole Earth Catalog.* *CoEvolution Quarterly* also carried forward much of the *Catalog*'s thematic formatting: every number of *CoEvolution Quarterly* also began with a section on "Understanding Whole Systems" or just, on occasion, "Whole Systems." *CoEvolution Quarterly*'s cast of contributors included Gregory Bateson, Paul Ehrlich, Howard Odum, Aldo Leopold, William Irwin Thompson, Wendell Berry, John Todd, Marshall

McLuhan, Ursula Le Guin, Ivan Illich, E. F. Schumacher, Gary Snyder, Margaret Mead, Garrett Hardin, Carl Sagan, Ramón Margalef, Roy Rappaport, Hazel Henderson, Kenneth Boulding, Peter Warshall, and Donella Meadows, among many others. This section of *Gaian Systems* samples selected contents of *CoEvolution Quarterly* as they relate to the neocybernetics of the Gaia hypothesis.

THE GAIA HYPOTHESIS

The front cover of the first number of *CoEvolution Quarterly* premediated the unfolding of Gaia's coevolutionary cosmology (Figure 3). While its point of self-reference remained the human observer, in this image the Earth seen from space has been exploded into a planetary or cosmic visage made up of microbes and galaxies and things in between.

Initially, Lovelock published his own preliminary Gaian work in specialized scientific periodicals.[20] Similarly, when *Science* and *Nature* rejected Lovelock and Margulis's first collaborative articles, they placed them in *Tellus, Icarus,* and *Origins of Life.*[21] And there the whole matter might have rested, buried in relative obscurity, were it not for *CoEvolution Quarterly*'s putting the Gaia hypothesis on the cover of its summer 1975 number. When Brand expressed interest that spring, Margulis welcomed the opportunity, as we learn from the draft of a letter to Lovelock written onboard a flight to a speaking engagement in St. Louis, "where I have to discuss the origin & evolution of everything in about ½ hour":

> Good news— . . . I've spoken today to Alan Ternes, editor of *Natural History* (a classy glossy job with a circulation of 370,000). He's apparently a friend of Stewart Brand, editor of the *Co-evolution Quarterly.* Brand, who has been pressuring me mightily, claims his mag. has a circulation of only 17,000. They apparently are in agreement that *Nat. Hist.* will publish the Gaia II & that appearance (even prior appearance) in *Coev. Q.* will not jeopardize a full article in *Nat. Hist.* . . . [Brand] is claiming that his journal is responsible and responsive, refuses to compartmentalize science and that my accusation that he's into food faddism & astrology is totally unfounded. At any rate, what he wants from us is permission to excerpt apparently nearly all Gaia II with the statement that [it] is from a full article coming out in *Nat. History.* I told him that I could not give him permission unilaterally but must consult you. Since he now has a definite commitment

Supplement to the Whole Earth Catalog

The **CoEVOLUTION** Quarterly

$2
1974
Spring

FIGURE 3. *The first* CoEvolution Quarterly, *front cover, spring 1974.*

from Ternes at *Nat. Hist.* and since after reading CQ I find myself sympathetic to his goals, I would hope you will agree to this plan.[22]

As matters turned out, "Gaia II" got the royal treatment in the low-circulation, low-budget countercultural journal *CoEvolution Quarterly* but short shrift from the glossy mainstream magazine *Natural History.* It did appear there, over a year later, not expanded but condensed and reframed, as "Is Mars a Spaceship, Too?"[23] *Natural History* returned the Gaia hypothesis to the context of alien life detection then on the public's mind due to the imminent arrival on Mars that summer of two Viking landers. Margulis and Lovelock's concluding remarks put the best face on a high-circulation debut somewhat muffled for general consumption. Turning Lovelock's original logic around, they noted that if the landers did find life on Mars (if that planet is, like Earth, "a spaceship"), that would disprove his circumstantial Gaian claim that the presence of active life must leave a detectable signature in the atmosphere of a planet that possesses it: "Failure of the Viking mission to find life on Mars will not prove the existence of Gaia, but it will add support to the hypothesis. Most scientific experiments are designed to disprove a hypothesis; when they fail the hypothesis is thereby strengthened. At great cost and effort, a rare planetary experiment for the Gaia hypothesis is now speeding toward a conclusion" (90). As anticipated, the Viking's non-result left the Gaia hypothesis intact.

However, *CoEvolution Quarterly* had welcomed the Gaia hypothesis a year earlier with no such muted exposition, becoming the first non-scientific journal to treat the topic. Lead-authored by Lynn Margulis with what is still a distinctly specialized treatment, Brand lets "The Atmosphere as Circulatory System of the Biosphere: The Gaia Hypothesis" sprawl over ten pages in the midst of graphics, diagrams, data tables, and multiple excerpts from previously published articles. It anchors an "Understanding Whole Systems" section that also includes Earth-seen-from-space articles by Carl Sagan and former astronaut Rusty Schweickart and a glowing review of Margulis's first book, *Origin of Eukaryotic Cells,* by Beat poet Michael McClure. A substantial extract from Ramón Margalef's 1968 text *Perspectives in Ecological Theory,* featuring "The Ecosystem as a Cybernetic System," resonates with the Gaia article that precedes it.[24] The article itself led its reader into the topic with a reproduction of Sachs von Lewenheimb's 1664 engraving *Oceanus Macro-Microcosmicus* (procured by Margulis), embellishing

William Harvey's description of the circulatory system of the body with worldly imageries of rivers, oceans, and winds, as an analogy for the Gaian role of the atmosphere ensuring the circulation of nutrients and flushing of wastes for the planetary "body." The article's assorted illustrations include what is surely an editorial image—that is, not one that Margulis herself provided for publication—of an archaic sculpture of the Earth goddess as a popular hook for the scientific message. And so began the history of cross-currents in the reception of Gaia between its authors' quests for scientific bona fides and its lay enthusiasts' desire for mythic resonances.

Margulis and Lovelock's sober exposition of the Gaia hypothesis breaks the argument down into elemental ecological cycles (the carbon cycle, the nitrogen cycle, etc.) and their relation to living systems, emphasizing the complexity and multiplicity of the phenomena under review. These early Gaia arguments do not bear on the biosphere as a whole or the planet altogether but specifically on the planetary atmosphere enveloping the biosphere as both source and sink for metabolic processes. For instance, with the key biological elements (carbon, nitrogen, oxygen, hydrogen, sulphur, and phosphorus),

> cycling times must be short because biological growth is based on continual cell division that requires the doubling of cell masses in periods of time that are generally less than months and typically, days or hours. On lifeless planets there is no particular reason to expect this phenomenon of atmospheric cycling, nor on the Earth is it expected that gases of elements that do not enter metabolism as either metabolites or poisons will cycle rapidly.... Because biological solutions to problems tend to be varied, redundant, and complex, it is likely that all of the mechanisms of atmospheric homeostasis will involve complex feedback loops.[25]

Brand's headnote wryly observed that for some persons, such serious scientific fare might appear anomalous in his venue, to the detriment of its authors: "Margulis and Lovelock will doubtless take some flak for appearing in suspect company—condom evaluations, poetry, and such." Nevertheless, its inclusion in *CoEvolution Quarterly* was an inspired intervention for all concerned. The authors' own headnote began: "We would like to discuss the Earth's atmosphere from a new point of view— that it is an integral, regulated, and necessary part of the biosphere"

(30). In the article proper, they restate the main point: "The purpose of this paper is simply to present our reasons for believing the atmosphere is actively controlled" (32).

The issue of Gaia's material circulation frames their cybernetic hypothesis regarding its operational circularity leading to self-regulation effecting the homeostasis of atmospheric composition. The "control" of the atmosphere was the hypothesized emergent outcome of a closed loop of biogeochemical cycles held in homeostasis at the planetary level, that is, feedback-regulated throughout much of geological time. The notion that this took place both "by and for" the biota was a classic piece of Gaian hyperbole that seems by now perhaps less hyperbolic.[26] However, the abiding picture here is that Earth's atmosphere, the repository for the gaseous wastes circulating in and out of organic capture and use, has evolved along with the life that has pumped it up. In the Gaian view, the maintenance of the atmosphere in viable proportions of oxygen and nitrogen is not the abiotic lucky happenstance of traditional geology but a systemic outcome operationally integrated with and regulated by the coupling of life and Earth. The stable persistence of these proportions over eons of fluctuations depends on living, largely microbial processes coupled to geological dynamics.

Such a properly granular and tentative account of the Gaian system's production of atmospheric self-regulation can always give way to the smooth discursive space in which a holistic vision covers over the numerical immensity of biological forms and concrete complexity of geological cycles. Unsurprisingly, Brand's headnote to this article also seized the "whole Earth" potential of the argument: Gaia "treats the anomalous Earth atmosphere as an artifact of life and comprehends the planet itself as a single life" (31). Brand's editing reinforced this mode of appreciation a page later by interpolating underneath the main article an excerpt from "The Quest for Gaia," a previously published piece coauthored by Lovelock and Sydney Epton. Bundled together here are the extreme claims of the early Gaia hypothesis—cooperation, totality, optimality, "control" transferred from the abiotic environment to the biota—that ruffled the established science of that moment:

> The atmosphere looked like a contrivance put together co-operatively by the totality of living systems to carry out certain necessary control functions. This led us to the formulation of the proposition that living matter, the air, the oceans, the land surface were parts of a giant

FIGURE 4. *The* Next Whole Earth Catalog, *back cover, fall 1980 (detail).*

system which was able to control temperature, the composition of the air and sea, the pH of the soil and so on so as to be optimum for survival of the biosphere. The system seemed to exhibit the behavior of a single organism, even a living creature. (32)

However, for all the cultural play between cybernetic specificity, holistic generality, and archetypal profundity in the Gaia conception, it may now be evident why the Gaia discourse brilliantly mediated through this 1975 extravaganza would find a dedicated audience among the countercultural intelligentsia gathered by *CoEvolution Quarterly* and the wider Whole Earth network. Its promissory countervision of life taking care of business in its own house put to shame the "selfish gene" of the same moment.[27] Five years later, on the back cover of the *Next Whole Earth Catalog*, an oversized return to the compendious catalog format, the name of Gaia is now affixed to an image of Earthrise (Figure 4).[28] Brand's caption summed up the popular Gaia notions of totality and singularity that would take root in ensuing decades: "The Gaia

Hypothesis, as proposed by the British scientist James Lovelock, suggests that the Earth's atmosphere and oceans are maintained as highly sophisticated buffering devices by the totality of life on the planet. The whole Earth, in other words, may function as a single self-regulating organism."

"ECOLOGICAL CONSIDERATIONS FOR SPACE COLONIES"

Princeton physicist Gerard K. O'Neill's proposals for space habitats in high orbit and their memorable depictions by NASA artists were contemporaneous with the introduction of the Gaia hypothesis. *CoEvolution Quarterly*'s publication of Margulis and Lovelock's "The Atmosphere as Circulatory System" in the summer of 1975 immediately preceded a fall number devoting the first thirty pages to O'Neill's proposals for high-orbital space colonies around the Earth. O'Neill's speculative technological constructions presented images of environmental closure that translated Gaia's terrestrial implications into idealized visions of sustainable habitats.[29]

The planetary imaginary of the Gaia concept and the microplanetary imaginary of the massive space habitat called out to each other under the trope of Spaceship Earth. With the Apollo program now wound down, O'Neill envisioned a "high frontier" beyond JFK's New Frontier. He had already been working at this project for several years with modest preliminary support from NASA. *CoEvolution Quarterly* published a long transcript of his congressional testimony seeking a major NASA commitment. Six months later still, in its spring 1976 number, *CoEvolution Quarterly* devoted eighty pages to the controversy that erupted over its positive presentation of O'Neill's vision as a potential form of countercultural commune in the sky. Brand worked the space-colony debate for *CoEvolution Quarterly* content by pitting post-psychedelic space-oriented technophiles such as himself against Whole Earth–identifying environmentalists and green technophobes. Numerous supporters and detractors responded, including Buckminster Fuller and Paolo Soleri, novelists Ken Kesey and Wendell Berry, the poets Gary Snyder and Richard Brautigan, astronaut Rusty Schweickart, cultural observer William Irwin Thompson, scientists Lynn Margulis, Paul and Anne Ehrlich, and Carl Sagan. A year later, Brand gathered, republished, and expanded these materials as the freestanding paperback volume *Space Colonies*.[30]

In the midst of these developments, Margulis gathered with a band of fellow ecologists to take up the issue, and *CoEvolution Quarterly* co-published their two-page multiauthored position statement.[31] Brand's headnote to "Ecological Considerations for Space Colonies" identified Margulis and two of her colleagues from the Woods Hole Oceano-graphic Institution on Cape Cod as the prime movers of the piece. It began: "There appears to be growing interest in the possibility of estab-lishing large space colonies capable of supporting hundreds or thou-sands of people in isolation from the earth for long periods. . . . Such colonies would present extremely difficult biological and ecological problems. These should be addressed at the very outset if any serious effort toward designing satellites or colonies on celestial bodies other than the earth is to proceed" (96). These cautionary sentiments are as cogent now as they were in 1976, but you would hardly know that from the coverage of Jeff Bezos's revival of O'Neill's colonies as a scheme wor-thy of serious reconsideration. Both of Gaia's authors were cited in "Eco-logical Considerations," for which text Margulis may have been largely responsible. Although this article never mentions the Gaia concept by name, it purveys a broadly Gaian sensibility applied to vexed issues in the capacity of closed environments to maintain habitability.

"Ecological Considerations" conveys the tight conceptual coupling we have noted between ecosystem ecology and the Gaia hypothesis. The basic proposition of such engineered habitats, once their shells are set in place, is the artificial but viable replication of the Earth's own "eco-system services" within a materially closed vessel—in other words, the design and production of miniature Gaias (Figure 5). Margulis and Sagan would later observe that sending forth such constructions from the Earth was like Gaia itself bearing designed offspring through tech-nological mediation: "The generation of new 'buds,' materially closed living systems, within the 'mother biosphere' resembles the structure of a fractal. Closed ecosystems are not artificial at all, but part of the natural processes of self-maintenance, reproduction, and evolution."[32] However, the opening passage of "Ecological Considerations" also rang the sobering note of Gaian complexity, the exceedingly recondite dy-namics by which viable planetary regimes arose and maintained op-erations, the wicked difficulty of achieving working replicas of living ecosystems in artificial form. This text summarized O'Neill's proposal as the aim "to build a new meta-stable ecosystem, complete with biotic

FIGURE 5. *Artist's rendering of a large materially closed ecology,* CoEvolution Quarterly, *spring 1976.*

resources and closed cycles for other essential resources, and capable of supporting man over long periods"; however, "No such system has ever been constructed on earth. The probability that such a system can be built and maintained indefinitely at present seems remote. It seems especially remote when we realize that we have no background in the analysis of the problem" (96). Sustained research into the Gaia hypothesis would amount precisely to an "analysis of the problem," but at that moment such efforts were confined to disparate and inconclusive research projects by ecosystem ecologists at the scale of terrariums and greenhouses.

O'Neill's own promotional materials provided scant recognition of this reasoned but downbeat assessment, assuming on the contrary that the interior ecologies of miniature self-supporting worlds fit for long-term human habitation were engineering problems to be mopped up once the structures were hurtled into orbit. In an interview with Brand under the title "Is the Surface of a Planet Really the Right Place for an Expanding Technological Civilization?" O'Neill explained the reason why it was not. Why, he asked, just when we've perfected the rockets to lift us out of Earth's gravity hole, should we want to drop down another one?

> The classic science fiction idea of colonization is always you go off and you find another planetary surface, like the moon or Mars. . . . They're the wrong distance from the sun. . . . The sort of analogy I like to use nowadays is to say that, "Here we are at the bottom of a hole which is 4,000 miles deep. We're a little bit like an animal who lives down at the bottom of a hole. And one day he climbs up to the top of the hole, and he gets out, and here's all the green grass and the flowers and the sunshine coming down. And he goes around and it's all very lovely, and then he finds another hole, and he crawls down to the bottom of that hole.["] And if we go off and try to get serious about colonizing other planetary surfaces, we're really doing just that.[33]

Here is a curious sort of planetary imaginary in reverse. On the surface of Earth, gravity has been holding us down, but we can now be done with it by going off into space. Free space is not only there for free, it's also free of gravity, whose absence we can turn to our advantage in a high-orbital environment. In his article that opens the *Space Colonies* volume, "The High Frontier," O'Neill declared that "the L5 Earth-Moon Lagrange libration point . . . could be a far more attractive environment for living than most of the world's population now experiences."[34] But, he cautioned, we lose that advantage if we just capitulate to another, alien gravity hole. In contrast with life in a high-orbital space colony—the virginal pastoral Eden of an idealized outdoorsy Earth with "the green grass and the flowers and the sunshine coming down," life on Earth has always been mired at the bottom of a deep hole. The ironies within O'Neill's preposterous turns of phrase underscore the larger problems with his scheme. These were abundantly documented in the "Ecological Considerations" statement as well as the critical commentaries Brand published alongside O'Neill, which we will take a closer look at in chapter 7. It is simply that the range of knowledge needed to engineer materially closed ecologies making the proposed space colonies even temporarily habitable without constant resupply did not exist then, nor does it now. Sagan and Margulis would speculate a decade later that "the full scientific exploration of Gaian control mechanisms is probably the surest single road leading to the successful implementation of self-supporting living habitats in space."[35] But in 1975, O'Neill took that implementation for granted.[36]

HEINZ VON FOERSTER

The early pages of *CoEvolution Quarterly* contain a number of items associated with Brand's resident expositor of *Laws of Form,* the spirited cyberneticist Heinz von Foerster. It was von Foerster who coined the phrase "second-order cybernetics" and developed its initial formulations.[37] In the early 1970s, with a strong push from the *Whole Earth Catalog*'s construction of the countercultural sensibility, key lines of cybernetic thought were morphing from grim control theory into playful explorations of the paradoxes of cognition. Von Foerster factored into systems discourse the logical binds of self-reference as a positive force, precisely as a cybernetics of cybernetics, that is, as a creative program to observe (at "second-order") cybernetics' own processes of system observation. His encyclopedia article on cybernetics later presented this as "the logic of autology, that is, concepts that can be applied to themselves."[38] A veteran of the Macy conferences and formidable authority in cybernetic matters, von Foerster was also a gracious and consequential facilitator of collegial connections within the systems counterculture.[39]

Let us come back to the composite text of the maiden appearance of the Gaia hypothesis in *CoEvolution Quarterly.* Running under a portion of "The Atmosphere as Circulatory System of the Biosphere" was a substantial excerpt from the Gaia article lead-authored by Lovelock, "Atmospheric Homeostasis by and for the Biosphere: The Gaia Hypothesis." The content of this republished sidebar, subtitled "Gaia and Cybernetics," was concerned more precisely with those larger regions of systems theory extending the discourse of entropy from thermodynamics and statistical mechanics to information theory.[40] The Gaia concept grew out of Lovelock's seminal innovation with regard to life detection at the planetary level: instead of going to great lengths to make contact with planetary surfaces, one could analyze their atmospheres at a distance for signs of an entropy reduction. Here, too, Lovelock and Margulis located Gaia in relation to the entropy concept, as that complicated quantity offered a "first cautious approach to a classification of life, . . . as follows: 'Life is one member of the class of phenomena which are open or continuous reaction systems able to decrease their entropy at the expense of free energy taken from the environment and subsequently rejected in a degraded form'" (36). The excerpt in question went on to give several mathematical formulas treating measures of both thermodynamic and informatic entropy.

As it happened, however, the next number of *CoEvolution Quarterly* devoted a full page to an extended letter to the editor from von Foerster. Titled "Gaia's Cybernetics Badly Expressed," it asserted defects in the mathematical treatments of entropy offered in "Gaia and Cybernetics." Von Foerster's objections were technical and did not touch one way or another on the main theme toward which Lovelock and Margulis were pointing their invocations of entropy, the properly cybernetic issue of the discrimination of boundaries between systems and their environments. It is by determining such boundaries that one can measure, in the current instance, entropy reductions within living systems and entropy increases without. I will dilate on this discussion for a moment, as we will be coming back to the issue of Gaia's boundaries in chapter 8. Lovelock's particular focus on entropy reduction as a systemic marker for life processes runs deeply through his development of the Gaia concept. *CoEvolution Quarterly*'s republished excerpt "Gaia and Cybernetics" returns to this topic:

> When the whole assembly of life is so seen it is clear that the true boundary is space. The outgoing entropy flux from the Earth indeed from Gaia "if she exists," is long wavelength infra-red radiation to space. This then, is the physical justification for delineating the boundary of life as the outer reaches of the atmosphere. There is also to a lesser extent an inner boundary represented by the interface with those inner parts of the Earth as yet unaffected by surface processes. We may now consider all that is encompassed by the bounds as putative life. Whether or not Gaia is real will depend upon the extent to which the entropy reduction within a compartment such as the atmosphere is recognizably different from the abiological steady state background.[41]

Von Foerster's critique left these cogent matters, which were consonant with his own treatments of entropy and self-organization, unremarked.[42] Instead, he offered a parting shot at Lovelock's participation in the widespread reification of "information," a concept by then mainstreamed in cybernetical deployments of information theory, treating these quantities as real rather than virtual, that is, as if information theory measured physical substances rather than statistical differences.[43] This conceptual issue gave von Foerster an opportunity to rehearse his own conversion to the constructivist perspective unfolding at that moment in 1975 from Humberto Maturana's biology of cognition, from the theory of autopoiesis, and from von Foerster's own second-order cybernetic revisions of epistemological ideas:

> I would hope that we shall never tire of reminding ourselves and each other that "complexity," "disorder," "entropy," "information," "order," "organization," "simplicity," etc., are not names for properties of things, but those for properties of descriptions, or—if you wish—are names reflecting properties of the observer (describer), his vocabulary, his natural or chosen limits of discrimination, etc., in short, his idiosyncrasies at the time of his observation.[44]

This is a pure statement of neocybernetic epistemology. Von Foerster's lively response to Margulis and Lovelock's article in *CoEvolution Quarterly* did not dismiss the main ideas of the Gaia hypothesis. His tone was collegial and colloquial. However, he wrote, "I am unable to (re)-construct the proper representation of these equations' intent." There does appear to have been transmitted mistakes of mathematical expression in the excerpt from the "Atmospheric Homeostasis" article in *Tellus*.[45] Von Foerster continued, "Moreover, sitting in the boondocks I have no way to find out who is to be charged with these booboos: CQ who misprints Lovelock and Margulis; Lovelock and Margulis who misquote Denbigh (1951) and Evans (1969); or Denbigh and Evans who misunderstand. But this is not my job." However, despite the disagreements over philosophical semantics von Foerster noted above, his critique also affirmed that "I found Lovelock's and Margulis's ideas too important to see them becoming vulnerable because of deficiencies of a different kind. As a comment on their—or anybody else's—classification of Life I suggest that you reproduce 'Autopoiesis: The Organization of Living Systems, its Characterization and a Model.'" Clearing side issues away from matters of true import, von Foerster's commentary is to my knowledge the first and original suggestion of a relation between the Gaia hypothesis and the theory of autopoiesis as a noninformatic description of the recursive organization and operational closure of cognitive systems.[46]

FRANCISCO VARELA

CoEvolution Quarterly continued to introduce its readers to the newer cybernetics beyond the seminal instance of Gregory Bateson's ecology of mind. The summer 1976 number features back-to-back interviews. One is Brand's own lengthy interview with Bateson and his first wife, Margaret Mead, delving into their participation at the fabled Macy conferences on cybernetics.[47] Preceding this item is "On Observing Natural Systems," a fine interview conducted by Donna Johnson with the co-author of the concept of autopoiesis, Francisco Varela. Von Foerster had recommended Varela, the brilliant young collaborator of his friend and colleague Humberto Maturana, to Brand, who arranged for his debut among the California systems intelligentsia. In his headnote to the interview, Brand situated Varela's work by rehearsing von Foerster's trademark epistemological themes and stock articulations of the distinction between first- and second-order cybernetics: "I share the opinion of Ludwig Wittgenstein, Gregory Bateson, G. Spencer Brown, Heinz von Foerster and others that failure to understand self-reference is the poison in the brain of most Western misbehavior, public and personal. In his recent landmark paper, 'A Calculus [for] Self-Reference' and in this interview, Francisco is helping build what von Foerster calls 'a cybernetics of *observing*-systems,' which is the rest of the story after 'the cybernetics of *observed*-systems'—feedback, goal-seeking, and such."[48]

"On Observing Natural Systems" encapsulated this very discourse, and in particular the confluence of Spencer-Brown's *Laws of Form* with von Foerster's cognitive constructivism, which would later prove seminal in particular for the full development of Luhmann's systems theory.[49] The following excerpts give the flavor of Varela's conversation. After some preliminary clarifications, his interviewer probed the discussion of "whole systems" further by asking, "So studying the organization of a whole system is studying the nature of its self-reference?"

> Varela: That's it. That is, the kind of self-referential organization that has provided the stable properties that it shows. And this is what gives the system its nature. When you have a closed interaction of chemical productions, you can have a cell, and not before that. When you have a closed interaction of descriptions, you can have self-consciousness, and not before. When you have a closed interaction of species, you have an ecological system, and not before. That is, the closure, the

self-referential-ness, seem to be the hinges upon which the emergent properties of a system turn. (27)

In the following chapter we will look further into the history of contacts between Varela and Margulis. One could speculate on that basis that his formulation of "a closed interaction of species" may well have infiltrated her formulation of autopoietic Gaia as "the organismal-environmental regulatory system at the Earth's surface, comprised of more than thirty million extant species."[50] At the least, it is reasonable to think that Varela's neocybernetic style of systems theory first and effectively came to Margulis's attention in this *CoEvolution Quarterly* interview, which was published precisely one year after the *CoEvolution Quarterly* debut of her lead-authored article on the Gaia hypothesis. And regarding the epistemological discourse at the nexus of Maturana, von Foerster, and Spencer-Brown, just as von Foerster had insisted that descriptions primarily reflect the properties of the observer, Varela went on to underscore "the full importance of introducing the observer into the observation":

> Whatever we purposely distinguish will reveal not only the properties we are looking at but the fact that we are doing these interactions out of our own properties, that is, the properties we discover in systems will depend on our own properties. In its purest form, that means that whatever description we do of the world will be based on the act of splitting it apart in different ways. And the way we see the world, therefore, reveals what is our choice of cleavage, as it were, and that there are many of them precisely because there are many observers. (29)

The prior epistemological distinction between subjects and objects becomes relative to an observer's choice. *Laws of Form* conveyed an equivalent point: the cognitive paradox of self-reference cuts the ground out from under the pretense of unmediated objectivity. Presumably it was Brand who placed a sidebar next to the text of "On Observing Natural Systems" citing the same passage from *Laws of Form* that von Foerster had quoted in his review in the *Whole Earth Catalog*: "We may take it that the world undoubtedly is itself (i.e., is indistinct from itself), but, in any attempt to see itself as an object, it must, equally undoubtedly, act so as to make itself distinct from, and therefore false to, itself. In this condition it will always partially elude itself."[51] Let us apply this epistemological statement to the observation of Gaia. The Gaian instance would be the final horizon of this epistemological reckoning: we cannot really fathom Gaia as a planetary system without looking, self-referentially, at ourselves, a part of Gaia, looking at Gaia. Objectivity is surpassed by participation. Moreover, Spencer-Brown's epistemological critique of holism informs Varela's constructivist looping of the properties of the observer into the attributes of their descriptions. It bears repeating that the partial Earth of our discussion in chapter 3 reappears here, hiding in plain sight in the glare of Whole Earth discourse.

In the summer of 1976, concurrently with the publication of "On Observing Natural Systems," Bateson and Brand organized a "Mind-Body Dualism Conference," also attended by von Foerster and Varela.[52] Varela returned to the pages of *CoEvolution Quarterly* that fall with "Not One, Not Two: Position Paper for the Mind-Body Conference," material he would streamline for incorporation into his first book, *Principles of Biological Autonomy,* published three years later.[53] In six packed pages of small type, "Not One, Not Two" addresses the conference theme of Cartesian dualism by formulating "Star cybernetics." "Star is (can be taken to be) a compact expression to signify a broad paradigm encompassing that series of convergencies rightly demanded by Bateson" in his invitational paper to the conference.[54] Varela lists this series as:

cybernetics↔*epistemology*↔*evolution*↔*ethics*↔*cognition*↔*ecology*[55]

I will draw out Varela's cybernetic reasoning in this article up to the point that its relevance to Gaian thought becomes clear. A decade later Varela will address Gaia theory directly.

My exposition of Varela's Star cybernetics greatly condenses the detail of the argument and extracts it from Varela's own highly formalized presentation. So then, the formula for what Varela calls "the Star* statement" is:

"the it/the process leading to it" (62)

This formula has not two but three parts—the components on either side of the slash and the "whole system" computed by the integration of those components. How then does one create a Star statement in any given instance? Varela explains:

> Take any situation (domain, process, entity, notion) which is holistic (total, closed, complete, full, stable, self-contained). Put it on the left side of the /. Put on the right side of it the corresponding processes (constituents, generators, dynamics). For example:
>
> being/becoming   environment/system
> space/time   context/text...
>
> In each case the dual elements become effectively complementary: they mutually *specify* each other. There is no more duality in the sense that they are effectively related; we can contemplate these dual pairs from a metalevel where they become a cognitive unity, a second-order whole. (63–64)

Star cybernetics flirts with philosophical holism but complicates it with a standard technique for resolving logical impasses: if one's view of a solution is blocked at one level, one adds another dimension to the view of the situation. The shift to a metalevel creates the conceptual space needed to sublate that obstruction. As we will see in a moment, placing affairs in a Star statement adds a second level to an observation. By contrast, Varela goes on, traditional idealism tends to stay on one level:

> In what I call the classic or Hegelian paradigm, the notion of dualities is tied to the idea of polarity, a clash of opposites:

> The basic form of these kinds of dualities is *symmetry*: Both poles belong to the same level. The nerve of the logic behind this dialectic is *negation*; all pairs are of the form *A/not-A* (e.g., *+/-, oppressor/oppressed*). (64)

Fixing negative relations between polar opposites, classical dialectics tend to block the view beyond the opposition proper. Impasse is built into this style of observation, and in any real-world situation, solution by the synthesis of antithetical entities is difficult if not impossible. However:

In our (shall we say) cybernetic or post-Hegelian paradigm, dualities are adequately represented by *imbrication* of levels, where one term of the pair emerges from the other.

The basic form of these dualities is *asymmetry*: Both terms extend across levels. The nerve of the logic behind this dialectics is *self-reference*: pairs of the form: *it/processes leading to it.* (64)

Let us formulate an instance corresponding to the theme of the meeting at which Varela was delivering this talk. If we map the mind–body dualism onto what Varela calls the classic or Hegelian paradigm, we get the Cartesian split. Negation rules, and self-division holds the field. However, if we map it as a Star statement, supplementarity absorbs negation, and self-reference heals self-division. In this form, both mind and body refer to the same self. The unity of their distinction emerges on the metalevel already implicit in the differentiation of levels constelled by the Star formation. Varela comments: "Pairs of the form Star *bridge* across one level, and this crossing is operational. They mutually specify each other" (64). Here is the strong neocybernetic note: "this crossing is operational" because both terms are now seized not as entrenched opposites but in their operational interdependence. Elsewhere I have called the logical modality of Star cybernetics *double positivity,* as in the imbricated or embedded and mutually supplementary relations of co-emergence between an autopoietic system and its environment. The system generates a boundary within which it carries out its dynamics and from which base of emergent possibilities it defines its environment not by negation but by operational distinction.[56] A Gaian observation of double positivity would be that from the moment of Gaia's coalescence, all earthly niches within the Gaian sphere have co-emerged from or in structural relation to the operations of living systems in the same measure that living beings have held their terms of existence in correlation with the contingencies, affordances, and limitations of their enabling environments.

In the fullness of "Not One, Not Two," Varela works the Star paradigm through numerous examples, but let us conclude with one that brings us back to "natural systems," Brand's "organic cybernetics," and our wider neocybernetic consideration of the Gaia concept:

It is, of course, the case that when we look to natural systems, nowhere
do we find opposition apart from our own projections of values. The
pair predator/prey, say, does not *operate* as excluding opposites. Both
generate a whole unity, their ecosystemic domain, where there is
complementarity, mutual stabilization, and benefits in survival for
both. So, although we can project values to the opposites predator/
prey, the effective duality is a larger one, of Star form: *ecosystem/
species interaction.* (64)

One of the oldest and most pointless debates over the Gaia hypothesis
concerns its supposed contradiction of competition as a primary evolutionary driver. Gaia is said to flout the supposed competitive rule that
would map evolutionary interactions among individuals and species
onto the classic oppositional paradigm in the form *survivors/nonsurvivors*
in the struggle for life. However, instead, let us map Gaia onto Varela's
Star formation: Gaia would be *it,* and the recursive ecosystemic couplings of organisms to each other and to their dwelling places would
be *the processes leading to it.* Geohistorically, Gaia would arrive toward
the end of the Archean eon as the planetary metalevel arising from the
aggregative dynamics of living systems (Margulis's "symbiotic Earth")
coupled to the coevolution of environments compounded from abiotic
processes and the residues of life. There is no final competition in *these*
couplings, only *imbrication, complementarity,* and *mutual stabilization*
between Earth and life processes and *benefits in survival* on the living
side of the Gaian consortium.

   Putting Varela's Star cybernetics next to Bruno Latour's critique of
the cybernetic approach to Gaia, one might deem Varela's Star statements to be a consummate diagram of the way cybernetic discourse
conceptualizes organization on multiple levels. Does this also or necessarily produce unwarranted totalizations of heterogeneous elements?
Some of Varela's statements in this early draft of his later work are
certainly equivocal. This is especially so in his remark, quoted above,
that once one observes dualities in Star formation, "There is no more
duality in the sense that they are effectively related; we can contemplate
these dual pairs from a metalevel where they become a cognitive unity,
a second-order whole." Were this to be read as full-blown holism, then
Varela would be implying the possibility of an observer that could actually place itself in a terminal position to cognize simultaneously the

whole composed of a system *and* its environment, a move that would supersede the distinction between and so totalize system-environment differentiations. However, such a reading would need to dismiss the way Varela frames his scheme within a constructivist epistemology that adheres to Spencer-Brown's postulates regarding the inevitably partial, hence nonholistic nature of any possible observation.

Let us look more closely at the way Varela has constructed the Star statement regarding this particular duality. It reads "environment/system" and not the other way around. Here the environment is the "it" that emerges from the cognitive processes of the system that constitutes its alterity. In this framing, in good constructivist fashion, epistemology precedes ontology. Knowing produces being, a formulation that underscores how self-reference lays down the bridge between the imbricated levels implicated here. Moreover, seen in a biotic autopoietic view in which any living organization knows its world in the form of its own elements, Varela's Star statement "environment/system" encapsulates the core proposition of the Gaia hypothesis that *living systems produce their environments*. Let us now briefly observe this situation through the complementary conceptuality of another observer who operationalizes Spencer-Brown's *Laws of Form*, the theory of system differentiation in Luhmann's presentation. What we can then say for the Gaian instance is that Gaia is a system that "uses itself as environment in forming its own subsystems."[57] In other words, just as Varela diagrams the situation, the differentiation of levels works in both directions at once. And in that case, Varela's proposal does not erase dualities through holistic mergers but calls instead for complementary pairs that mutually specify each other once a bridge arises allowing the closure of their recursion. The crucial factor is thus not the multiplication of levels per se but the maintenance of differentiations such that events of systemic emergence have some environmental space in which to unfold.

Multiple orders emerge from the various neocybernetic schemes of the systems counterculture. From Fuller to Varela, with stops for von Foerster, Spencer-Brown, Bateson, and Lovelock and Margulis, these thinkers point their accounts of system dynamics toward material cycles, recursive forms, and circular operations that close the loop around preexisting elements such that their consortia bootstrap new orders of coherence and bring new beings and processes into the world. In the *Whole Earth Catalog,* Fuller set forward *synergy* to mean the "unique behavior of whole systems, unpredicted by the behavior of their respective subsystems' events,"[58] while in *CoEvolution Quarterly*, Varela observed how "the simultaneity of interactions . . . gives whole systems the flavor of being what they are."[59] The main participant in these conversations responsible for giving "whole systems" a consistently holistic construction was the editor, Stewart Brand, adept at crafting popular mottos to ease the approach to the complexity of the systems discourses he curated. *CoEvolution Quarterly*'s framing operations certainly inflected its readers' reception of the Gaia concept. Nevertheless, Brand's curatorial instincts were keen, and the actual substance of both the neocybernetic discussions and the Gaia discourse Brand put forth to his readership may be more accurately described as, at the least, postholistic. In such neocybernetic descriptions, the situation of cognitive processes on multiple levels follows from the multiplicity as well as the finitude of observers.

DAISYWORLD

James Lovelock's statements regarding the cybernetics of Daisyworld provide a foil for our discussion of the neocybernetics of Gaia. Daisyworld is a model of automation rather than autopoiesis. It marks an increase in the sophistication of Lovelock's cybernetics but not a revolution in his cybernetic thinking. Lovelock declined the opportunity to engage with autopoietic systems theory. Similarly, after the arrival of chaos theory in the 1980s, on those occasions when he narrates Gaia in relation to the newer computational systems discourses—dynamical systems theory applied to population dynamics, the discourse of emergence in complex adaptive systems—he keeps them separate from his conceptuality of cybernetics proper. Cybernetics for Lovelock is set at control systems engineering and, notwithstanding occasional perturbations, stays at that steady conceptual state. His invention of the Daisyworld simulation brings out his ties to computational platforms for cybernetic model building. Daisyworld—a computer model of homeostatic self-regulation in the coupled interaction between an idealized biota and its virtual environment—is the culmination of Lovelock's heuristic exploration of Gaian parameters through cybernetic models. Moreover, its appearance in *CoEvolution Quarterly* under the modest title "Daisyworld: A Cybernetic Proof of the Gaia Hypothesis" is the climax of that periodical's sponsorship and spawning of the seminal forms of Gaia theory, from the physiological Earth of "The Atmosphere as Circulatory System" in 1975 to the Daisyworld of 1983 modeled in differential equations run on a desktop.[60]

CoEvolution Quarterly's "Daisyworld" article is a revised republication of a paper in which Lovelock had recently announced Daisyworld to the participants of a meeting held in the Netherlands in June 1982, the Fourth International Symposium on Biomineralization. This interdisciplinary topic is of strong Gaian interest. It concerns the sedimentary precipitation of the postbiotic geosphere through various metabolic processes that evolve in the biosphere—lithification, calcification, silicification, and so on. Lovelock and Margulis appeared together in the section of the meeting on "Global Recycling and Biomineralization." Margulis and her coauthor situated Gaia in relation to biomineralization with a presentation on "Microbial Systematics and a Gaian View of the Sediments."[61] But Lovelock had something different in mind. He recycled the title of his paper here, "Gaia as Seen through the Atmosphere," whole cloth from his 1972 paper of the same name, in which he first published his theories about a self-regulating biosphere under the name of Gaia. Perhaps by this refrain Lovelock signaled another formative moment for Gaia theory, for his paper centered on a first rollout of the Daisyworld model to a specialist audience, accompanied by various atmospheric considerations.[62]

Lovelock begins by taking stock of the fortunes of the Gaia concept in its first decade. He refers to the paper coauthored with Margulis, "Atmospheric Homeostasis by and for the Biosphere: The Gaia Hypothesis," published in *Tellus* in 1974. Margulis gets full credit for her part in developing the effective public debut of the Gaia hypothesis. Reviewed ten years later, however, Lovelock concedes, Gaia's turn on the scientific stage has been inconclusive at best. By now the extent and tenor of scientific resistance to Gaia has taken shape. Lovelock notes similar fates for pre-Gaian observations of the active coupling of life to Earth processes. But he sees in this history of dismissal no insuperable counterargument but rather a larger philosophical disinclination to accept the premises of cybernetic reasoning. As with "earlier attempts to unify the biological and geochemical approaches to understanding the Earth, the gaia hypothesis has tended to be ignored rather than criticized by geochemists, almost as if Aristotle still ruled and anything moving towards a circular, even a nonlinear, argument was forbidden. Gaia which uses the circular reasoning of cybernetics was taken to be teleological" (15).

Lovelock's presentation of Gaia has always rested on cybernetic control theory for the most plausible account of the Earth's evident maintenance of planetary conditions within the small range of geochemical states suitable for life's habitation. The Earth's abiotic history of massive perturbations from volcanic outgassings and planetesimal impacts, as well as life's own generation of atmospheric disequilibria between persistent levels of reactive gases, in particular, methane and oxygen, could well have driven the climate out of life's habitable range on a number of occasions. Thus, some mode of active control must account for the relatively steady state of Earth's planetary regime. The "presence of a control system, Gaia" (17), maintains the biota's active participation in these global processes of self-regulation. However, as Lovelock also concedes, regarding the precise mechanism by which this planetary coupling rises to operational efficacy, "It is not immediately obvious how such a course of events could lead to planetary homeostasis" (17). Moreover, responding directly to W. Ford Doolittle's critique of Gaia published in *CoEvolution Quarterly* in 1981, Lovelock notes the objection there that "the biota have no capacity for conscious foresight or planning and would not in the pursuit of local selfish interests evolve an altruistic system for planetary improvement and regulation" (17).[63]

Lovelock invents Daisyworld as a concrete rebuttal to those particular charges that the Gaia hypothesis *overanimates* (in Latour's idiom) the world with the imputation of conscious intentions, teleological aims, or some sort of planetary self capable of either selfish or altruistic ends. At the same time, he combats the critique of his application of control theory in the original hypothesis precisely by doubling down on the cybernetics of Gaia. He cannily notes how the logic of recursion goes against the linear grain of grammatical, narrative, and mathematical syntax:

> The sequential logic of descriptive writing is not designed for the concise explanation of control systems with their inherent circularity recursiveness and non-linearity. Even the formalism of mathematics loses its elegance when an attempt is made to describe a simple nonlinear control system such as, for example, an electrical water heater controlled by a bimetallic strip thermostat. I have chosen therefore to

present a simple model of an imaginary planet whose temperature is regulated at a biological optimum over a wide range of solar radiation levels as a working example of a Gaian mechanism. (17)

Daisyworld's nonlinear equations model a rudimentary planetary system, steadily forced, as is ours, by a sun gradually gaining in luminosity, whose biota are minimally composed of black daisies that thrive in cool conditions and white daisies that thrive when it's hot. Seeded with both varieties, Daisyworld starts out cool and is then externally forced toward warmer conditions. Due to their different albedos, or indexes of reflectivity, the two kinds of daisies feed back upon their climate to different effects. The low-albedo black daisies heat the planet by absorbing the sun's rays, while the high-albedo white daisies cool the planet by reflecting that same radiation back to space. The black daisies thrive at first as the initially cool conditions suppress the growth of the white daisies. But as the black daisies proliferate, the planet warms up enough to favor the spread of white daisies and suppress the growth of black daisies. The rising tide of white daisies diminishes the black daisies while also reflecting heat away from the planet. These counter-effects settle down or regulate the positive amplification between the sun and the black daisies that had been driving up the temperature. Pushing back on the absorption of solar forcing, Daisyworld as a whole maintains its virtual climate at a steady level for as long as it can. It does so automatically, with no teleological impetus but only the mutual interplay of negative and positive feedbacks: "No foresight or planning is required by the daisies, only their opportunistic local growth when conditions favor them."[64] The black and white daisies model the mutual coupling of two Gaian feedback loops, either of which can exert a negative—that is, regulatory or stabilizing—effect on the other to achieve and conserve a virtual homeostasis, up until the model's solar forcing becomes too great for the system to control. Driven past that tipping point, Daisyworld's life goes extinct.

A technical article published later that year with his student Andrew Watson as first author presents a full mathematical treatment of the original Daisyworld. Watson and Lovelock write there: "The daisyworld equations form a system of non-linear, multiple feedback loops. The analysis of such systems is not a trivial problem, even for the highly simplified situation on daisyworld. Some information on the steady state behavior of the equations can, however, be obtained without a dispro-portionate amount of mathematical effort."[65] In the *Biomineralization* paper, Lovelock provides a set of what will become iconic figures, graphs generated by the Daisyworld model that show how this mathematical parable of a biosphere–geosphere coupling automatically computes "a stable point around which the daisies can successfully homeostat the temperature over a wide range of luminosities."[66] The larger import of the Daisyworld parable in Lovelock's presentation is that this simula-tion of the postulated existence of a planetary control system models the stability and resilience of the biosphere even when confronted with massive shocks to the system. Daisyworld also models the limits of these natural controls once planetary variables pass beyond the range of possible regulation.

I will not engage the debate over the validity or usefulness of Daisy-world.[67] Suffice it to note, along with Timothy Lenton and his colleagues, that climate modelers and other researchers took up Daisyworld with some enthusiasm but did not thereby need to commit themselves to a full-bodied acceptance of the Gaia hypothesis. Daisyworld provided the Gaia hypothesis with computational interest at an opportune mo-ment, fostering the sense that in some fashion it did constitute a "proof" of Gaia's operation as a planetary control system:

> Although Daisyworld was presented as a "parable," the model is so elegant, and so many studies have followed up on it, that it might have created a false impression of the likely nature of global regula-tory mechanisms and their relationship with individual-based natu-ral selection. Daisyworld is a special case in that traits selected at an individual scale also lead to global regulation. The microevolutionary dynamics are therefore stabilizing, addressing the persistence of regulation and illustrating a key feature of any plausible regulation mechanism—but providing no explanation for how or why a biota with these properties would arise.[68]

But as Lovelock himself made entirely clear when first presenting Daisy-world to his scientific peers, what Daisyworld models is a cybernetic *ex-planation* for Gaian dynamics without thereby proving their existence.

Coming back to Daisyworld's popular debut in *CoEvolution Quarterly*, I would speculate that the actual immodesty of the claim made in its title, "Daisyworld: A Cybernetic Proof of the Gaia Hypothesis," was Stewart Brand's editorial fabrication. Nonetheless, Lovelock had already put the cybernetic language and its accompanying polemic into the prior *Biomineralization* article we have just been examining. For its *CoEvolution Quarterly* appearance, Lovelock only needed to extract that prior paper from its professional context through some rewriting of the opening section. This gave him (presumably) the opportunity of *CoEvolution Quarterly*'s nonspecialist venue to rehearse his observations about the state of cybernetics and its ongoing marginalization in the academy. Lovelock has never sounded more like a card-carrying member of the systems counterculture than in these remarks that seem made to his current editor's order.

The *CoEvolution Quarterly* article gained popular appeal with an informality and candor not present in the professional paper: "It is now just over ten years since Lynn Margulis and I published our first paper on the Gaia Hypothesis. You may be wondering what has happened in the meanwhile. You will have noted that the idea does not yet seem to have set big science on fire."[69] He again notes how "such names as Redfield, Hutchinson, and Sillen," whose statements, according to Lovelock, anticipated Gaian ideas, were not heeded. Again, Lovelock lays his own participation in this unfriendly reception on the fear of cybernetics: "One of the extraordinary things about science is that whilst it swallows the intricacies of relativity and of genetics, it has never been comfortable with whole systems; witness the unpopularity of cybernetics. How many universities, I wonder, have departments of cybernetics?" (66–67). He again pegs unreasoning metaphysical adherence to inherited models of Aristotelian causality as the culprit in Gaia's cool welcome. For most scientists, "The circular and recursive logic of whole systems is alien to them. This is especially true of geologists, geochemists, biochemists, and exobiologists who might otherwise have been interested in Gaia. It is true that engineers and physiologists are enlightened by their professional need to lift themselves from the narrow trough of linear thought. Unfortunately they tend to keep the conspicuous advantages of whole systems thinking to themselves" (67).

For the soft-spoken Lovelock, these remarks are particularly acerbic. Lovelock will often tweak the scientific academy from his hard-won position as a scientist whose independent status removes the duress of conformity within a university department or corporate laboratory.[70]

His critical tone and pro-whole systems polemic could also bear the mark of *CoEvolution Quarterly*. Moreover, it may be that they also come with a taste of the Lindisfarne Association, which had inducted Lovelock and Margulis together in 1981. We will examine this intellectual context more closely in the following chapter. Meanwhile, we should note that while she would dutifully expound it on later occasions, Margulis herself did not sign on to Daisyworld as a way of doing Gaian science. Her Gaian applications of autopoietic systems theory read like a counterstatement on her part with regard to the direction one could take the "circular and recursive logic of whole systems" that Lovelock sees in Gaia. In this pursuit, a decade later Margulis would cultivate the Lindisfarne ethos for her own round of polemical interventions on behalf of autopoietic Gaia. We will catch up with that story in chapter 6.

CHAPTER 5

The Lindisfarne Connection

Stewart and I engaged . . . in this East Coast/West Coast thing, with the roots in the systems stuff of Gregory Bateson and the ecological movement. And the meetings on both coasts began to be part of a larger ecology of mind, to use the name of Gregory's bestseller. And so we knew about one another, we worked together, and a lot of the same players are all in the same meetings. So there's a lot of overlap. Even though the styles of Lindisfarne and the *Whole Earth Catalog* are so different, I still published in the *Whole Earth Review,* and there is a commonality that we were participating in, but we were articulating very different styles.

—WILLIAM IRWIN THOMPSON, personal communication

Gaia is something that acts as a domain within which we move, and in that sense it extends beyond us and doesn't depend on language; but we can't separate that kind of outerness or exteriority from the fact that we are inside the domain, involved in the kind of circularity that brings us forth, as Francisco would put it. It is this pattern of circularity that is, I think, really at the heart of the idea of emergence, whether we are talking about Gaia, the cell, mind, or language.

—EVAN THOMPSON, *Gaia 2: Emergence*

William Irwin Thompson

The Gaia hypothesis and the autopoiesis concept entered the world separately in the early 1970s. Within a few years they intersected in the pages of *CoEvolution Quarterly.* Now we will explore the further mingling of their conceptual histories as their respective theorists developed significant collegial relations. Beginning in the 1980s, Lynn Margulis gave the concept of autopoiesis a prominent profile in her evolutionary and Gaian discourses. How did this come about? When *CoEvolution Quarterly* brought out its article on the Gaia hypothesis, the systems counterculture took note. For the further cultivation of these intellectual contacts, however, the networking around the Gaia concept accomplished by Brand and *CoEvolution Quarterly* shifted over to a consequential but more recondite thought collective, William Irwin Thompson's Lindisfarne Association.

Cultural historian, essayist, and poet are some of the ways to indicate the intellectual and creative personae of Lindisfarne's founder, but planetary visionary does more justice to the spirit of his efforts. Thompson brought literary and anthropological depth as well as cybernetic acumen to the mythic resonances of the Gaia hypothesis. Along with its occasional but decisive presentations of Gaia and neocybernetics, *CoEvolution Quarterly* also gave intermittent notice to Thompson and his activities. For instance, it published a speech Thompson delivered at an open-air event in 1978 called the Whole Earth Jamboree, held to commemorate the tenth anniversary of the *Whole Earth Catalog*. In 1972, Thompson's memoir of countercultural questing in the 1960s, *At the Edge of History,* and the *Whole Earth Catalog* were both finalists in the Contemporary Affairs category of the National Book Award. The *Whole Earth Catalog* won: "Since then Bill has done other good books," Brand explained, "but he has been most active as the cofounder and head of Lindisfarne Association, which has put together a remarkable number of people and events somewhat more private but in many ways very similar to this."[1] My own approach to the Gaia concept hews to a relatively nominalist effort to distinguish mythic and scientific domains of interest. However, regarding the mythopoetic depths the name of Gaia carried into scientific precincts, Thompson has noted, "Before Gaia was a hypothesis she was a goddess, so what more appropriate area could there be for an exploration of myth and science?"[2] Thompson treats Gaia first and foremost as an archetype, of which its theory is a recent secular derivative. His contribution to Gaia discourse is to bring the deep Gaian imaginary forward alongside rigorous rehearsals of Gaia theory and neocybernetic systems theory.

Thompson established the Lindisfarne Association around the same time that Brand started up *CoEvolution Quarterly*. Lindisfarne's first incarnation was as a residential community with "seminars, workshops and lectures in philosophy, the physical sciences, Jungian psychology and world order models"; in addition, "To bring the centralizing spiritual and planetary vision of Lindisfarne into sharp focus, William Irwin Thompson, Lindisfarne director, will offer a course on the Transformations of Human Culture."[3] Such ecumenical purposes ran more or less parallel with the ecological vision of *CoEvolution Quarterly*, particularly in its support for systems philosophers such as Bateson, von Foerster, and Varela. Brand's talk at an early Lindisfarne meeting is published in a Lindisfarne publication that includes articles by Bateson, Jonas Salk, Paolo Soleri, E. F. Schumacher, John Todd, and Lewis Thomas.[4] For its part, *CoEvolution Quarterly* left an appreciable imprint on the Lindisfarne group. Thompson relates that it was in the pages of *CoEvolution Quarterly* that he first became aware of the authors of the Gaia hypothesis and the work of Varela.[5] Between 1975 and 1978 the Lindisfarne Association went from its initial Fishcove campus in Southampton, Long Island, to a new headquarters in the Chelsea district of New York City. Among its other activities, it ran a fellow-in-residence program. The first Lindisfarne residential fellow was Gregory Bateson in 1976, while completing his last book, *Mind and Nature: A Necessary Unity*. In 1977 and 1978, Francisco Varela followed Bateson as fellow-in-residence completing his first book, *Principles of Biological Autonomy*. When the residential component of Lindisfarne waned at the end of the 1970s, Thompson annually drew a movable conference from an expanding roster of Lindisfarne Fellows.[6]

Erich Jantsch may well have given Lynn Margulis an initial clue concerning the "autopoietic Gaia system," if she happened to attend to his *The Self-Organizing Universe* in the year of its release.[7] However, her effective encounters with neocybernetic theory came from Lindisfarne Association meetings. In the summer of 1981, Margulis attended her first Lindisfarne Fellows event, co-organized by Thompson and Varela, and also attended by Humberto Maturana, Heinz von Foerster, Henri Atlan, and James Lovelock. Further Lindisfarne gatherings included a May 1988 Fellows meeting held in Perugia, Italy, at which the conversation again centered on the scientific convergence of Lovelock, Margulis, and Varela. From these two occasions in particular, Thompson developed the Gaia-themed essay collections *Gaia: A Way of Knowing* and *Gaia 2: Emergence* and the monograph *Imaginary Landscape*.[8] Drawn from the Perugia meeting, *Gaia 2* also documents Varela's most detailed extant commentary on the Gaia concept. Thompson suggested in that volume that expounding the Gaia concept through autopoietic systems theory marked a shift between first- and second-generation cybernetic thinkers. In these works he also elaborated provocatively on Gaia's political, economic, and cognitive implications. These Lindisfarne documents are perhaps the most important Gaia discourses of the 1980s outside of Lovelock's and Margulis's own writings. However, they arose in a private context yielding niche publications that never enjoyed wide distribution. Their importance now also lies in what they reveal regarding the common context within which both Lovelock and Margulis cultivated separate strands of Gaia discourse after the 1970s.

In an article on the trajectory of Varela's scientific career, the philosopher Evan Thompson begins with a personal reminiscence that records his first encounter with Varela at a 1977 Lindisfarne Fellows meeting organized by his father, with Bateson as chair: "I was not quite 15 years old; Francisco was almost 32. At that time Francisco was known within the circle of second-generation cybernetics and systems theory for his work with Maturana on autopoiesis and for his 'calculus of self-reference.' But outside this circle he was known for an interview and a paper that had appeared about a year earlier in *CoEvolution Quarterly*."[9] We examined those texts in the last chapter. So here is a direct transfer, with Bateson and Varela at the core, from the residually technophilic *CoEvolution Quarterly* milieu to the spiritual and intellectual ambience of Bill Thompson's Lindisfarne project.

This was a different atmosphere altogether, with a charismatic but determinedly non-authoritarian central figure and a well-articulated mission concerning the pursuit of a planetary culture through interweavings of science, art, and spirituality. In this regard, Bateson was a towering but transitional figure. He summed up the great wealth of insights to be won from an information-theoretic mode of cybernetic thinking, but also some of its limitations.[10] Thompson considered Varela the forerunner of a new phase of cultural transformation, with a constructivist practice strong enough to bring forth what would soon be specified as a Gaian world: "When I read Varela's paper on non-dualism, 'Not One, Not Two,' . . . I knew that I wanted Varela to succeed Bateson as our second scholar-in-residence. . . . Varela's non-dualism seemed to get at the heart of my discomfort with Gregory's dualism of object and

information, *pleroma* and *creatura,* mind and nature."[11] Thompson himself would envision a mode of thought in which not just mind and nature but myth and science were also amenable to mapping with Varela's
Star cybernetics so as to enter a state of differentiated non-opposition.
Referring to Varela's neocybernetic rejection of "computationalism"—
that is, his critique of the mainstream cognitive science of that moment
in which the mind is considered as a linear information processor and
thought is taken to be a computation upon objective representations—
Thompson describes the coalescence of the Gaian planetary vision at
Lindisfarne in the 1980s:

> When Lindisfarne's ecological world view was enhanced by the
> critique of Computationalism in the "Embodied" cognitive science
> of Francisco Varela in 1977, and then by the Gaia evolutionary theory
> of James Lovelock and Lynn Margulis in 1981, it began to be obvious
> to all of us that a new science was showing its face at Lindisfarne and
> that just such a science was as critical to the process of planetization
> as any esoteric philosophy of the past. . . . I tried to make this imagi-
> nary landscape visible to all, for I believed this new world view held
> out our best hope for effecting the transition from a disintegrating
> industrial civilization to an emerging planetary culture.[12]

At the same time, one of the guiding principles of the Lindisfarne Fellowship was the positive maintenance of dissensus or ideological non-
commonality among its variously scientific, literary, activist, and con-
templative members. In an explicit retort to Jürgen Habermas's notion
of rational consensus, Thompson nicely put the rationale for this pursuit of radical diversity: "A World should not be seen . . . as an organization structured through communicative rationality, but as the cohabitation of incompatible systems by which and through which the forces of
mutual rejection serve to integrate the apparently autonomous unities
in a meta-domain that is invisible to them but still constituted by their
reactive energies."[13]

Gaia: A Way of Knowing

On November 19, 1980, Maribeth Bunn wrote to Lynn Margulis on behalf of William Irwin Thompson, then living in Switzerland:

I am taking the liberty of passing on the particulars of the 1981 Fellows' Conference for Bill since I know he is anxious to have confirmations of dates etc from all participants. The Conference is entitled "Biology and the New Image of Humanity." The theme was selected in consultation with Dr. Francisco Varela, a Lindisfarne Fellow whose new book *Principles of Biological Autonomy* has just recently been published. Bill has also invited your colleague, Dr. Lovelock, who has agreed to participate in the conference. The dates for the meeting are June 4–7, 1981 and will be held at the Lindisfarne Fellows House/Wainwright Center at Green Gulch Farm, San Francisco Zen Center. . . . The traditional format of the Fellows' gathering is such that each speaker makes a presentation (not necessarily a formal paper—often a simple sharing of new works, ideas or spin-offs from earlier presentations at the meeting) from 40 minutes to an hour and this is followed by general discussion.[14]

Margulis accepted the invitation as well. Thus began a long-standing relationship with manifest consequences for the evolution of Margulis's Gaia discourse in particular and biological exposition in general. Thompson would draw the bulk of his first Gaia-themed essay collection—*Gaia: A Way of Knowing—Political Implications of the New Biology*—from the 1981 conference on "Biology and the New Image of Humanity." His preface to this volume concludes: "The Gaia hypothesis alone would not be enough to express the way of knowing or the politics of life. With the atmospheric chemistry of Lovelock, we have the macrocosm; with the bacteriology of Margulis we have the microcosm, but moving between the macrocosm of the planet and the microcosm of the cell is the mesocosm of the mind. It is here in the cognitive biology of Maturana and Varela that knowing truly becomes the organization of the living that brings forth a world."[15]

Thompson's introduction to the volume reprises his introductory remarks at the 1981 Lindisfarne meeting drawing the conceptual interconnections of the systems counterculture into an overarching view. He begins by noting the absence of Bateson, who would die of lung cancer within a month of that meeting. Bateson has been a crucial participant in wider cybernetic transformations, one of the key thinkers "responsible for opening up new paths in cybernetics, epistemology, and self-organizing systems biology."[16] Thompson indicates the biopolitical substance of his own integration of neocybernetics and symbiogenetics under the sign of Gaia. Our symbiotic integration with the Earth is *not* universal merger but rather—as implicitly modeled on the eukaryotic consortium detailed in Margulis's work on cell evolution—dynamic differentiation leading to greater coevolutionary complexity. "The fundamental principle that I see coming out of this new mode of thought is that living systems express a dynamic in which opposites are basic and opposition essential. One cannot say that the ocean is right and the continent is wrong in a Gaian view of planetary process. What this means for me is that the movement from archaic industrial modes of thought into a new planetary culture is characterized by a movement from ideology to an ecology of consciousness" (27). Again, reading this through Varela's Star schema, whereas industrial modes of thought fell under the dialectics of negation and antithesis, the ecology of consciousness (Bateson's "mind") suitable for a planetary culture will maintain multiply positive, complementary differentiations, in which both "ocean" and "continent" take their mutual places in the Gaian system.

The first six chapters of *Gaia: A Way of Knowing* comprise an international tour of biological systems theory, in the following order: Bateson (posthumously), Varela, Maturana, Lovelock, Margulis, and Atlan.[17] Varela contributed "Laying Down a Path in Walking," a neocybernetic critique of adaptationism in evolutionary theory and representationalism in neuroscience. Varela's "nonrepresentationist" position is a further application of epistemological constructivism: "To understand the neural processes from a nonrepresentationist point of view, it is enough just to notice that whatever perturbation reaches from the medium will be in-formed according to the internal coherences of the system. Such perturbation cannot act as 'information' to be processed. In contrast, we say that the nervous system has *operational closure,* because it relies on internal coherences capable of specifying a relevant world."[18] Maturana's closely related contribution was "Everything Is Said by an Observer," a nontechnical rehearsal of the main arguments of *Autopoiesis and Cognition* for how a closed system nonetheless maintains cognitive interaction with its environment. A living system is "a closed system, a system which only generates states in autopoiesis. . . . So in the interaction of a living system and its medium, although what happens to the system is determined by its structure, and what happens to the medium is determined by its structure, the coincidence of these two selects what changes of state will occur."[19]

The inventors of the Gaia hypothesis then followed the architects of autopoiesis. Lovelock's "Gaia: A Model for Planetary and Cellular Dynamics" sketched the history of the Gaia hypothesis and reviewed the major lines of argument for it. One is thermodynamic: "When the air, the ocean, and the crust of our planet are examined in this way, the Earth is seen to be a strange and beautiful anomaly," due to its extreme disequilibrium of energy potentials.[20] The other line is cybernetic, for which the logic is lucid, even if the experimental practice is vexed. Taking Gaia to be a system, one then assumes its components to be integrated into that system's productions and so open to investigation not as freestanding phenomena but as components of coordinated functions. For instance, Lovelock and his colleagues "think now that an otherwise enigmatic, apparently wasteful process of the biosphere—that of producing methane, only to have it flow up into the atmosphere where it is oxidized, apparently doing no good—is in fact part of a feedback loop concerned with the regulation of oxygen" (91–92). The "cellular dynamics" of Lovelock's title carried the discussion into Margulis's domain of expertise. Some of the geological promoters of the Anthropocene seem to have missed Lovelock's point formulated in the early 1980s in consonance with Margulis's discourse on the microcosm: "When we talk about life or the biosphere, we tend to forget that procaryotes, simple bacteria, ran a successful biosphere and represented life on Earth for nearly 2 aeons (two thousand million years). They are still today responsible for a great deal of the running of the present system" (95). No matter how heavily humanity's activities may weigh at the surface of the planet, the absolute viability of the biosphere will continue to rest on the prosperity of the microbes.

Margulis picked up that theme in "Early Life: The Microbes Have Priority" with a synopsis of her account of evolutionary phylogeny. The collective chronicle of life's evolutionary history, she explained, was in the process of reinstating increasingly detailed sketches of its previously missing opening chapters. This expanded narrative undercut prior notions, imported from wishful political agendas, that the evolution of life has been "'progressive,' leading to 'higher' and therefore better life forms. One must realize that even three billion years ago, neatly functioning atmospheric cycles were modulated by organisms.... If the

vast stretch of pre-Phanerozoic time once seemed uneventful, it was because we lacked the tools to examine it."[21] Margulis's article in this first
Lindisfarne collection focused on earlier work within her dedicated
specialization of microbial evolution, just prior to its strong neocybernetic turn throughout the 1980s and 1990s.

The most sustained discussion of autopoietic systems theory in *Gaia:
A Way of Knowing* was not biological but socioeconomic. Thompson's own
article, "Gaia and the Politics of Life," proceeded to a second section,
"Toward an Autopoietic Economy," which treated the late-modern proliferation of "shadow economies" as emergent autonomous formations
redolent of the Gaian interconnectedness and evolutionary mobility of
microbial symbioses.[22] More broadly, from the new systems biologies
documented in this volume, Thompson set forth a cultural synthesis of
mythopoesis and science in distinctly neocybernetic terms: "What I am
offering in this book is not so much a description of some scientific theories but an unfoldment in which *the observer of the scientific observer
changes the science of the scientist.* The literary writer, the poet, becomes
possessed by science, and in reflecting the work back to the scientist,
the scientist sees his image transformed."[23] Thompson's second-order
observation acted not as a mere reception and sorting operation but as
a new construction in its own right, specifically as a determination of
cultural values proper to the artist's role. Chapter 8 will discuss more
fully how Thompson's formulations of the "politics of life" and *Gaia politique* leap out to us now as precisely *biopolitical.* Thompson got there
avant la lettre of current biopolitical thought by adding the element of
mindfulness in the discourse of autopoietic cognition to the recursion
of Gaia upon the biota.

Gaia 2: Emergence

The contributors to the second Gaia volume, *Gaia 2: Emergence—The
New Science of Becoming,* comprise a cross section of the Lindisfarne
Fellowship in the later 1980s. Lovelock, Margulis, and Varela deliver the
first three chapters, immediately followed by Evan Thompson's presentation of material from his collaboration in progress with Varela on
the volume "Worlds without Ground: Cognitive Science and Embodied
Experience."[24] In addition to William Irwin Thompson, the other contributors are physicist Arthur Zajonc; botanist and geneticist Wes

Jackson, founder of the Land Institute; John and Nancy Jack Todd, found-
ers of the New Alchemy Institute and Ocean Arks; developmental systems theorist Susan Oyama; and their Italian hosts, philosopher Mauro
Ceruti and editor Gianluca Bocchi. However, between the early 1980s
of its predecessor and the late 1980s recorded by this second volume,
a new set of systems discourses—dynamical systems theory and complexity theory, chaos theory for short—had hit with substantial impact.

In his introduction for *Gaia 2,* Thompson stood back from these newer
sciences of emergence to envision Gaia as a world myth or global imaginative structure for a new planetary culture in the making. "Ironically
enough," he noted in the introduction,

> it was the physicists, the highest of the high priests of matter and
> scientific materialism, who enabled us to break out of the grip of
> the positivists. . . . As Heisenberg said, "We do not have a science of
> nature, we have a science of our descriptions of nature." Now in our
> contemporary passage from mimesis to autopoiesis, the video art-
> ists, such as Gene Youngblood and Bill Viola, are as fascinated with
> the world of Maturana and Varela as once the Renaissance Italian
> artists were fascinated with perspective and motion in the shift from
> religious faith to scientific observation.[25]

Thompson carries over the fascination with autopoiesis to the fractal
imaginary of chaos theory and the recognition of complexity. To get a
grip on the unruly multiplicities of systems and their interrelations, "one
jumps to a higher level to transpose the behavior into an image" (16).
Here Thompson virtually works Varela's Star cybernetics toward his poi-
etic ends of generating cultural information through images and models.
The biocognitive theory of autopoiesis can then carry that mental image
(the "it") back to its embodied conditions (the "processes leading to it"),
back to the domain of affect. In the modeling practices of René Thom
or Ralph Abraham, "complexity begins to be resolved by the imagination and experienced in a very direct, human, and embodied way" (16).
Multiple realizations of this way of seeing and feeling create a "planetary
culture" that "is essentially a complex ecology of multiple cognitive domains" (17). Perhaps a "phase-portrait" is now possible by which to grasp
such contemporary experience in its "geometries of behavior" (21):

What is the geometry of behavior, what is the shape of the thing I
see dimly out there at the edge of my understanding on the horizon
between perception and imagination? What is the phase-space of
the atmosphere? What is the phase-space of the Self in the immune
system? ...

I have asked Lovelock, Margulis, and Varela to join us because I
think that the shape of the atmosphere in the work of Lovelock, the
shape of the planetary bacterial bioplasm in Margulis, the shape of
the immune system in the work of Varela, and the shape of the world
economy all reveal homeomorphic phase-portraits, and, I believe,
this new shape of things to come is one of the emergent properties of
our new mathematical and narrative imagination, of a mentality that
has outgrown the Galilean world view of modernism. (21–22)

The answer toward which all of these questions and intuitions are
pointed is, of course, an updated version of the Gaia concept. Thompson's
Gaia at this moment is a dynamical systems model for a noospheric
contemplation of the Gaia of Lovelock and Margulis: "Gaia is the phase-
space of our planet, and a phase-portrait of the geometry of its behavior
would not produce the familiar billiard ball, but a complex topology of
permeable membranes" (23). Margulis's symbiotic microcosm was especially conducive to this line of imagery, in which biotic membranes
proliferated alongside and within the substantial compost of the Earth:
"A bacterium is not an object; as a temporal flow in a bioplasm, it is a
phase-space that interacts with the oceans and the atmosphere ... in-
terpenetrating the geophysical processes studied by Lovelock" (24). In
the microcosm Margulis describes, nascent microbes are always delivered into communities and so are entrained with the ecology of their
colony. This is already Gaia at the micro-dimension. "The more one tries
to envision these little creatures acting in concert, the more they seem
as if they were the antibodies of the planet maintaining a stable identity through time" (24). We will come back to the discourse of immunitary Gaia in chapter 8. For now, Thompson's introduction to *Gaia 2*
closes by evoking "the shape of Gaia" in "these new phase-portraits of
the geometry of behavior of emergent domains" (28). If we model Gaia
in this way, "we will see the organisms extruding their environments in
a fluid process of natural drift" (28).

Immediately following Thompson's geometrical élan, the imperturbable Lovelock presented "Gaia: A Planetary Emergent Phenomenon." His standard reflection on the story of Gaia theory's scientific fortunes as of 1988 is gamely inflected now toward the meeting's theme as spelled out by Bill Thompson in its program: "Because humanity is experiencing the emergence of a new planetary culture, the *avant-garde* imagination in both art and science is fascinated with the whole problem of 'Emergence.'"[26] Here I will draw out Lovelock's paper in particular, for it specifically sets up the crowning feature of this volume, a forty-page transcript of the meeting's concluding symposium, "From Biology to Cognitive Science."[27] Lovelock's "Gaia: A Planetary Emergent Phenomenon" transmits the tenor of *Ages of Gaia*. Its exposition is steady and understated. For instance, in its basic statement of the Gaia hypothesis, the term "tolerable" shows up where "optimal" had once been: "The Earth's surface environment is, and has been, actively regulated at a state *tolerable* for the biota by the biota" (my italics).[28] In this and other ways, "The Gaia hypothesis has matured over the past fifteen years and can now be more clearly stated as a theory that views the evolution of the biota and of their material environment as a single tight-coupled process, with the self-regulation of climate and chemistry as an emergent property" (30). For Lovelock, this is the "real Gaia," which must now be distinguished from "a taxa of parasites and inquilines" (31), that is to say, from impostors living commensally in the dwelling of another beast altogether. Lovelock wends his way through an apt critique of adaptationism, by way of Darwin's pre-Gaian assumptions, to one of his finest poetic images for Gaia as a product of life and death: "In his time, of course, Darwin did not know, as we do now, that the air we breathe, the oceans, and the rocks are all either the direct products of living organisms or else have been greatly modified by their presence. In no way do organisms just 'adapt' to a dead world determined by physics and chemistry alone. *They live with a world that is the breath and bones of their ancestors* and that they are now sustaining" (32; my italics).

Lovelock couches some of his presentation through concepts of emergence in Varela's particular idiom for system dynamics, in remarking on the evolution of Gaia as the "emergence of a domain" in which "important properties, such as climate and chemical composition, are seen as an emergent consequence of this evolutionary process" (33). He also recurs to the holistic idiom of the earlier cybernetics: "Gaia would be expected to be emergent, that is, the whole will be more than the sum of the parts" (33). Nonetheless, Lovelock devotes the bulk of his discussion to a roundup of Daisyworld research. By now, Daisyworld has spun off a number of variations, such as "a model that included ten different-colored daisy species, their albedos ranging in evenly spaced steps from dark to light" (37). He highlights the following points. Unlike those population models of multiple species in competition that tend toward chaotic bifurcations, Daisyworld is stable, nonchaotic. Moreover, "The stability of Daisyworld is even more remarkable since no attempt was made to linearize the equations used in the model. Not only is the model naturally stable but it will resist severe perturbations, such as the sudden death of half or more of all daisies, and then recover homeostasis when the perturbation is removed" (37). Daisyworld models the emergence of the self-regulation that maintains that stability, for like Gaia, "The Daisyworld thermostat has no set point, instead the domain always moves to a stable state where the relationship between daisy population and planetary temperature and that between temperature and daisy growth converge. The emergent system seeks the most comfortable state rather like a cat as it runs and moves before settling" (38). Lovelock's paper concludes by turning back to Gaia theory in general, to underline several key connections. Evolution and emergence intersect as "Living organisms have to evolve with their planet to the stage of emergence when they are able to regulate their planet, otherwise the ineluctable forces of physical and chemical evolution would render it uninhabitable." Gaia is "a single indivisible domain" that "depends upon coherent coupling between the evolution of the organism and the evolution of its material environment" (41).

"FROM BIOLOGY TO COGNITIVE SCIENCE"

Bill Thompson called to order the Perugia meeting's concluding event, a "General Symposium on the Cultural Implications of the Idea of Emergence in the Fields of Biology, Cognitive Science, and Philosophy," with a request for Varela to speak to Lovelock's paper, in light of Varela's "particular work in developing the concepts of 'autopoiesis' and 'autonomy.'"[29] Varela obliged with a lengthy assessment and critique of Lovelock's Gaia theory at the point to which it had then arrived, especially with the addition of the Daisyworld computer models.[30] Let us examine Varela's tour de force of scientific conversation for its neocybernetic perspective on Lovelock's cybernetic orientation. He prefaced his remarks with praise for Gaia as "one of the great ideas of twentieth-century science" (210), with which he was in basic agreement. However, he had several points to make. His first concerned Lovelock's continued use of phrasings that hypostatize the "life" of Gaia. Varela suggested that the more general systems concept of autonomous function derived from operational closure would better suit the Gaian occasion:

> Jim has made it very clear . . . that Gaia cannot be described as other than having the quality of life. . . . But it seems to me that this difficult issue can perhaps be helped and clarified by making a distinction. . . . It is the difference between being alive, which is an elusive and somewhat metaphorical concept, and a broader concept, which is perhaps easier to tackle, that of autonomy. The quality we see in Gaia as being living-like, to me is the fact that it is a fully autonomous system . . . whose fundamental organization corresponds to operational closure. (211)

We should note that Varela himself, with regard to Gaia, does not move from the idiom of autonomy to that of autopoiesis. This is not precisely the autopoietic Gaia of Margulis, although she was in the audience to which Varela presented these statements. I would submit that his description of systemic autonomy is abstract enough to contain autopoietic Gaia and continues to hold as well for my own treatment of metabiotic Gaia.

For instance, according to Varela, Gaia's self-constitution produces a systemic identity: "It is this quality of self-identity that I see in Gaia. So, operational closure is a form, if you like, of fully self-referential network constitution that specifies its own identity. . . . Autonomy, in the sense of full operational closure, is the best way of describing that living-like quality of Gaia, and . . . the use of the concept of autonomy might liberate the theory from some of the more animistic notions that have parasitized it" (211). As we traced through the neocybernetic content of *CoEvolution Quarterly,* Varela's treatment of autonomous systems was a particular enlargement of the discourse of autopoiesis. Instead of further straining after biotic or living referents or analogues, his remarks suggested, one could bring the Gaia concept under the description of operational autonomy in metabiotic systems. And although Varela would not have put it precisely this way, the recognition that there are nonliving or metabiotic modes of autonomy based on operational closure resolves the central problem with the overly "strong" form of the Gaia concept, the form in which it becomes prey to parasitizing by "animistic notions." Autopoiesis broadly considered also describes a general mode of systemic self-reference, *one* form of which is biological.

Varela's next point moved from the nature of the worldly Gaia to the artifact that is Daisyworld. Varela's neocybernetic thrust concerned his perception of Daisyworld's inability to model the complex adaptability of Gaia's ongoing emergence as a globally distributed network of systems, a planetary network that, like an immune system, continues to learn on the job. For Varela, Daisyworld's heuristic limitations derive from its conceptual origins in feedback engineering. In chapter 4 we cited Varela addressing this same criticism regarding the first cybernetics' residually linear or input/output orientation: "Early cybernetics is essentially concerned with feedback circuits, and the early cyberneticists fell short of recognizing the importance of circularity in the constitution of an identity. Their loops are still inside an input/output box."[31] In Daisyworld, too, the operational outcome of its feedback loops, like that of a thermostat, may be "linear" in the sense of hovering homeostatically around a single stable point.

> Daisyworld, in the best tradition of feedback engineering, which Jim has referred to, is not the same thing as a fully plastic network, that is, a network which has some way of changing itself. . . . Here there is a distinction between a single, linear feedback mechanism, or circumstance where you have one, two, or three feedback loops, and a network. A whole bunch of feedback mechanisms added together does not amount to the same thing as a network, for a network has a distributive quality and has its own dynamic. . . . So I propose, I hope not too boldly to its own inventor, that the best model for Gaia is not one of the old tradition of feedbacks added together, but one of a fully distributed network. . . . In the same way that you will not get a cell by just adding together the regulatory circuits of enzymes and substrates, you will not get Gaia out of the regulatory circuits of Daisyworld. I believe that one will not have a fully convincing argument for Gaia until the full plastic network qualities of Gaia become apparent. For then, you see, you will actually be able to put your

finger on the learning capacity of Gaia to show just how it becomes adaptive. (212)

I take Varela to be saying that Daisyworld is insufficient to model the autonomy of "real Gaia" in its self-constitution of a systemic identity. Such an "argument for Gaia" would have to show not just relatively stable properties in a homeostatic sense but also the ability to regenerate the definitive emergence of its unity as a system even as it evolves or "drifts" through a history of different configurations. Daisyworld is "a group of equations" (213), but it is not a system in its own right, at least not in the strong sense of an ensemble possessing a self-constituting closure of operation.

After a fair amount of further exchange between Varela and Lovelock, Thompson's evenhanded moderation articulates the distinction we have been working to clarify, between the classical homeostatic paradigm on the one hand and neocybernetic constitutive recursion on the other. Speaking directly to Varela, Thompson remarks: "I see your comment on Jim's talk as a generational development. The first and founding generation of cybernetics . . . gave us basic concepts for systems guidance and correction, the feedbacks you're talking about. Now your generation comes along with its connectionist language . . . or your own 'autopoiesis,' and says, 'Our generation wants to take it another step, from feedback to the metadynamics of the system as a learning one'" (214–15). Turning to Lovelock, Thompson solicits his response and frames a possible rebuttal of Varela's critique: "Of course, part of the force of Daisyworld is that it comes at complexity through simplicity, that it serves as a parable. Do you feel, Jim, that the metaphoric force of your argument is lost if the simplicity of feedback is immersed in the complexity of networks?" (215). Lovelock accepts the gist of Thompson's suggestion. He then graciously asks Varela to assist his understanding by expounding his sense of the topic that has been hovering over the meeting as a whole:

So my first question is about "Emergence." I love this word, emergence, it's a beautiful word and it really means something to me, but first I want to take it down to the very lowest level. One of the simplest cybernetic devices ever made was Watt's steam engine governor. Do you know the thing? . . . Now, what I want to ask is: Is a cybernetic device as simple as that—one that shows emergent properties? (215–16)

What ensues is a succinct but major discourse through which Varela develops a pointed articulation of his current neocybernetic practice as a systems thinker. First of all, "If we are going to make sense of emergence as something interesting, then we have to distinguish between an emergent property and an emergent domain" (216). Take Watt's ur-cybernetic steam engine governor. Here is a simple mechanical feedback circuit in which changes in the steam engine's output of work modulate the centrifugal force acting upon rotating weighted balls connected to a valve that regulates the engine's throttle. This mechanism indeed "exhibits an emergent property" embodied in its homeostatic governance of the steam engine's operation. However, according to Varela, this is "not an emergent domain. Why? Because to me an emergent domain is one that creates or specifies or gives rise to a new identity or a class of things" (216). Varela employs Lovelock's mechanical example to segue from designed to natural systems, at which point neocybernetic conceptuality kicks in with its specific grounding in living systems. The primary example Varela offers of an emergent domain is the organic cell, for which "our little notion of autopoiesis . . . specifies the circular mechanism" (217). Varela succinctly rehearses some of the metabolic intracellular functions by which a

> bootstrapping activity gives rise to these coherent unities that,
> in my view, are the minimal living structure, which is why, with
> Humberto [Maturana], I wanted to give this activity a specific name,
> autopoiesis, for they are self-producing in that specific sense of being
> buckled together. There is no set-point here. There is no sense of just
> a one-dimensional property arising, but what arises, what emerges
> is a class-identity, and that gives rise to an emergent domain, which
> is life. And that, you know, is pretty dramatic. Now my feeling is that
> with Gaia it's much the same sort of thing. (217)

Varela's autonomous Gaia would be a planetary bootstrapping of biotic and abiotic components, a self-constituting domain of self-maintaining operations, a meta-domain arising out of life in worldly context, whose own emergence as an autopoietic domain has now been drawn up into the "metadynamics" of the Gaian system. This brings Varela to a certain codification of his current version of NST:

> I would agree with Bill and would call this second generation cybernetics; it's this post-cybernetics type of work that I'm very fond of. . . . The main difference, as I see it, is that in order to analyze this class-identity in a way that is not loops upon loops but has this quality of the emergent domain, one has to add the quality that I tried to capture in my presentation, and that is the metadynamics. Metadynamics is the system described not just in a way that explains how the parts begin to relate and give rise to the whole as more than the sum of its parts, but also in a way that explains the process by which the whole knows how to change itself in such a way so as to maintain that quality, that emergent property. (217–18)

Varela's statement shows that one way beyond holism is *through* the whole to a differentiated systematicity by which "the whole knows how to change itself" and so enters time, becomes part of a sequence, part of an inner-outer history emerging from moment to moment. Another way is to follow the resonance of emergence toward embodiment, "to pursue the metadynamics down to a technical level of connectionist, distributive patterns." Varela continues: "The exciting thing is that in some cases, such as with the immune system, you can take it down to some levels of precision, even mathematical precision: Whether such a metadynamics can be equally explicit and precise, say, in a Gaian context is what brings us all together to think and explore" (229).[32]

The symposium continued to range widely over Gaia discourse, the theory of autopoiesis, and the concepts of emergence and systemic level. However, the passages just reviewed strike me as the most developed and consequential formulations of NST within the Lindisfarne milieu proper. Margulis's recorded contributions to the symposium did not break new ground for her, nor did she venture to launch her own formulations of autopoietic Gaia into this conversation dominated by the co-inventors of Gaia and autopoiesis, respectively. As we will study in the next chapter, she chose other forums for her own autopoietic discourses. But it seems certain that the Lindisfarne connection had a lot to do with their conception.

CHAPTER 6

Margulis and Autopoiesis

The big trouble in biology is directly related to the big trouble
in our social structure and its priorities. This is a big subject.

—LYNN MARGULIS, "BIG TROUBLE IN BIOLOGY"

The word *autopoiesis* did not appear in the 1981 edition of Lynn Margulis's
primary evolutionary text within her own discipline, *Symbiosis in Cell
Evolution.*[1] However, autopoietic theory would become a fixture in her
subsequent texts. For instance, multiple entries for "autopoiesis" and
"autopoietic systems" appear in the next (1993) edition of *Symbiosis in
Cell Evolution.*[2] Margulis now gave prominence to the concept of autopoiesis not only in regard to Gaia theory but also as running the ecological gamut from the cell to the biosphere. Francisco Varela provided a
sketch if not a blueprint for this development. Bootstrapping the concept of autopoiesis to her work on symbiosis and environmental evolution, Margulis extended Lovelock by taking the science of Gaia on a
distinctly neocybernetic path.

Margulis herself occasionally used a first-order idiom that restricted
the sense of cybernetics to mechanical or computational applications.[3]
On these occasions she may have been responding to some degree to
Lovelock's move into computer modeling, pursuing the Gaia concept
through, to pick up Stewart Brand's phrasing in *II Cybernetic Frontiers,*
the "machine cybernetics" of Daisyworld. Margulis herself would remain committed to the "organic cybernetics" of Maturana and Varela's
original theory of autopoiesis. The concept of autopoietic Gaia was a
way to keep her symbiotic planet anchored to the biota. On other occasions, especially in writings coauthored with Dorion Sagan, she would
entertain ideas of machine evolution. She nevertheless found conceptual ways to bring these speculations under an autopoietic and Gaian

description. Her symbiotic conception of autopoiesis more often held her Gaia concept in a biocentric posture; nonetheless, her theorizations traced for Gaia a metabiotic course beyond strictly organic status. Lovelock himself would strengthen his description of Gaia as a natural system arising from the coeval coupling of biological and geological dynamics, rightly insisting on keeping Gaia theory in due balance between the material-energetic and organic components of its operations.[4] I will venture to follow both leads and extend the sense of autopoietic Gaia as denoting a more comprehensive, metabiotic coupling capable of coordinating both geological and technological dynamics with the operational closure of living systems.

Lovelock's one mention of autopoiesis that I know of occurs at the end of *Ages of Gaia*. It is drawn not from Maturana and Varela but from Erich Jantsch's construction of autopoiesis in *The Self-Organizing Universe*: "The tightly coupled evolution of the physical environment and the autopoietic entities of pre-life led to a new order of stability."[5] For his part, Jantsch made an early contribution to the autopoietic approach to Gaia by taking autopoiesis back to the abiotic nexus of dissipative structures as described by the physical chemist Ilya Prigogine, and then forward once more to Gaia as a singular superorganic system.[6] Given that the form of autopoiesis can be construed as a theory of minimal life emerging from prebiotic autocatalytic processes, Jantsch proceeded to backdate the evolution of autopoiesis from biotic cells to abiotic chemical reaction systems: "In the more than 3000 million years before the appearance of the first multicellular organisms, three main levels of autopoietic existence appear: dissipative structures, prokaryotes and eukaryotes. In macroevolution, however, the identification of autopoietic levels is more difficult. Nevertheless it seems that the prokaryotes are matched on the macroscopic branch by the autopoietic Gaia system."[7] Jantsch alluded here to Margulis's serial endosymbiosis theory: the nucleated eukaryotic cell evolved from viable mergers of its bacterial precursors. While Jantsch may have stretched the idea of autopoiesis thin over prebiotic areas of application, his was nonetheless a seminal grasp of its possibilities as a unifying concept within systems theory. And while Heinz von Foerster was likely the first to put the two concepts side by side, perhaps suggesting but not saying outright that Gaia itself was autopoietic, Jantsch may have been the first person to state an autopoietic conception of the Gaian system.

The onset of autopoietic discourse in Margulis's texts of the 1980s and 1990s also coincided with the 1981 launch of Margulis's writing collaboration with her first son, the science journalist and author Dorion Sagan.[8] Margulis and Sagan distributed autopoietic and Gaian themes across a series of noteworthy documents for which it is sometimes difficult, especially at the outset, to construct a definitive chronology. It is also often challenging, in their coauthored writings, to determine the tilt of authorial provenance in any given passage. Lead-author status in the byline does not necessarily indicate consistent authorial priority, and one cannot assume perfect unanimity of judgment in this collaboration of two very different and equally strenuous thinkers. This volatile mix of melded authorship appears to have been a deliberate textual strategy allowing both voices to jostle and provoke each other in a kind of mutual endosymbiosis.[9] However, with regard to the topic of concern in this chapter, the evidence suggests that the regular introduction of autopoietic theory into their coauthored texts was due largely to Margulis's particular intellectual commitments.

The texts I discuss in this chapter were begun as early as 1981 and first published between 1986 and 1991. I will proceed in what seems to me the most likely sequence of composition based on a progressive refinement in expression and coherence around the core topic of autopoiesis.[10] Aimed at a general audience, *Microcosmos* bears the marks of a first collaboration: its seams occasionally show, especially in the final chapter, which brings autopoiesis together with the Gaia concept. The more technical university-press book *Origins of Sex* has the steadier exposition; it begins with a brief chapter detailing the concept of autopoiesis in relation to the self-production of living systems and in distinction from genetics and processes of reproduction, and goes on to introduce an important formulation of "component autopoiesis." Tying these texts together, a popular article lead-authored by Sagan, "Gaia and the Evolution of Machines," published in 1987, features the most vigorous speculative extension of autopoietic Gaia to date. The content of this article corresponds to a chapter of Lovelock's own major text under development throughout this period, *Ages of Gaia*, in which machine themes also receive speculative treatment. These convergences all draw attention to the way that the Gaia concept consistently hails its cybernetic origins and so calls upon its theorists, Margulis in particular, to consider its neocybernetic reformulation as autopoietic Gaia in relation to nonautopoietic systems. A few years later, Margulis returns on her own to the conceptual intersection of autopoiesis and Gaia in a linked pair of academic articles that constitute her most concerted treatments of the autopoietic idea in the immediate vicinity of Gaia theory.

Microcosmos

The consolidation of Gaia theory by the early 1980s informs this text's evolutionary narrative. First of all, the concept of the microcosm is already planetary in scope. For instance, Margulis and Sagan's account of Darwinism shifts the accent from the competition for survival of individual organisms to life taken altogether: "The view of evolution as chronic bloody competition . . . dissolves before a new view of continual cooperation, strong interaction, and mutual dependence among life forms. Life did not take over the globe by combat, but by networking."[11] Moreover, the early bacteria took over around three billion years ago with a wide-open evolutionary lottery driven primarily by the recombinatory dynamics of lateral gene transfer: "The result is a planet made fertile and inhabitable for larger forms of life by a communicating and cooperating worldwide superorganism of bacteria" (17). This "superorganism" is not yet Gaia; it is rather the microcosmic underpinnings thereof, the "planetary patina" of the prokaryotes (126). Expounding Sonea and Panisset's bacteriology, Margulis and Sagan reason that "if, indeed, all strains of bacteria can potentially share all bacterial genes, then strictly speaking there are no true species in the bacterial world. All bacteria are one organism, one entity capable of genetic engineering on a planetary or global scale" (89).[12] Here and elsewhere in *Microcosmos,* Margulis and Sagan purvey a sort of biotic holism alongside an early draft for a presentation of autopoiesis.

Autopoiesis enters *Microcosmos* in the context of the origin of life. Margulis and Sagan assemble a picture of life's origin as a phenomenon bootstrapped from the self-organization of Earthly chemistry over an eon of prebiotic time: "From dissipative structures to RNA hypercycles to autopoietic systems to the first crudely replicating beings, we begin to see the winding road that self-organizing structures traveled on their journey to the living cell" (57). Although *Microcosmos* never explicitly cites it, Jantsch's amalgam of Prigogine and Maturana and Varela in *The Self-Organizing Universe* rises to the surface here. The definition of autopoiesis advanced in *Microcosmos* blends the argumentative framing of Maturana and Varela's original version and Jantsch's looser, more extended treatment. For instance, here is Jantsch: "Autopoiesis refers to the characteristic of living systems to continuously renew themselves and to regulate this process in such a way that the integrity of their structure is maintained."[13] Compare a similar concentration on integrity of structure in the initial statement on autopoiesis in *Microcosmos*:

> To be alive, an entity must first be *autopoietic*—that is, it must actively maintain itself against the mischief of the world. Life responds to disturbance, using matter and energy to stay intact. An organism constantly exchanges its parts, replacing its component chemicals without ever losing its identity. This modulating, "holistic" phenomenon of autopoiesis, of active self-maintenance, is the basis of all known life. All cells react to external perturbations in order to preserve key aspects of their identity within their boundaries. (56)

Both of these statements address the self-maintenance of the autopoietic identity, but "holistic" in scare quotes leaves the recursive or self-referential form of autopoietic dynamics unremarked.

In *Microcosmos*, the manner of autopoietic recursion becomes marginally explicit in the final chapter, "The Future Supercosm." Speculations about the possibility of taking terrestrial life successfully into extraterrestrial environments elicit this work's most extensive rehearsal of the Gaia concept. The passage in question offers a sketch of the mechanistic paradigm also under Lovelock's critique and which the neocybernetic view will supersede. Classical physical views of living dynamics based on the science of Descartes and Newton were linear rather than recursive. This linear hangover remains the case with a lot of the machine cybernetics and information theory then fashionably being applied to living systems through "computer-age analogies: amino acids are a form of 'input,' RNA is 'data-processing,' and organisms are the 'output,' the 'hard copy' controlled by that 'master program,' that 'reproducing software,' the genes" (264). Autopoiesis enters this section of *Microcosmos* precisely to rebut such bioinformatic computationalism: "We have held to a somewhat different and more abstract view. . . . Life, a watery, carbon-based macromolecular system, is reproducing autopoiesis. The autopoietic view of life is circular" (264).

"Life . . . is reproducing autopoiesis" is an interestingly compressed formulation. But how *is* the "autopoietic view of life . . . circular"? Gaia makes its entrance at this point to provide the terms needed to resolve the sense of these conceptual constructions, but just barely. It arrives with this introduction: "The freelance atmospheric chemist James Lovelock sees life best represented by a self-supporting environmental system which he calls Gaia" (265). What Lovelock's Gaia as an "environmental system" has just brought with it is the currently missing component of its proper construction, an importation of atmospheric and climatic materiality into the discussion. This Gaia abides the superorganicism still at large in Margulis and Sagan's current formulations: "Gaia, the superorganismic system of all life on Earth" (265). Autopoietic Gaia shares at first these biotic phrasings: "According to Lovelock's idea, which he calls the Gaia hypothesis, the biota itself, which includes *Homo sapiens,* is autopoietic. It recognizes, regulates, and creates conditions necessary for its own continuing survival" (266). Margulis and Sagan project their current theorizing of the microcosm in relation to "the biota itself" as an autopoietic system. At the same time, the finer part of this construction is its salient observation of autopoietic Gaia as a system of planetary cognition through environmental recognition.

With the following statement, however, Margulis and Sagan interlock Gaian and autopoietic circularity, in the appropriate rhetorical form of a discursive chiasmus—a figure of speech in which the order of words or phrases in parallel clauses is repeated but reversed, by which syntactic turn the statement circles back on itself. Here autopoietic Gaia finds its verbal form through the depicted reciprocity of a systemic coupling of life and Earth: "On earth the environment has been made and monitored by life as much as life has been made and influenced by the environment" (265). Let us quickly compare to this hard-won Gaian construction in *Microcosmos* one of Margulis's own formulations of Gaia over a decade later in *Symbiotic Planet,* in which the biotic bias at large in *Microcosmos* is largely corrected: "The sum of planetary life, Gaia, displays a physiology that we recognize as environmental regulation. Gaia itself is not an organism directly selected among many. It is an emergent property of interaction among organisms, the spherical planet on which they reside, and an energy source, the sun."[14] This passage retains a summative biotic formulation but then grounds it in cosmological processes. The rejection of an organismic description is another way of saying that Gaia, unlike any literal organism, does not reproduce and so leaves no progeny that may be more or less "fit" for natural selection. As such, this statement retains a vestige of autopoietic conceptuality even while "autopoietic Gaia" does not rise to utterance in this passage of Margulis's memoir. Rather, in this statement, life, Earth, and sun have fallen into place as the "coupled system" of Lovelock's mature idiom. Margulis follows with an expression that nicely reinforces this formulation: "Gaia is the regulated surface of the planet incessantly creating new environments and new organisms. . . . Less a single live entity than a huge set of interacting ecosystems, the Earth as Gaian regulatory physiology transcends all individual organisms" (120). In my idiom, Gaia's operational closure around the flow of solar radiation drives a cyclical interplay between life and Earth that is metabiotic in the final instance.

Origins of Sex

The opening chapter of *Origins of Sex,* "What Is Life? DNA, Autopoiesis, and the Reproductive Imperative," is still a highly serviceable inscription of autopoiesis as the primary process of self-production into a basic exposition of cellular and molecular biology. As we noted in chapter 3, when Varela, Maturana, and Uribe introduced the concept of autopoiesis in the 1970s, it came as a conceptual retort to an overemphasis on molecular genetics as the sole driver of evolutionary variation. Margulis and Sagan are now at pains in this text both to honor that prior call to account for living organization and to integrate that corrective supplement into the standard presentation. *Origins of Sex* makes room for autopoiesis by qualifying the function of metabolism. Autopoiesis names the principle of the imperative for continuous self-production, and "metabolism is the mechanism of autopoiesis."[15] There are not one but two intimately intertwined living imperatives. In the first instance, autopoiesis must accomplish the operational continuity of a living system in its own right and at least until the arrival of an organism's reproductive capability. Then, on that basis, for life to continue beyond the finite duration of that cellular or organismal self, reproduction must realize genetic continuity across generations. Meanwhile, despite anthropomorphic tales to the contrary, genes have no such desires. Macromolecules "are indifferent to existence," Margulis and Sagan insist: "chemical systems have no priorities" (12). The introduction of autopoiesis into these descriptions also displaces the function of reproduction from its stereotypical presentation as the supreme expression of living beings and brings it back into relation with its inexorable preconditions. Nevertheless, at the same time, the authors render autopoiesis in its necessarily contingent relations to genetics and reproduction:

> Autopoiesis occurs, then, to maintain an organism during its own life, but by itself autopoiesis does not guarantee that an organism will show genetic continuity or that the characteristics of any given organism will persist faithfully through time. The process that ensures genetic continuity is reproduction. But autopoiesis remains the primary process. On the one hand, without it the organism would not survive to reach the stage at which reproduction becomes feasible. On the other hand, autopoiesis does not depend on reproduction, at least within a single generation. (13)

The Gaia concept does not fit the topic of *Origins of Sex* and does not make an appearance there. Instead, Margulis and Sagan develop their account of serial endosymbiosis in the evolution of the eukaryotic cell in relation to related theories about the evolution of sexual reproduction, which threshold occurs well after eukaryosis has been stabilized and the earliest protists, the eukaryotic microbes, enter the world. They also make some observations that bend in the direction of the higher-order ecologies for which the Gaian system would be the final iteration. The chapter "Meiosis and Cell Differentiation" begins: "A central thesis of this book is that the eukaryotic cell is homologous to a community of microorganisms" (170). It characterizes the "eukaryotic individual" as the systemically integrated sum of its bacterial precursors, whose separate genetic residues are not entirely bound within the nucleus but are also distributed throughout the cell and its organelles and bound up by its outer membrane:

> All eukaryotic individuals must reserve, in a form capable of continued reproduction, their genetic components, the remnant bacteria in the combined form of the nucleocytoplasmic, mitochondrial, plastid, and undulipodial genomes. If we accept the cell as a microbial community, the germ plasm is equivalent to component autopoiesis: a

complete set of heterologous genomes and their protein synthetic systems contained within a membranous package—not the nuclear membrane but the plasma membrane. (175–76)

This articulation of "component autopoiesis" arrives once the discussion shifts from the opening chapter's rehearsal of the minimal imperatives of a basic prokaryotic being such as a bacterium to the more demanding processes needed to maintain the eukaryotic cell's integration and synthesis of its heterogeneous genetic inheritance. Moreover, "We can apply the principles of community ecology directly to the development of the individual" (176). If we return Margulis and Sagan's cellular extension of community ecology back to its primary reference, then we already inhabit a theory of the ecosystem for which "component autopoiesis" drives the necessarily higher-order forms of community self-production and self-maintenance. Ecological communities themselves have shifting but relatively stable identities. Site-specific, they endure and mature. However, even while their living components carry on their reproductive ways, insofar as they also maintain their individual autopoieses and contribute them to a group dynamic, such communities do not reproduce their organization in the form of progeny. Rather, their continuity over time must emerge from the composite maintenance of the ecosystemic consortium. We are thus a step or two closer to a description of the operational sense of autopoietic Gaia as a self-producing but nonreproducing entity.

"Gaia and the Evolution of Machines"

"Gaia and the Evolution of Machines" refines an autopoietic treatment of Gaia theory in relation to the technosphere first sketched out in a letter Margulis sent Lovelock in December 1985.[16] This article appeared in the *Whole Earth Review,* the immediate successor of *CoEvolution Quarterly,* now editorially retooled for the rise of cyberculture and the postecological ambience of the world after *Neuromancer.* This was an appropriate venue for an article addressing the decade-long interest of the Whole Earth network in Gaia discourse, while updating that discussion with the latest in machine cybernetics. Sagan and Margulis began by deferring priority to Lovelock, as Margulis herself did not herself participate in the original gestation of Gaia's first descriptions.

Nevertheless, they now draw the idiom of their own description from the autopoietic language developed in *Microcosmos* and *Origins of Sex.* They underline Gaia's systemic self-maintenance as logically prior to its maintenance of planetary variables, in that it is the "self-maintaining properties of cells, organisms, communities and ecosystems" that "can be extrapolated to the atmosphere and surface sediments of the planet Earth" (15). To this implicitly autopoietic Gaia concept they now add the planetary accretion of the technosphere: "Not only are members of the more than 10 million existing species components of the Gaian regulatory system but so are our machines. Here we argue that although not by themselves alive, like viruses and beehives, machines are capable of reproduction, mutation and evolution. That is, even though they are not autopoietic, machines do evolve" (15). The Gaian matrix absorbs the technosphere within the finality of its operations.

Gaia now constitutes the metabiotic matrix within which autopoietic and nonautopoietic systems couple their distinct operations. Let us recall the opening passages of "Autopoiesis: The Organization of Living Systems": "Reproduction and evolution are not constitutive features of the living organization. . . . [A]ll biological phenomenology, including reproduction and evolution, is secondary to the establishment of this unitary organization" (187), that is, secondary to the establishment of the autopoietic organization. Sagan and Margulis place their current argument on the same discursive tracks, but they repurpose Maturana and Varela's logical architecture. They endorse the autopoietic organization as the prime criterion of living systems. Then they take reproduction, genetic mutation, and evolution—the same history-bound and ontogenetic qualities that, according to the autopoietic critique, mainstream biology has misplaced as the prime criteria of life per se—and transfer them to the epiphenomenal and metabiotic realm of machines, of designed technological systems. In other words, while granting living systems' exclusive title to autopoietic self-production, Sagan and Margulis put the mechanistic side of modern biological theory back where it belongs, on the description of machines.

Later in the article, borrowing facets of the autopoietic conceptuality developed in *Origins of Sex,* Sagan and Margulis term the particular quality of this active matrix of coordinated operations *consortial* and apply it to the "community ecology" of both the biological individual and the Gaian consortium:

The consortial quality of the individual preempts the notion of independence. For example, what appears to be a single wood-eating termite is comprised of billions of microbes, a few kinds of which do the actual digesting of the cellulose of wood. Gaia is the same sort of consortial entity but she is far more complex. Consortia, associations, partnerships, symbioses, and competitions in the interaction between organisms extend to the global scale. Living and nonliving matter, self and environment are inextricably interconnected. (16)

These Gaian formulations of systemic interdependence maintain the operational differentiation of biotic, abiotic, and metabiotic domains. Margulis and Sagan leverage the operational closure of autopoietic form to discern the mutual specification of differential systemic operations at the ecological scale. Biotic autopoiesis—that is, organic life—still takes precedence in their account, even as it indicates, in the coupling of non-autopoietic components to autopoietic operations, the mutual dependence of contemporary humanity and its technosphere. Over and above the reproduction of Gaia's living components, human life and machine reproduction are now interdependent as well:

Although there is an ineffable continuum between the living and the nonliving, we are beginning to understand the functions and organizations that are common to living entities. Living systems, from their smallest limits as bacterial cells to their largest extent as Gaia, are *autopoietic*: they self-maintain. As autopoietic systems they are bounded—they retain their recognizable features even while undergoing a dynamic interchange of parts.... Autopoiesis is a prerequisite to reproduction.... Components of autopoietic systems reproduce. The reproduction of autopoietic systems depends on the autopoiesis of the components of such systems. (18)

This passage affirms and extends the idea of "component autopoiesis" introduced in *Origins of Sex* in order to account for the autopoietic form of the prokaryotic consortium that becomes the eukaryotic cell. Now they take that idea up to the Gaian instance at the zero degree of its dependence on the reproductive continuity of its autopoietic components. At the same time: "Machines reproduce. Alone, they do not self-assemble. They do not self-maintain: machines alone are insufficient *parts* of autopoietic systems. Despite our machineless past, however, our autopoiesis now depends on machine organization in much the same way that cells of our body depend on human organization (anatomy and physiology)" (19). The interdependence between contemporary human life and machine reproduction, over and above the reproduction of Gaia's assembly of components, presses the purview of autopoiesis beyond its biotic base. Taking the premises of the machine issue when framed by the concept of autopoiesis to their logical conclusions, Sagan and Margulis independently retrace the metabiotic course of that concept's neocybernetic development toward the self-producing operations of technological society and the sociotechnical reproduction of communications. Moreover, "The reproduction of technological societies and their components is part of the autopoiesis of the biosphere" (19). And,

> From a biospheric view, machines are one of DNA's latest strategies for autopoiesis and expansion. The classification of machines as non-autopoietic and nonliving does not negate the fact that they reproduce, and reproduce with mutation, as avidly as viruses. Like beehives, termite mounds, coral reefs, and other products of the activity of life, machines—if indirectly through DNA and RNA—make more of themselves. Through us they make other machines. (19)

In what may well be an oblique satire of Richard Dawkins's selfish-gene concept, this passage completes the transfer of neo-Darwinism's biological priorities to the machinic phylum. We turn away from the blind dispersion of bodily phenotypes to the designed evolution of machines. The Gaia concept dismissed by Dawkins now presides over "DNA's latest strategies for autopoiesis." Genetic determinism implodes within a creative and combinatory metabiotic biosphere now in the process of sending out mechanical spores to other planets. In this vision of a multiply coupled, autopoietic Gaia, humans are variously entrained parts of the technosphere, but that network or grid is itself a nonautopoietic part of this Earth's biosphere and as such takes part in its incessant modification, not just of the geosphere, but also, incrementally, of the cosmic environment:

> The Viking Lander on the surface of Mars does not maintain its own structure or actively preserve its boundaries. Alone, lacking com-

munication, it is no longer autopoietic. But from 1975 to 1982, when all of its communication with the Earth was halted, even the Viking Lander was part of an autopoietic system. Machines, by themselves on Mars, are not autopoietic. Machines tended by their workers form part of the autopoietic systems of their makers. (19)

The autopoietic discourse in "Gaia and the Evolution of Machines" anticipated the Anthropocene technosphere by several decades. Margulis and Sagan fastened upon a mode of systems description founded on the self-produced, membrane-bounded operational closure of living systems, on the cellular organization, in the midst of those systems' capacities in the fullness of evolutionary time to arrive at higher-order autopoietic consortia of pre-evolved components. Gaia itself exhibits these improbable but evolutionarily successful metabiotic couplings of living ecologies, geological formations, and technological systems across its planetary interface; mutual feedbacks of living and nonliving processes that continuously remix the system; and deeply interfolded differential effects of Earth and life processes, including minds and societies. Gaia has always partaken of the air, the rocks, and the oceans. In the Anthropocene epoch it also wraps itself globally around technological processes and productions. The Gaian discourse of the technosphere indicates the need to maintain the biospherical bona fides of machine beings.

Replacing Neo-Darwinism

When Margulis returns on her own to the matter of autopoietic Gaia, her target is the school of neo-Darwinism, the biological orthodoxy compounded from the "modern synthesis" of Mendelian inheritance together with Malthusian competition as updated by molecular genetics. In this account, evolutionary variation derives largely from random genetic mutations that serendipitously improve the survival prospects of their phenotypes once in a blue moon and so get passed on to progeny. This has always been a very thin reed on which to hang a progressivist account of macroevolution. Around 1989–90 Margulis composed two essays connected by textual overlap, a shared sense of cultural distress, and a dramatic increase in institutional animus. She delivered the first of these, "Kingdom Animalia: The Zoological Malaise from a

Microbial Perspective," within the biological academy to a plenary session on "Emerging Systems: Molecules, Genes and Cells" at the centennial meeting of the American Society of Zoologists.[17] Soon thereafter, the article "Big Trouble in Biology: Physiological Autopoiesis versus Mechanistic Neo-Darwinism" appeared in one of John Brockman's Reality Club trade publications with an extraordinary diatribe against the biological establishment in particular and big science in general.[18] In both articles, Margulis speaks out from a position I would specify as the Lindisfarne variant of the systems counterculture.

In the latter essay, several references to William Irwin Thompson underscore the countercultural valence of her arguments. However, tracing the affect of "Big Trouble in Biology" to the Lindisfarne milieu shows problematically for Margulis in particular. For Thompson, the Lindisfarne Association was strictly a private intellectual gathering funded by private sources for, in the sense given by the Austrian cultural critic Ivan Illich and cited by Thompson, "counterfoil research—research aimed at questioning the assumptions on which society operates, in concrete terms."[19] To pursue that agenda free of institutional entanglements, Thompson turned away from an academic career in which he had already earned early promotion to full professor, departing the Departments of English at MIT and then York University by 1973. For his part, Lovelock has prided himself on a career in science that depended on neither corporate nor university employment. His position as a "Visiting Professor in Cybernetics at Reading University," advertised in various biographical sketches, appears to have been a fabrication assisted by in-house colleagues there so that Lovelock could publish in scientific journals that otherwise would not consider work by anyone lacking an academic credential. However, Margulis had students to employ and a university lab to run. Even with her move from Boston University and appointment in 1988 as Distinguished University Professor at the University of Massachusetts, Margulis was already feeling the pinch of diminishing funding for her field of "organismic" research, let alone for the more unconventional transdisciplinary agenda of her own scholarship. She did not keep the matter of that professional discontent to herself.

"KINGDOM ANIMALIA"

In "Kingdom Animalia," Margulis informs an audience of dedicated zoologists celebrating the one hundredth anniversary of their professional association that their views on animal biology participate in an intellectual disorder rooted in the overextension of insular and philosophically misguided concerns to life altogether and the planet in general. The original published version immediately establishes this combative tone with a remarkable "Synopsis" that begins with two images of bodily trauma: "Pain and cognitive dissonance abound amongst biologists: the plant–animal, botany–zoology wound has nearly healed and the new gash—revealed by department budget reorganizations—is 'molecular' vs. 'organismic' biology. Here I contend resolution of these tensions within zoology requires that an autopoietic-gaian view replace a mechanical–neodarwinian perspective."[20]

The article proper starts with a preliminary rehearsal of how to break out of the battered institutional and conceptual shackles of the plant–animal contrast. From her microbial perspective, animals are "embedded in the context of their microbial predecessors. They are not 'superior,' or 'higher' forms of life to be contrasted with the 'lower' animals and 'higher' plants. Rather, animals are peculiar, if familiar, descendants of coevolved microbial communities" (862). This classic Margulis move prepares one to carry their evolutionary vision back to the Archean scene before the emergence of species specificities, at which point later differentiations among animals, plants, and fungi dissolve into the microcosmic commons from which all eukaryotic cell forms first arose through the symbiogenetic mergers of the ur-bacteria. The autopoietic form of the living is rooted here with the origin of life altogether. Gaia first arises within an entirely prokaryotic Earth. From this microbial vantage, Margulis now shifts the discussion to autopoietic Gaia as representing the nonzoocentric worldview that must supersede modern biology's prostration before the mechanistic philosophy enshrined in neo-Darwinism's gene-centered doctrines. Margulis's current treatment of autopoiesis is worth drawing out at some length, because it provides a professional audience with an unusually high level of physiological and biochemical detail.[21]

What becomes particularly clear in this presentation is how the original theory of autopoiesis is rooted in physiology rather than molecular genetics. This orientation is patent in the original conceptuality put forth by Varela, Maturana, and Uribe: "Consider for example the case of a cell: it is a network of chemical reactions."[22] Margulis lays out her autopoietic perspective now with increased biochemical detail.

The section of "Kingdom Animalia" titled "Autopoietic Gaia to Replace Neodarwinian Mechanics" begins:

> Autopoiesis, a term invented by Maturana and Varela (1980) and elaborated by other authors (Fleischacker, 1988) refers to the living nature of material systems. Well within the materialist view that recognizes the physical-chemical composition of organisms, autopoiesis refers to the self-making and self-maintaining properties of living systems relative to their dead counterparts. Autopoietic, unlike mechanical, systems produce and maintain their own boundaries (plasma membranes, skin, exoskeltons, bark, etc.). Autopoietic systems incessantly modulate their ionic composition and macromolecular sequences (i.e., amino acid and nucleotide residues in their proteins and nucleic acids). (865–66)

This autopoietic materialism is reinforced by the concreteness of its physiological examples (the "plasma membranes" of cells, the "bark" of plants) and by the biochemical specifics related to the "incessant modulation" of intracellular processes. As Margulis now shifts to a definition of Gaia, she carries over the systems inflection of cellular chemistry to Gaia's own "autopoietic" regulation of planetary chemistry: "Gaia is defined as the large self-maintaining, self-producing system extending within about 20 kilometers of the surface of the Earth. The Gaia hypothesis states: the surface sediments and troposphere of the Earth are actively regulated by the biota (the sum of the live organisms) with respect to the chemical composition of the reactive elements (e.g., H, C, N, O, S), acidity (e.g., H^+, OH^-, CO_3^-, HCO_3^-), the oxidation-reduction state and the temperature" (866).

I will stay on this course of chemical interest for the moment. For alongside the polemical content, which will only intensify in "Big Trouble in Biology," Margulis carries the theory of autopoiesis farther up the phylogenetic line from the idealized autopoietic cell of Maturana and Varela, and this physiological specificity makes the ultimate destination of autopoietic Gaia that much more plausible. This long passage offers several striking examples of salutary concreteness in presenting a "chemically self-conscious, autopoietic point of view" and also provides a blunt reminder of how the "autopoietic imperative" mixes life and death together in the chemical matters of Gaia's persistence over the eons.

Why do Pacific salmon swim upstream to die in the area where they themselves spawned? A neodarwinian uses military or economic terms, tending towards an explanation in terms of "reproductive strategies," of offspring outcompeting others with fewer genes in common. In the autopoietic point-of-view, attention is paid to the chemical components of the fish. For example, that the dead bodies of the upstream adult salmon provide phosphorus for the diatoms that, during the next season, serve as food for salmon fry.

Another example: Why do small quantities (less than or equal to 0.5 ml inocula) of certain bacteria added to fresh growth solution not grow whereas larger ones (greater than or equal to 1.0 ml) grow well? The observation that death of the organisms comprising nearly the entire inoculum provides conditions for growth of the few remaining bacteria is described as pure "altruism" (and thus rejected) by neo-darwinians. From the chemically self-conscious, autopoietic point of view it is sufficient to recognize that component lipids and other compounds shed by a large inoculum provide sufficient ambient conditions, probably including food, for the initial growth of at least a few of the bacteria in pure culture. At least ten orders of birds contain species in which parents or nestmates eat their offspring. Cell death, tissue resorption and cannibalism are common means for the auto-poietic imperative of replacement of molecular components. (868)

The coevolution of meiotic sexuality and programmed death specific to the kingdom Animalia would appear to motivate this particular emphasis on the place of mortality in a zoologically fleshed-out autopoietic conception. Margulis the autopoietic materialist places the zoological interest in metazoan sex and death against the sublime Gaian backdrop of deep evolutionary time extending six times farther back than the origin of animals per se. During all that time, it has been Gaia that has enjoyed a form of autopoietic immortality, a deathlessness of incessant operation: "All organisms are part of a single continuous bounded autopoietic system that has never been breached since the origins of life in the Hadean or Archean eon. While portions of the system (cells, individuals, populations, species) are always losing autopoietic properties, the entire system itself persists. Death must co-occur with life. Failure to retain autopoietic properties is death—and death by loss of components, desiccation, disintegration, and atrophy is intrinsic to the continuity of life" (879). Moreover, by insisting on an *autopoietic* Gaia theory, Margulis develops all of this conceptual imagery as a neocybernetic antidote to the zoological malaise of neo-Darwinian genetic parochialism and planetary small-mindedness. However, she reserves her most stringent animus against mainstream neo-Darwinism for a sister paper to this one, aimed at a wider if still elite intellectual audience.

"BIG TROUBLE IN BIOLOGY"

How are we to understand such a vehement critique of the author's own scientific discipline and academic institutions? Is her tirade the idiosyncratic outcome of strictly personal stresses, or a rejoinder to the usual accumulation of workplace or disciplinary resentments? Does this essay vent the justifiable anger of a female genius who has finally had it up to here with the daily indignities of a masculinist academic culture within a patriarchal society? Even if there were some truth to these rationales, settling for any of them would be a patronizing response denying wider validity to her extraordinarily deep and specific arguments. And what if she is right? What if her critique of the philosophical aberrations of mainstream Anglo-American science was on the mark in 1990 and is still largely correct thirty years later? Perhaps Margulis's uniquely refined scientific awareness of planetary dynamics enabled her to see farther into and cry out sooner about the global crisis that is crashing down all around us now. What if her arduously acquired access to alternative ways of scientific seeing and knowing gave her sufficient perspective and courage to expose the deformity of some of modern Western culture's misguided verities, no matter how ceremoniously they have been wrapped in the robes of science?

The primary destructive verity in her field, the "big trouble," is what she considers the indoctrination of biologists with the conviction that the aim of their science is to reduce the phenomena of life to nonliving mechanisms. The doctrine to be resisted states that "life is a mechanical system fully describable by physics and chemistry. Biology, in this reductionist view, is a subfield of chemistry and physics."[23] As Margulis will detail in a later iteration of the Reality Club publications, biologists in general suffer from P.E., "physics envy . . . a syndrome in which scientists in other disciplines yearn for the mathematically explicit models of physics."[24] In biology proper, the reigning disciplinary expression of such "physicomathematics envy" (214) is the set of neo-Darwinist doctrines that encourages her colleagues and students to ditch fieldwork and intimacy with the physiologies of embodied living beings for mathematical and computational approaches to molecular genetics and population dynamics. "Hence biologists receive Guggenheim Fellowships for calculations of the evolutionary basis of altruism or quantification of parental investment in male children, while the tropical forests are destroyed at the rate of hundreds of acres per day" (213). It is not just a matter of academics preferring ivory-tower offices with supercomputer interfaces to strenuous confrontations with real-time environmental depredations. It is also that, falling in line with such bloodless dogmas, biologists have been giving away the store, ceding knowledge of life to people who *really* have no idea what life is yet are "their supposed superiors: physicists, chemists, and mathematicians" (215). And because, "like monasteries of the Middle Ages, today's universities and professional societies guard their knowledge" (213), her biological colleagues may not even know that there actually are respectable alternative ways of conceiving their disciplinary objects and doing their science. The trouble is that they have been too carefully guarded against heretical doctrines that open out onto alternative cosmologies. Too few biologists, Margulis laments, know that "a life-centered alternative worldview" even exists, "called 'autopoiesis,' which rejects the concept of a mechanical universe knowable by an objective observer" (214).

However, as the auto mechanic said to the owner of the broken-down car, "There's your trouble!" Scientists are not supposed to challenge the popular notion of their unique possession of the truth about nature. Proper scientific mechanists guarantee to deliver the kinds of ontological finalities that funders prefer to anticipate—well, eventually, once a complete data set—say, the "human genome"—has been assembled and computed upon. The theory of autopoiesis comes with no such 100,000-mile guarantees. It promises only a universe in unpredictable flux on the edge of whatever it takes to maintain living self-productions contingent upon the integrity of operational closures. Moreover, autopoietic theory is wedded to constructivist epistemologies that have withdrawn refuge in positivism from mainstream science's conceit of objectivity. An autopoietic universe is not an absolute totality or a grand unified anything. As it comes into being out of a pre-living world, an autopoietic universe does not supersede but supplements the physicochemical cosmos with processes of life and cognition that are possible only for embodied living beings, among which scientists may be placed, incapable of bodiless omnipresence and absolute knowledge.

In its primer on autopoiesis (considered strictly in its original biotic sense), "Big Trouble in Biology" builds upon a smaller table in "Kingdom Animalia," now giving "Criteria of Autopoiesis" with an expanded breakdown of "Properties of Autopoietic Systems." Six properties in all—identity, unitary operation, self-boundedness, self-maintenance/circularity, external supply of component raw materials, external supply of energy—are unfolded with regard to aspects such as "Boundary Structure Produced by System" and, perhaps uniquely in the literature of autopoiesis at this moment, "Examples of Biochemical/Metabolic Correlates" (216). Margulis fleshes out biotic autopoietic materiality with regard to such factors as nucleic acids, fatty acids, multienzyme-mediated networks, lipogenesis, polymerization, and so forth. In the following passage she works through an autopoietic framing of metabolism, placing the familiar workings of cell biology and physiology into the encompassing logical framework of autopoietic organization and self-production:

> Autopoietic systems metabolize, whereas nonautopoietic systems do not. Proteins, viruses, plasmids, and genes are all components of live material. When contained within the boundaries of animal, plant, or other cells, they may be required to sustain cells or organisms and their autopoietic behavior; yet proteins, viruses, plasmids, and genes, intrinsically incapable of metabolism, are never autopoietic in isolation. Metabolism includes gas and liquid exchange (breathing, eating, and excreting, for instance); it is the detectable manifestation of autopoiesis. Autopoiesis determines physiology and hence is the imperative of all live matter. Autopoietic entities, that is, all live beings, must metabolize. These material exchanges are the *sine qua non* of the autopoietic system, whatever its identity. (217)

The bottom line of Margulis's autopoietic critique is that the mechanistic worldview sees no fundamental distinction between living and nonliving entities. As we noted earlier, for all his systemic orientation, Lovelock's cybernetics have always countenanced this mechanistic orientation: "The only difference between non-living and living systems is in the scale of their intricacy, a distinction which fades all the time as the complexity and capacity of automated systems continue to evolve.

Whether we have artificial intelligence now or must wait a little longer is open to debate."[25] Fifty years later, Lovelock's vision of the coming Novacene era as a time of "electronic life" makes his mechanistic roots entirely clear.[26] For Margulis, autopoietic Gaia contests this mindset even as it resides within her longtime collaborator. Meanwhile, the mechanistic worldview still finds a place for "intelligence" within its universe, somewhere or another, artificial or otherwise, ideally constituted as an informatic patterning with no particular location and rising above the contingencies of living bodies.

Whatever you may think of these scenarios, they are the manifest repercussions of specific scientific and technological *thought collectives* granted ideological dominance in modern culture. The concept of the thought collective was propounded by the Polish microbiologist Ludwik Fleck. "Big Trouble in Biology" registers the uptake of Fleck's sociology of science within Margulis's polemical armory. She introduces Fleck's work midway through this article with a harrowing biographical sketch of his survival of the Holocaust, spared, like the chemist Primo Levi, due to his utility to the Nazis and impressed into service in Auschwitz and Buchenwald manufacturing typhus vaccine for German troops. According to Margulis's account, Fleck sent bogus vaccines to the front while holding back "the real vaccine, in exceedingly short supply, to protect himself, his family, and friends. Surrounded by lives in daily danger, Fleck paid close attention to how easily scientists and technicians mentally imbibe the prevalent 'common myth'" (222). The implication is that professional ideals will eventually defer to the demands of personal survival and succumb to overriding ideological forces. However, Fleck developed his sociological theory of scientific facts prior to these dire experiences. He originally published *Genesis and Development of a Scientific Fact* in 1935.[27] Brought into English translation in 1979, Fleck's sociology of scientific thought collectives anticipated by over a generation the rise of the sociology of scientific knowledge in the 1970s and 1980s, for instance, in a work such as Latour and Woolgar's *Laboratory Life: The Construction of Scientific Facts,* also published in 1979.[28] Margulis offers this synopsis of Fleck's sociological analysis:

> The theory claims that all "scientific facts" are merely consensuses among socially interacting "card-carrying" scientists.... "[T]he fact" is a product of a complex social process beginning with individual observation or measurement and terminating with the integration

of a stylized "true statement" into the knowledge of the society at
large.... [C]ertain words and phrases become banners for the immediate identification of scientific friend or foe.... [S]ocial activities ...
cement into cohesive groups otherwise unruly scientists and technicians.... "[T]hought-collectives"—are then recognizable.... [T]he
thought-collective achieves the status of "professional tribe," as do
today's Neo-Darwinists, whose members are bound together by many
ties, including those of common scientific language. (223)

Using Fleck's theory of thought collectives to identify and characterize the biological tribe against which she is waging a mental fight,
Margulis bootstraps an astonishing sociological and ideological critique
of her own field. "Why do members of the Neo-Darwinist social group
dominate the biological scientific activities in U.S. and other English-
speaking academic institutions? Probably there are many reasons, but
a Fleckian one is that the Neo-Darwinist mechanistic, nonautopoietic
worldview is entirely consistent with the major myths of our dominant
civilization" (225–26). At this point Margulis makes the first of several
references to William Irwin Thompson's 1981 mythopoetic analysis of
cultural evolution, *The Time Falling Bodies Take to Light.*[29] The current
showdown between the planetary culture envisioned in the Lindisfarne
ethos and, in Thompson's words, our "materialistic civilization that is
concerned almost exclusively with technology, power and wealth" elic-
its from Margulis some of the edgiest, most politically trenchant commentary anywhere in her writings:

> A world philosophy based on the recognition of the autopoietic and
> nonmechanical nature of life *must* upset the believers in the funda-
> mental myths of our technological civilization. In the world of the
> Native American, humanity belongs to the earth; in the world of the
> money machines, the earth belongs to humanity. In the autopoietic
> framework, everything is observed by an embedded observer; in the
> mechanical world, the observer is objective and stands apart from
> the observed. (226–27)

I suspect that Neo-Darwinists, upon observing physiology and contemplating autopoiesis, suffer cognitive malaise. Their mathematized
formulations systematically ignore physiology, metabolism, and

biological diversity; they fail to describe the incessant, responsive, reciprocal effects of life embedded in environment. Suffering philosophical distress, physics-worshiping Neo-Darwinists must reject autopoiesis and its attendant life-centered biology with the same zeal with which the Spanish true church, guarded by its Inquisitors, rejected the mescal- and peyote-eating religions of the Native Americans. (228)

Gaia arrives in "Big Trouble in Biology" as the destination of these meditations on *embeddedness*. For the theory of autopoiesis, the self-reference of cognition is axiomatic: observers are always already embedded in their observations, whether or not they observe this to be the case or are aware of their existential and epistemological status in this regard. For its part, as Margulis telegraphs the matter, life itself is "embedded in environment." Following out this autopoietic logic, living (autopoietic) systems are ineluctably in cognitive relation to the environments from which they emerge. Living beings are environmentally-embedded observing systems, period. The recursive self-constitution of living systems feeds back into the incessant recursive reconstitution and thus evolution of their environments. This is autopoietic Gaia. Gaia's components generate a continuous set of fractal or infinitesimally differentiated planetary loops: this would be autopoietic Gaia as planetary strange attractor in William Irwin Thompson's dynamical-systems image of Gaia's phase-space. In whatever way one works out the construction of this observation, the passage into the autopoietic worldview is strongly mediated by the very kind of immersive experience of planetary connection that the bloodless biology against which Margulis inveighs denies to its disciples, who thus become its dupes:

Who are the victims of these latter-day religious wars for the souls of the biological science practitioners? Primarily graduate students, young investigators, and teachers, in whom direct observations of life and experience in the field often foster an expansive autopoietic attitude. The study of physiology and immersion, especially in tropical nature, tends to lead students to a perception that the living planetary surface behaves as a whole (the biosphere, the place where life exists on the Earth). Yet the Academy guards, using Neo-Darwinism as an inquisitory tool, superimpose a gigantic super-structure of

mechanism and hierarchy that protects the throbbing biosphere
from being directly sensed by these new scientists—people most in
need of sensing it. The dispensers of the funds for scientific research
and education and other opportunity makers herd the best minds
and bodies into sterile laboratories and white-walled university
cloisters to be catechized with dogmatic nonsense to such an extent
that many doctoral graduates in the biological sciences cannot distinguish a nucleic acid solution from a cell suspension, a sedimentary from an igneous rock, a kelp from a cyanobacterium, or rye from
ergot. (228–29)

This plea on behalf of her students as well as her colleagues is one
expression of the ethical imperative that drives "Big Trouble in Biology"
and that orients her radical attitudes altogether. An exposition of the
Gaia hypothesis proper arrives after this quintessentially Gaian evocation of the need for initiation into immersive experiences leading to "a
perception that the living planetary surface behaves as a whole." Gaia's
rehearsal provides one last hook on which to hang neo-Darwinism in
effigy. Then an ecological fatalism descends on Margulis's denouement,
anticipating that our mechanistic civilization is beyond self-reform.
Nothing in the past thirty years has done much to dispel this gloomy
view. Our cultural transformation will have to be apocalyptic, some
form of global convulsion: "Among academic biologists inside the con-
vent walls, Neo-Darwinist reductionism will prevail until the sudden-
ness of a new planetary culture replaces the technological civilization
to which Thompson refers. Only after the new civilization binds us con-
sciously to our nonhuman planetmates, especially the truly productive green ones, can the physiology of autopoietic visionaries replace
the mechanics of the Neo-Darwinists inside the academic cloister"
(229). In these sweeping passages the neocybernetics of Gaia rises to
a consummate expression of the philosophical stance of the systems
counterculture.

PART III
GAIAN INQUIRIES

CHAPTER 7

The Planetary Imaginary

The planetary imaginary is constituted whenever found or made images of worlds living or otherwise are bodied forth in some workable medium and taken up into popular or artistic images, journalistic or fictional accounts, or other currents of communication. The planetary imaginary nurtures intuitions of the actual Earth's complex operations. It inspires new techniques for reflecting those processes. Critical readings may discern in such Earthly connections worldly value commitments. For instance, with the first images of the Earth sent down from space, a technocultural event of image production radically shifted the planetary imaginary toward an epoch of cosmological self-recognition. The face of the Earth seen from space became a mirror in which we see ourselves reflected and transformed. The image of Earth seen from space had previously been a speculation, a science fiction trope, or a measured projection—a geographical artifice, such as a globe. Gaia was conceived as a planetary system when the gaze taking Earth as its object could shift from fictive agencies to human observers applying technical prostheses. Gaian science began as technical images of our own planet viewed against its cosmic background fell into place back on Earth. A NASA ATS-III weather satellite (Figure 6) transmitted the first whole Earth image in color (Figure 7) on November 10, 1967.

In "Cyborgs and Symbionts," as if speaking from a posthuman future, Donna Haraway noted the emphatic play of cultural mediations that arise when "the signals emanating from an extraterrestrial perspective, such as the photographic eye of a space ship, are relayed and translated through the information-processing machines built by the members of a voraciously energy-consuming, space-faring hominid culture that called itself Mankind."[1] Haraway made these comments while observing the connections between NASA's space program, the Gaia hypothesis, and the emergence of a global culture forever transformed by having

NASA G-66-3652

FIGURE 6. *The ATS-III weather satellite orbits the Earth. NASA.*

these instruments and images of cyborg cognition: "I do not think that most people who live on earth now have the choice not to live inside of, and not to be shaped by, the fiercely material and imaginative apparatuses for making 'us' cyborgs and making our homes into places mapped within the space of titanic globalizations in a direct line of descent from the cybernetic Gaia seen from NASA's fabulous eyes" (xix). Astronautic, computational, and communications technologies combined in this moment to mediate and publicize unprecedented images of the planet, spurring speculations over previously unimagined processes of the Earth system.

The spaceborne images of Earth from the mid-1960s onward were already prompting such an epochal reimagining of our planet as a whole system prior to the arrival of the Gaia hypothesis as a published scientific proposition in the early 1970s.[2] As the idea of Gaia was gestating in Lovelock's early work on planetary atmospheres, images of the Earth taken from space were powerfully shaping systems thinking through a Whole Earth imaginary with an expanded sense of planetary tempo-

FIGURE 7. *The 1967 ATS-III videograph placed on the front cover of the* Whole Earth Catalog *for spring 1969. NASA.*

rality. To see the Earth not merely as a physical object but as a living system of cosmological duration was also to see it in persistent if not eternal operation. Here was an entity not just captured in a cosmic pose but also grasped at a particular moment in its cosmic evolution. The planetary imaginary may communicate facets of complex systems, actualize something potential but as yet nonexistent, perhaps a previously unrecognized Earth dynamic. It may depict the wild proliferation of microbial life in the Archean eon or forecast the coming evolution and passing establishment of new ecologies. It sketches matters whose full extent cannot be contained in a single view or grasped at any single scale—for instance, the Gaian consortium itself conceived as a coupled complex of geobiological systems evolving over planetary time. Gaia's disparate discursive and visual mediations have constituted a body of texts for which the Earth seen from space remains the foremost representation.[3] In the fall of 1968, the first *Whole Earth Catalog* placed the first spaceborne whole Earth image (Figure 7) on its front cover. As we noted in chapter 4, in the spring of 1969 it repurposed the Earthrise photograph (Figure 2) in the same manner. The *Whole Earth Catalog* cultivated its consummate suite of cybernetic discourses alongside the planetary imaginary crystallized by Earthrise.

However, even as it seemed that the whole Earth was now in view, it also became more clear now just how much one still couldn't see. Like the full moon but more so, the whole Earth has not one but many dark sides. Even spaceborne images of Earth have no way of showing the profusion of geobiological cycles in operation under and amid the clouds that proclaim its fitness for life but shroud the view. These dynamics must still be painstakingly assembled, then visualized in a diagram or a narrative. Coupled to an astounding array of terrestrial and orbital geosensors, Earth system science now provides the data to produce evidentiary replicas of specific Earth system dynamics. Nevertheless, the entirety of those constructions will still place only partial figures in motion on a dark and moving background. Despite our abiding desire to transcend all limiting horizons, in the era of Gaia the planetary imaginary partakes of the permanent partiality of our knowledge of a world in process toward a future we cannot model with any certainty.

Dune

The planetary imaginary is not the global imaginary. An actual planet is a substantial cosmological phenomenon. If it possesses life, a planet potentially bears ecological processes capable of constituting its own observers. In contrast, a globe models a planet's surface with demarcations typically corresponding more to parochial human interests than to natural features. For instance, near the beginning of Frank Herbert's classic science fiction novel *Dune,* the stock villain of the piece, Baron Vladimir Harkonnen, is introduced at first behind and then alongside a massive globe.

> It was a relief globe of a world, partly in shadows, spinning under the impetus of a fat hand that glittered with rings. . . . The fat hand descended onto the globe, stopped the spinning. . . . It was the kind of globe made for wealthy collectors or planetary governors of the Empire. . . . Latitude and longitude lines were laid in with hairfine platinum wire. The polar caps were insets of finest cloudmilk diamonds.

The fat hand moved, tracing details on the surface. "I invite you
to observe," the basso voice rumbled. . . . "Nowhere do you see blue of
lakes or rivers or seas. And these lovely polar caps—so small. Could
anyone mistake this place? Arrakis! Truly unique. A superb setting
for a unique victory."[4]

This oblique view of the planet Dune presents it, so to speak, in luxurious effigy. Expressing a feudal will to exploit Arrakis for political and
economic gain and personal glory, the Baron voices a global imaginary
that responds not to planetary matters, their processes and their interrelations, but to maps, grids, and lines of control. In contrast, the
planetary imaginary corresponds to a necessarily partial view of interconnected territories. The former is political and economic, all too
human; the latter is ecological and geomorphic, more than human. And
whereas you can spin a globe so as to have no dark side—no shadow of
the unknowable, at least within its limited notion of the known world—
you cannot really spin an actual planet such as Earth. Its intricacies will
finally surpass our technological as well as our epistemological grasp.
As a speculative faculty, the planetary imaginary is a significant mode
of human self-observation that undercuts or goes beyond its own capacity to observe. The planetary view is always partial, and so decenters
the human in relation to its worldly situation.

Moreover, just as *Dune* depicts both global conflicts and planetary
coordinations, the modern era is one not just of accomplished globalization but also of ongoing planetization. The globalizing process we are
still in the midst of operates an instrumental humanism driving com-
modity extraction amid political and economic consolidations, negoti-
ating corporate and cultural differences and the mobility of their interrelations relative to a humanity still splintered into feudal or ideological
tribes. In contrast, planetization readjusts the human view of the Earth
as an ultimately incalculable and uncontrollable system. Meditations
upon ecology and systems theory, for instance, by Edgar Morin, Peter
Westbroek, Michel Serres, Peter Sloterdijk, and Bruno Latour, are setting promising outlines for a philosophy adequate to planetization.[5]
For another instance, as we have seen, through his own writings and
direction of symposia sponsored by the Lindisfarne Association beginning in the 1970s, William Irwin Thompson has led important conversations on the formation of a planetary culture.[6]

As planetization emerges in the work of such thinkers, various strands of systems theory have intertwined with varieties of ecological thought to theorize the embeddedness yet noncentrality of the human within wider natural or cosmological schemes. For instance, Gregory Bateson's *Steps to an Ecology of Mind* has been a seminal manifesto for such a systemic viewpoint. In a passage noted by James Lovelock in *Ages of Gaia,* it concedes its status as metaphysics: "The cybernetic epistemology which I have offered you would suggest a new approach. The individual mind is immanent but not only in the body. It is immanent also in pathways and messages outside the body; and there is a larger Mind of which the individual mind is only a subsystem. This larger Mind is comparable to God and is perhaps what some people mean by 'God,' but it is still immanent in the total interconnected social system and planetary ecology."[7] Bateson developed a conceptual shift from ecology as a natural-scientific metadiscipline on a par with cybernetics and specifically focused on the interrelations of life and environment, to ecology as a mobile figure for any situation of interdependent system-environment complexity.[8]

Dune is historically and conceptually concurrent with this philosophical turn. One of the first great American ecological novels, serialized starting in 1963, published entire in 1965, arriving in the vanguard of 1960s counterculture, Herbert's fiction presents mind expansion and alternative communities in a context of planetary environmental concerns. Both *Steps to an Ecology of Mind* and *Dune* participate in, as well as further promote and refine, a larger body of ecological discourse that comes into specific conversation with the first cybernetics that emerges in the later 1940s and gathers intellectual and cultural momentum coming into the 1960s.[9] As both are being written, the wider cultural reception of cybernetic discourses—captured in phrases and concepts such as whole systems, self-organizing systems, informatics, computation, communication, artificial intelligence, noise, entropy, and synergy—is reaching a critical mass. In their own ways, both *Dune* and *Steps* mark two highpoints of this particular cultural crest. Concurrently, many other thinkers are crossing over from mainstream scientific agendas to the systems counterculture, or in Bateson's phrase, "the new epistemology which comes out of systems theory and ecology."

The discourses of cybernetics and ecology combine to form ecosystem ecology, another key conceptual incubator for the planetary imaginary that first comes to a head in the 1960s. The historian Joel B. Hagen notes that the "ecosystem is an intuitively appealing concept for most ecologists, even for those critical of the way ecosystem ecology has developed as a specialty. It is the only ecological concept that explicitly combines biotic and abiotic factors and places them on roughly equal footing."[10] In *Symbiotic Planet,* Lynn Margulis gives an economical reformulation of this incisive ecosystemic conception of the relation between abiotic and biotic elements, physical flow and biological cycle: "Sunlight moves through life, empowering cyclic work."[11] *Dune* also depicts ecosystem ecology's physicalist orientation to "abiotic factors" in relation to living systems through an appendix to the narrative proper, titled "The Ecology of Dune." This paratext contributes the backstory of the first imperial planetologist of Arrakis, Pardot Kynes. Sketching a planetary image closely akin to Alfred Lotka's "Mill-Wheel of Life" in his *Elements of Physical Biology,* "To Pardot Kynes, the planet was merely an expression of energy, a machine being driven by the sun."[12]

In its global dimension, *Dune* depicts the machinations of royal families wrangling over the control and exploitation of melange—the precious spice as indispensable to this storyworld as coffee, tobacco, and alkaloids of coca are to ours, and extractable only from the sands of Arrakis.[13] The planetary imaginary of the narrative emerges alongside these tribal elements. It immediately tangles its ecological themes up with both modes. Just as the Harkonnens vie to control the spice trade, the ecological mentors of the indigenous Fremen focus on control of the planet's climate. These crosscurrents dovetail upon the protagonist Paul Atreides, who nonetheless struggles to achieve and sustain a planetary rather than a global attitude. He does so, it may be said, with the aid of *Dune*'s unique pharmacopeia, the mind-altering spice that blows in the wind, and the Water of Life the Fremen draw from the spice-producing sandworms. That is, while *Dune*'s global commodification is a scene of multiple contestations, Paul's visionary career modulates its planetary imaginary through the increment of racial and ecological mystery that is playing itself out in his person.

However, one of the most important scenes of self-realization in the story is not psychedelic but rather a sober moment registering connections among the mundane details of Fremen social ecology. The adolescent Paul wins a duel of honor and with that his defeated Fremen challenger's wife, Harah, for a consort. Harah then shows the newcomer through the sietch or cavern commune and environs of their under-ground society. As they pass along the corridors of the sietch, Harah has occasion to describe Fremen lifeways. These have already incorporated techniques imported to the naturalized inhabitants of Arrakis by the imperial planetologists, Pardot and Liet Kynes. First the father and then the son have crossed over to the side of the Fremen to guide their society through the intricacies of political resistance. Bringing first-world scientific and technological knowledge to their adopted native culture, they have also infused the Fremen with an ecological prophecy, the vision of a desert world rendered verdant by proper and deliberate environmental management. As Paul and Harah walk along, she informs him about the dew collectors: "Each bush, each weed . . . is planted most tenderly in its own little pit. The pits are filled with smooth ovals of chromoplastic. . . . It cools with extreme rapidity. The surface condenses moisture out of the air. That moisture trickles down to keep our plants alive" (335). When they pass a classroom, Harah explains how, now that both Pardot and Liet Kynes are dead, their ecological teachings are being kept alive by these lessons: "'Tree,' the children chanted. 'Tree, grass, dune, wind, mountain, hill, fire, lightning, rock, rocks, dust, sand, heat, shelter, heat, full, winter, cold, empty, erosion, summer, cavern, day, tension, moon, night, caprock, sandtide, slope, planting, binder.'" Eventually they reach Harah's living space, and as Paul hesitates at her threshold, he has an encompassing thought: "It came to him that he was surrounded by a way of life that could only be understood by postulating an ecology of ideas and values" (336).

The story has already established that the young nobleman Paul has received ecological training on a par with the Kyneses' science. So the implication is that, in this thought "postulating an ecology of ideas and values," Paul has added together his comprehension of the ecological significance of the dew catchers with the way that "sandtide, slope, planting, binder" concern the ongoing Fremen efforts to control the wayward tides of the dunes. Both are part of a wider plan to shape their environment. What Paul seems to realize is that the natural environment and the engineering techniques that would cultivate and control it, together with its inhabitants and the pedagogical program to inculcate them with an ecological sensibility that comprehends the Kyneses' environmental prophecy, are all elements mutually constituted within a specific interlocking of natural, psychic, and social systems.[14]

Steps to an Ecology of Mind

Dune anticipates *Steps to an Ecology of Mind* in remarkable ways. This circumstance suggests that the ecological sensibility Bateson hoped to cultivate in his later writings was a discursive variant of a planetary imaginary like the one Herbert exercised in a fictional narrative mode. *Steps* is a compilation of essays and articles written in the two decades prior to the book's publication in 1972. Bateson states in the foreword: "Broadly, I have been concerned with four sorts of subject matter: anthropology, psychiatry, biological evolution and genetics, and the new epistemology which comes out of systems theory and ecology" (xii). He treats ecology as a variety of the discourse of cybernetic systems theory broadly construed. As such, it can partake of the strategic generality of cybernetic formulations of information and communication. In particular, it can pass beyond strictly natural-scientific reference and still provide the conceptual foundation and systemic occasion for the immanentist metaphysics of his ecology of mind. The later papers of *Steps,* especially those clustered at the end, all written between 1967 and 1971, are largely and explicitly engaged in the project of redescribing and repositioning human culture and communication as parts of a mental or informational ecology that is planetary in scope.

Bateson proposes "a new way of thinking about ideas and about those aggregates of ideas which I call 'minds.' This way of thinking I call the 'ecology of mind,' or the ecology of ideas." His conceptual approach is not just new, he states further, but necessary, for it provides a vital but otherwise unavailable understanding of the manifold complexities of worldly relations extending from natural evolutionary forms to human cultural behaviors. "Such matters as the bilateral symmetry of an animal, the patterned arrangement of leaves in a plant, the escalation of an armaments race, the processes of courtship, the nature of play, the grammar of a sentence, the mystery of biological evolution, and the contemporary crises in man's relationship to his environment, can only be understood in terms of such an ecology of ideas as I propose" (xv). It is a noteworthy circumstance and striking resonance that, as composed in 1971 and first published in 1972, Bateson's ecophilosophical declaration in this passage repeats nearly verbatim the passage we just looked at from *Dune.*

On the one hand, for Bateson in 1971, the comprehensive interrelatedness of planetary matters could "only be understood in terms of such an ecology of ideas as I propose." And on the other, for Herbert, who composes this piece of text for his hero Paul sometime before 1965, the current, renovated culture of the Fremen "could only be understood by postulating an ecology of ideas and values," period. Now, Bateson's every use of the term *ecology* in *Steps* occurs in an article postdating *Dune*'s publication. I would explain this mutual resonance and wide overlap in literary and philosophical approaches not as the repetition of a source but as concurrent responses to the same cultural paradigm that by the 1960s names itself, simply, *ecology*. Both authors receive the American inflection on the twentieth-century shift in the discipline of ecology from a descriptive form of natural history to a theoretical systems science. Positioned outside the science of ecology proper, both are free to move its systemic descriptions of humanity's wider interrelations with its worldly environments toward their cultural implications. Or again, the planetary imaginary of both authors belongs to the same *ecology of ecology*. For instance, both draw from the particular line of ecosystem ecology. "The Ecology of Dune" makes this disciplinary identification explicit through one of Pardot Kynes's exhortations: "'The thing the ecologically illiterate don't realize about an ecosystem,' Kynes said, 'is that it's a system. A system! A system maintains a certain fluid stability that can be destroyed by a misstep in just one niche'" (482).

In such a passage, the planetary imaginary takes the form of a pedagogical exhortation to maintain an encompassing and properly systemic ecological vision. In light of these fictionalized problematics of scientific literacy, *Steps*'s final paper, "Ecology and Flexibility in Urban Civilization," contains a final section, "The Transmission of Theory," that offers another remarkable echo of *Dune*. In this striking meditation, Bateson notes that a "first question in all application of theory to human problems concerns the education of those who are to carry out the plans" (503–4). We saw the narrative of *Dune* depict a comparable concern when Paul, being guided by Harah, was first made cognizant of the Fremen's ecological planning, and then encountered the chanting children, a scene from the pedagogical regime designed to imprint Pardot and Liet Kynes's plan for eco-engineering the environment of Dune on its future executors. Bateson connects a related issue of ecological citizenship to the insight that the larger system constituted by the human social organism *plus* its worldly environment is traversed by mental and informational as well as physical and geobiological circuits. Regarding this systemic description, he asks:

> Is it important that the right things be done for the right reasons? Is it necessary that those who revise and carry out plans should understand the ecological insights which guided the planners? . . . The question is not only ethical in the conventional sense, it is also an ecological question. The means by which one man influences another are a part of the ecology of ideas in their relationship, and part of the larger ecological system within which that relationship exists. (504)

By such reasoning, *Dune* would also be a systemic element within the ecology of ideas that yields Bateson's *Steps,* a fictionalized ecophilosophy in resonance with the ecosystemic ensemble to which Bateson gives expression. In "The Ecology of Dune," we read further:

> "There's an internally recognized beauty of motion and balance on any man-healthy planet," Kynes said. "You see in this beauty a dynamic stabilizing effect essential to all life. Its aim is simple: to maintain and produce coordinated patterns of greater and greater diversity. Life improves the closed system's capacity to sustain life. Life—all life—is in the service of Life. Necessary nutrients are made available to life by life in greater and greater richness as the diversity of life increases. The entire landscape comes alive, filled with relationships and relationships within relationships." (477–78)

The text of *Dune* bootstraps a Gaian recognition out of ecosystem ecology just a few years *avant la lettre*. But whereas Kynes's fictive discourse may be taken to represent ecology proper and so to be contained within a geobiological frame, Bateson's ecology of ideas extends the concept beyond its home reference to the natural world. His reinterpretation of ecology as a systemic epistemology allows for variety regarding the level at which one draws the boundaries of the system one intends to observe and the environment constituted by that distinction. In this way, a planetary system composed from the co-operation of life *and* its ecological environments may be enlarged, in cosmological perspective, to encompass immanent Mind as well: "I now localize something which I am calling 'Mind' immanent in the large biological system—the ecosystem. Or, if I draw the system boundaries at a different level, then mind is immanent in the total evolutionary structure" (460).

This is the planetary imaginary doing philosophy, turning upon ecology to make the conceptual shift by which "the large biological system—the ecosystem" may be taken to include as well those meta-biotic matters—the mediations of consciousness and communication—for which life per se is the precondition if not precisely the operation. Nevertheless, granting this interdependent nesting of systemic conditions, Bateson presses its implications for cultural evolution to some profound conclusions. For one, the body of ecology about which it reasons conditions the ecological mind: "The problem of how to transmit our ecological reasoning to those whom we wish to influence in what seems to us to be an ecologically 'good' direction is itself an ecological problem. We are not outside the ecology for which we plan—we are always and inevitably a part of it" (504). In line with the second-order cybernetics about to emerge at Heinz von Foerster's Biological Computer Laboratory, this is a classic self-referential insight placing the observer within the system constituted by his or her observation. Bateson does not minimize the vertiginous quality of this realization: "Herein lies the charm and the terror of ecology—that the ideas of this science are irreversibly becoming a part of our own ecosocial system" (504).[15] However, if we are inescapably bound to the Earth's evolving conditions, then as we negotiate our own trepidation it is bracing to note how Bateson concludes with a sublime ethical edict not to get hung up on hasty activisms to the point of losing touch with the primary vision of planetary participation: "If this estimate is correct, then the ecological ideas implicit in our plans are more important than the plans themselves" (505). Whereas cultural strategies adapt to fluctuating social circumstances, the higher wisdom to be won from the hard science is what will continue to count. One could say the same about Gaia: the best planetary ideas outlast the cultural programs that come and go in their name.

The Ecology of *Neuromancer*

In 1984, *Neuromancer* hit, not like a bomb exactly, but more like a massive slow-release torpedo whose effects grew inexorably as the world increasingly realigned itself around the storyworld of the novel. William Gibson famously named the plane of virtual reality within his near-future storyworld, previously unimagined in such profuse detail, *cyber-space*. This appellation was surely a submerged repercussion of the systems counterculture as it was going into eclipse in the Reagan era. Cyberspace took place alongside a noir vision of a disnatured planet in which biological bodies are in submission to technological manipulations and digital prosthetics. Let us take a moment to contemplate this particular threshold in the cultural imaginary of informatic cybernetics. On the face of it, cyberspace and cyberculture would seem to be in full flight from Gaia, at least if one takes Gaia as a stand-in for some residual concern or reverence for the "natural world" and its geobiological processes. The seemingly unmediated neural immediacy of virtual reality is a "bodiless exultation" (Gibson, 6) in which organic imperatives are momentarily or permanently transcended. When, after a season of forced exile, Gibson's protagonist, Case, jacks back into cyberspace, he sees "hypnagogic images jerking past like film compiled from random frames. Symbols, figures, faces, a blurred, fragmented mandala of visual information ... flowered for him, fluid neon origami trick, the unfolding of his distanceless home, his country, transparent 3D chessboard extending to infinity" (52). Readers of *Neuromancer* tend to fixate on this digitally rendered environment, an exclusively optical expanse, "bright lattices of logic" seemingly detached from planetary contingencies.

Nonetheless, on occasion the text of the novel throws cyberspace into relief by referring to its flip side, its embodied preconditions. Following the passage just cited, we read: "And somewhere he was laughing, in a white-painted loft, distant fingers caressing the deck, tears of release streaking his face" (52). Between the digital virtuality of cyberspace and the actual place where Case is bodily present, we have the two ostensible ecologies of *Neuromancer*. One is a domain for the mind with a variable and malleable population of digital entities. The other is the degraded environment of the outer storyworld. Green metaphors are flipped on their head, as in the single appearance of the word *ecology* in the text: "The dubious niche Case had carved for himself in the criminal ecology of Night City had been cut out with lies, scooped out a night at a time with betrayal" (11), and here, with "the mall crowds swaying like windblown grass, a field of flesh shot through with sudden eddies of need and gratification" (46).

On the urban front, in ironic homage to Paolo Soleri's once-hopeful projects, cityscapes are "dominated by the vast cubes of corporate arcologies" (6). *Arcology* was Soleri's own contraction of *architecture* and *ecology* to name his earnest efforts since the 1960s to retool urban design. The text deploys the word to gain its tone but travesties its sense. In a similar fashion, Buckminster Fuller's retrofuturistic architectural signifier appears in "towers and ragged Fuller domes" (31), as when Case remembers from his adolescence "the rose glow of the dawn geodesics" (46). We are to think that cities absorbed into the Sprawl had once been encased against a noxious outer world in now-crumbling geodesic domes, failed Earthbound attempts to shield human populations away from a dying Gaia in closed artifical ecosystems. And *Neuromancer*'s occasional open vistas, redolent of "blasted industrial moonscape . . . broken slag and the rusting shells of refineries" (85), reprise the entropic horizons of J. G. Ballard and Philip K. Dick stories in the 1950s and 1960s. Mentioned once in passing are a "pandemic" rendering horses and presumably other large animals extinct and "the rubble rings that fringe the radioactive core of old Bonn" (92, 96), placing the Earthbound setting of the novel in the aftermath of a reactor meltdown or limited nuclear exchange.

However, beyond the initial opposition of these inner and outer storyworlds—one beckoning, the other broken—*Neuromancer* does travel to something like another planet altogether, when the storyworld shifts from the Earthbound settings of the earlier action to the high-orbital space colonies where the main and climactic events will be staged. Gibson brilliantly imagines these orbital settings and provides some significant accounting for their operation. But residing somewhat in the glare and shadow of the wider celebration and exaltation of cyberspace, *Neuromancer*'s high-orbital environments tend to slip by without special notice. In fact, they have a specific but largely unrecognized provenance that tethers *Neuromancer* back to the chapter of American technoculture that also features the earliest stirrings of the Gaia hypothesis. The planetary imaginary in *Neuromancer* renders artificial Gaias through a fictive repurposing of Gerard O'Neill's high-orbital space colonies.

Part 3 of the novel begins with the following incantation:

> Archipelago.
> The islands. Torus, spindle, cluster. Human DNA spreading out from gravity's steep well like an oilslick.

Call up a graphics display that grossly simplifies the exchange of data in the L-5 archipelago. One segment clicks in as red solid, a massive rectangle dominating your screen.

Freeside. (101)

To begin with, the L in the designation L-5 refers to the French mathematician and astronomer Joseph-Louis Lagrange. L5 is one of several "Lagrangian libration points," locations in space where the relative gravities of a two-body system—here, the Earth and the Moon—balance such that an object placed in orbit there, near the line of the Moon's own path around the Earth, will not decay but be stable indefinitely. All the satellites, shuttles, and space labs we are familiar with, the ones that fall back to Earth on occasion, are in low orbit. An object at L5 is in high orbit. And as O'Neill had declared to Stewart Brand, "the L5 Earth-Moon Lagrange libration point . . . could be a far more attractive environment for living than most of the world's population now experiences."[16] Indeed, if one reviews the text of *Neuromancer* with the space-colony issues of *CoEvolution Quarterly* in hand, the conclusion is inescapable. High orbit, L5, torus, spindle, cluster: it's already here in O'Neill, served up on *CoEvolution Quarterly*'s hip countercultural platter. Gibson liberally helped himself to these goodies and recycled NASA's gorgeous depictions straight into the high-orbital storyworld of *Neuromancer*.[17]

For instance, the narrator's statement "Human DNA spreading out from gravity's steep well like an oilslick" is an image whose originality and perversity, as far as I know, are all Gibson's. However, the embedded phrase "gravity's deep well" is an homage to O'Neill's gravity hole. An illustration for "O'Neill Space Colony, Model III (6.2 miles long—1.24 miles diameter) at the L-5 Lagrangian Point" lays out the main design that Gibson will adapt (Figure 8). Here is the Archipelago, made up of a constellation of such gigantic constructions, and here is a "blunt white spindle, flanged and studded with grids and radiators, docks, domes" (Gibson, 77). The NASA illustration showing a sample interior of a Model III O'Neill space colony (Figure 9) enters *Neuromancer*'s high-orbital settings when Case gets a briefing on Freeside's terrain: "Casinos here. . . . Hotels, strata-title property, big shops along here. . . . Blue areas are lakes. . . . Big cigar. Narrows at the ends. . . . Mountain effect, as it narrows. Ground seems to get higher, more rocky, but it's an easy climb. Higher you climb, the lower the gravity" (107).

FIGURE 8. *O'Neill Space Colony Model III at the L-5 Lagrangian Point. NASA AC75–1085.*

In *CoEvolution Quarterly*'s first space-colony number, Brand had already explained how the artificial gravity produced by the centrifugal rotation of the cylinder around its long axis fades to nothing at the center of the tip. NASA's artists also appear to have humored O'Neill's notion of "the green grass and the flowers and the sunshine coming down" in an image reproduced on the front covers of *CoEvolution Quarterly* for fall 1975 and of the freestanding *CoEvolution Quarterly* issue *Space Colonies,* envisioning a moist and verdant landscape basking in reflected natural sunlight and happily reproducing a recognizable terrestrial ecosystem (Figure 9). The artist renders an orbital San Francisco complete with bays, harbors, sailboats, rolling hills, seagulls, and a Golden Gate Bridge, all gently bent by the inside of the circular shell of the cylinder within which this promised land has been planted.

Neuromancer turns all this idealized landscape into cyberpunk by converting O'Neill's earnest and sunny orbital San Francisco into a debauched and sleepless resort: "Freeside is many things, not all of them evident to the tourists who shuttle up and down the well. Freeside is brothel and banking nexus, pleasure dome and free port, border town

FIGURE 9. *Endcap view of a cylindrical colony with suspension bridge. NASA AC75–1883.*

and spa. Freeside is Las Vegas and the hanging gardens of Babylon, an orbital Geneva" (101). Given the debased status of the Earth from which Freeside is an escape, it is no surprise that it, too, is no ecological show-case. Freeside is imagined largely in detachment from its basis in some ecosystemic coupling of living and nonliving components. Only in a sort of afterthought, which we will examine in a moment, does *Neuromancer* bring the ecological possibility of a space colony such as Freeside for-ward as an issue. That it does so at all suggests to me that Gibson has in fact attended thoughtfully not just to O'Neill's designs, which were available elsewhere, but also to the airing of the issues surrounding them in *CoEvolution Quarterly*'s space-colony volumes.

Gibson hides these ecological considerations in plain sight by embed-ding them into the construction of a different island in the Archipelgo, not the luxurious space resort of Freeside but the high-orbital Rasta-farian shantytown called Zion: "Zion had been founded by five work-ers who'd refused to return, who'd turned their backs on the well and started building. They'd suffered calcium loss and heart shrinkage before rotational gravity was established in the colony's central torus" (103). The torus is one of three forms O'Neill envisioned for space-colony habitats—cylinder, torus, and sphere. And Zion, in contrast to the stratified society of Freeside, is clearly a people's colony. With a mural showing "a painted jungle of rainbow foliage" (109), it houses an egalitarian population of space Jamaicans speaking patois, passing spliffs, and making a life in space while providing Case's crew with off-grid transportation and bodyguard services: "Case gradually became aware of the music that pulsed constantly . . . dub, a sensuous mosaic cooked from vast libraries of digitalized pop; it was worship . . . and a sense of community. . . . Zion smelled of cooked vegetables, humanity, and ganga" (104). In the ecology of *Neuromancer,* Zion suggests a nature preserve—a vestige of green space in the midst of largely abiotic built environments—through which an enclosed habitat could actually ensure its own viability for the long term.[18]

The ecological specifics come late in the novel. They are brief and would seem to be afterthoughts that the narrator sends through Case's mind while he is otherwise occupied with his current caper, the Straylight run. And yet their coming at the start of a new chapter subtly emphasizes them. Seemingly out of nowhere, the narrator explicitly addresses the ecologies of three different enclosed spaces distributed throughout Freeside and Zion. At the moment, Case and his crew are at the tip of Freeside's spindle with a "steep climb out of gravity," penetrating the Villa Straylight. In the midst of more pressing matters, it occurs to him that

> the Villa Straylight was a parasitic structure. . . . Straylight bled
> air and water out of Freeside, and had no ecosystem of its own. . . .
> Freeside's ecosystem was limited, not closed. Zion was a closed
> system, capable of cycling for years without the introduction of external materials. Freeside produced its own air and water, but relied
> on constant shipments of food, on the regular augmentation of soil
> nutrients. The Villa Straylight produced nothing at all. (225–26)

The likely sense of these ecological details for most readers would be the metaphor suggested between the Villa Straylight's materially parasitic relation to Freeside and the parasitic as well as self-consuming decadence of the Tessier-Ashpool clan immured there inside their private hive. But taking it at face value, this passage literally and cogently addresses the ecological status of its high-orbital environments, and it does this in terms that are fully developed in *CoEvolution Quarterly*'s space-colony debates.

These debates were wide-ranging.[19] For instance, as collected in *Space Colonies,* a number of commentators there worry the vexing sociological issues that would be involved in the constitution of a specific human population for a particular space colony. Gibson deflects that problem by setting forth Freeside as a tourist destination and Zion as an ethnically unified subculture. But in any event, such demographic problems do not seem insurmountable. Granting that engineering solutions enable high-orbit space colonies to be built in the first place, the more challenging issue confronting them is precisely the living ecology to be initiated and established there and then shared and maintained by their full ensemble of living inhabitants as a closed ecosystemic whole. The most learned commentators on this score, including some who had previously contributed to "Ecological Considerations for Space Colonies," were near unanimous in their conclusion that the current state of research and technique was not adequate to the ecological problems space colonies would confront.

Paul and Anne Ehrlich sent *CoEvolution Quarterly* a statement to this effect. Ehrlich is famous for authoring *The Population Bomb,* a much-debated popular text of 1968 warning of coming resource depletion on an overpopulated Earth. Those touting the space-colony program talked up its utility for siphoning off excess human population or providing a last refuge for a remnant of humanity after the garden is gone. The Ehrlichs acknowledged these arguments: "The prospect of colonizing space presented by Gerard O'Neill and his associates has had wide appeal especially to young people who see it opening a new horizon for humanity. The possible advantages of the venture are many and not to be taken lightly." Nonetheless, they were not convinced:

> On the biological side things are not so rosy. The question of atmospheric composition may prove more vexing than O'Neill imagines, and the problems of maintaining complex artificial ecosystems within the capsule are far from solved. The micro-organisms necessary for the nitrogen-cycle and the diverse organisms involved in decay food chains would have to be established, as would a variety

of other micro-organisms necessary to the flourishing of some plants. . . . Whatever type of system were introduced there would almost certainly be serious problems with its stability—even if every effort were made to include many co-evolved elements. We simply have no idea how to create a large stable artificial ecosystem.[20]

Another skeptical commentator was the distinguished design ecologist John Todd, who already had considerable practical and experimental experience on research questions that had occupied ecosystem ecology since the 1950s—the composition, construction, and testing of artifical environments open for energy flow but materially closed, creating manufactured replicas of solar-powered natural ecosystems. His lengthy critique displays the epistemological humility appropriate to the planetary imaginary:

After a decade of living intimately with designed ecosystems I am coming to know that nature is the result of several billion years of evolution, and that our understanding of whole systems is primitive. There are sensitive, unknown and unpredictable ecological regulating mechanisms far beyond the most exotic mathematical formulations of ecologists. When I read of schemes to create living spaces from scratch upon which human lives will be dependent for the air they breathe, for extrinsic protection from pathogens and for biopurification of wastes and food culture, I begin to visualize a titaniclike folly born of an engineering world view.[21]

In other words, we might well build and launch a titanic space colony only to see it shipwrecked before long by the failure of its internal ecology, its evolution toward a nonviable state. This scenario actually happened in 1992 with the most significant and large-scale experiment in closed ecological habitats to date, Biosphere 2. After seventeen months, the O_2 level of its closed atmosphere dipped precipitously and the habitat became unviable for its human inhabitants. The closed environment had to be opened up for the survival of its crew.[22] In *Neuromancer*'s not-so-distant near-future fantasy, only Zion with its "cooked vegetables . . . and ganga" has succeeded in becoming what any beyond-Earth habitat with staying power will need to be, a micro-Gaia, a self-sustaining, self-maintaining home for permanent residents, "a closed system, capable of cycling for years without the introduction of external materials."

Inside the Villa Straylight the Tessier-Ashpool clan aspires to be a closed system, but it has its ecology all wrong: it cannot recycle; it can only consume others as well as itself. Their generations alternately awaken and hibernate, while pursuing an exclusive purification of their own genetic line through cloning, cryogenics, inbreeding, and murderous infighting. They are a fit parody of Western humanity en masse when it dismisses its own coevolutionary embeddedness within diverse living environments that transcend human understanding and control. Zion intimates that a space colony could be viable, but only if it can work out how to take Gaia with it. If it can re-create a coevolutionary consortium of living systems, including human beings, resting on virtuous microbial foundations and environmentally interlaced with the necessary suite of elemental geochemical cycles, it just might be able grow its own long-term ecosystem. Placing Zion on the sidelines as a tenuous but possible lifeline, the text then projects its otherwise flagrant organic deficits elsewhere, upon Marie-France, Ashpool, and 3Jane's incestuous inward spiral. So the Tessier-Ashpool clan at the tip of the spindle also epitomizes both the detachment and the exclusivity of the space-colony idea when it is rendered as a purely human-engineering challenge and not a necessarily Gaian phenomenon, that is, rendered as a global but not a planetary phenomenon.

Finally, it is the same Tessier-Ashpool clan that has constructed the sentient artificial intelligences, the AIs that are pulling the puppet strings of Case and his colleagues. One of these cyberentities, Wintermute, succeeds in stage-managing its own digital evolution through the overcoming of human prohibitions against its intercourse and merger with its reluctant counterpart AI, named Neuromancer. But in the end, the AI element cannot be taken any more seriously than can the novel's visions of cyberspace, which realm is specified as the virtual environment within which the AIs carry out the term of their artificial being. And even granting these machine systems some future-cosmic plausibility, they too will still depend upon viable environments, no matter how galactic, planets of some sort, material and living environments that transcend the ultimate finitude of the AI's own systemic complexity. At any rate, the Gaian vision around the edges of *Neuromancer* embeds them within some more-diverse ecology that will determine the limits of their current and future possibilities. In this way the novel's most sublime, disembodied, digital, and cosmic apocalypse leads back home once again to the planetary imaginary *Neuromancer* keeps mostly under wraps—the ecological sensibility incubated by the systems counterculture of the 1960s and 1970s.

The New Earth and Its Universe

In 1970 the *Whole Earth Catalog* publicized an obscure thin volume pregnant with neocybernetic futurity, George Spencer-Brown's *Laws of Form,* with a substantial review written by Heinz von Foerster. We looked at this event briefly in chapter 4. I will reprise some of that discussion here. Von Foerster noted that "laws are not descriptions, they are commands, injunctions: 'Do!'" *Laws of Form* is, among other things, a manual in cosmogony. The form of laws in *Laws of Form* enjoins *performative* utterances, statements that create the state of affairs they declare to be the case. They do not describe; they create or bring forth. "The first constructive proposition in this book is the injunction: 'Draw a distinction!' an exhortation to perform the primordial creative act" (von Foerster, "Laws of Form"). For example, to say *Let there be light* is also to say *Let there be a distinction between light and something other than light—call it darkness.* Light must come forth in distinction from darkness; otherwise all would be light, in which brilliance nothing could be distinguished. Moreover, the first injunction to draw a distinction carries a second, unuttered injunction: *Indicate yourself by that act.* This second injunction, to declare your implication in your own acts of distinction and observation, is often unmarked; it is sometimes suppressed. Nonetheless, it is the law. Indicating the self-reference of distinctions in this manner complicates any description of the universe. The observer must be a part of the universe that it can know. Whatever is "out there," its observation also refers us to the contingencies of location and cognitive agency from which its being known arises. In reference to the nature of the universe, that point of location is sufficiently specified as "somewhere on Earth."

Let us bring these formal considerations to a planetary image created by NASA artists in 1986 to depict "cosmic evolution" as the purview of their program in astrobiology (Figure 10).[23] In this visual account, the

FIGURE 10. *"Cosmic Evolution," Exobiology program at NASA Ames Research Center, 1986. NASA.*

Big Bang is presupposed. As narrated in the text of NASA's caption, creation got going with "the formation of stars." In the beginning, cosmic evolution is preterrestrial and prebiotic. Then the Earth forms out of the solar disk. Life gains a situation in the oceans and sets forth along the terrestrial evolutionary path. Let us now turn the laws of form toward a neocybernetic depiction of this cosmic tableau. As a systemic production, observation is not just a state into which one enters but an ongoing operation that the generation of further distinctions must continuously carry forward. To observe the forms of observation depicted in the narrative of this image, let this macrocosm include as part of its entire description the form of the observer within the system distinguished by its observations. Let the observer enter the system of cosmic evolution precisely in the unfolding moments of creation, as the upsurge of creative distinctions that materialize as "the formation of stars, the production of heavy elements, and the formation of planetary systems." This observation of observation closes the loop connecting star-formation to ideation. We reboot our understanding of this evolutionary path as a recursive, self-producing process maintained by the continuous reentry of the form of distinction into the form of the distinguished. As we reenter the astrobiological gaze into the universe it contemplates in this fashion, its outward course bends back down the well of cosmic time to ride the waves of terrestrial evolution back up to its own spatiotemporal location on a mountaintop above the clouds. As Gaian beings, we are rooted into a mature planet of which we are a recent manifestation. In its natural arc, the astrobiological gaze returns to Earth in a more complex state than when it departed. In its return to Earth, the astrobiological gaze renews the Earth.

RETURN TO EARTH

Peter Sloterdijk's *Spheres* trilogy develops the longer history of the return to Earth as a form of modernity's planetary imaginary: "The cosmological process of modernity is characterized by the changes of shape and refinements in the earth's image in its diverse technical media."[24] Modernity accompanies "the downfall of heaven," the slow-rolling collapse of a classical picture of Earth conceived as entirely at rest beneath the heavenly motions of a divinely populated and immunitary dome of cosmic concern for human needs and desires. Our planet now traverses a decentered cosmic space while a globalized Earth continually works out new existential living arrangements. Sloterdijk projects these new thought patterns broadly across the modernizing West. For instance, by the latter part of the eighteenth century, the loss of a transcendent heaven over Earth has become "the transcendental turn" in the philosophy of Immanuel Kant with "the turn of the cognizer towards their own cognitive apparatus and the local cognitive situation. . . . The earth . . . is now the transcendental star that comes into play as the locational condition of all self-reflections. . . . As the star on which the theory of stars appeared, the earth shines with self-generated phosphorescence" (25).

Could Sloterdijk have formulated these striking descriptions of a self-reflexive intellectual modernity much before the 1970s? It was then that second-order cybernetics and social systems theory began to take up these themes of German philosophical idealism from Kant and Hegel through Husserl and Heidegger into the theories of observation and paradox cultivated by NST. Niklas Luhmann in particular, a frequent reference in Sloterdijk's texts, radicalized the theory of self-referential systems by crossing Husserl with the models of autopoiesis and cognition drawn from Humberto Maturana and Francisco Varela.[25] Sloterdijk's modernity is to some extent a back-formation from our own neocybernetic moment. The "thought figure" of the return to Earth from a point beyond the world is now "concrete as a movement in the physically real space: the astronaut Edwin Aldrin, who became the second human to set foot on the moon on 21 July 1969, shortly after Neil Armstrong, took stock of his life as an astronaut in a book with the title *Return to Earth*."[26] It may be only in the astrobiological era heralded by NASA, Earthrise, *CoEvolution Quarterly,* and the Gaia hypothesis that we have begun fully to consummate the return to Earth that Sloterdijk traces onward from the Renaissance of Copernicus and Galileo.

As the astronauts have testified, the observer of the Earth from space returns to Earth seeing differently, with an altered view that changes the system constituted by that altered state.[27] Latour has recently treated a version of this phenomenon in the way that Gaia theory has refocused thinking on Earthly matters of systemic complexity in planetary functions. This sudden or unexpected return *of* Earth *to* the center of things can be "dizzying":

> And our vertigo is much more pronounced than the one set off by
> Galileo when he described the Earth orbiting around the Sun. It
> took a good deal of imagination, in the seventeenth century, to be
> frightened by the "eternal silence of these infinite spaces," since
> in practice, on Earth, no one could detect the slightest difference
> between the heliocentric version and the geocentric version of ev-
> eryday experience. . . . But here, with Lovelock, it is very easy to feel
> the extent to which this new form of geo-centrism—I ought to say
> Gaia-centrism—has consequences! This time, we are not at all in the
> same world.[28]

As registered in Gaia theory at the leading edge of Earth system science and in many other non- and posthuman turns in current theory and philosophy, what is new about the Earth at this larger moment is its unanticipated resumption of a kind of universal centrality. The Anthropocene is a symptom, however feverish, of the way our new Earth has become peremptory in enjoining participatory rather than alienated stances of observation.

In the modernist telling, the heliocentrism of Copernicus was a revolutionary breakthrough to the cosmic reality of terrestrial insignificance. Now, however, Gaian thought has spiraled us toward a new Earth beyond that prior paradigm shift. On the one hand, in astrobiologist David Grinspoon's review of the usual heroic version of events, "Four centuries ago Galileo liberated us from the tiny prison of geocentrism, by revealing that Earth, which we've always called not just a world but the world, is only one of many planets. Our loss of a privileged place was compensated for by a massive enlargement of our universe."[29] But, on the other hand, in the current moment, "we cannot see nature clearly if we insist on ignoring our own growing role . . . for all we know we could be determining the future of all life" (209). Here is a cosmic cliffhanger: astrobiology anticipates that life in the universe is not unique to Earth, but it cannot confirm that status. It may then be that a presumptuous species of Earthlings is holding the fate of life in the universe in its all-too-human grip. If the fate of the Earth does have cosmological consequences, and if, due to our tampering with its conditions of habitability, our fate on this Earth is now in human hands, then at least for the time being, Earth has to come back to the center of our universe.

This is the new geocentrism—the "Gaia-centrism" noted by Latour. Cosmographer David McConville has developed a distinctly neo-cybernetic take on the new geocentrism as an observational practice. In "On the Evolution of the Heavenly Spheres" he describes a conceptual tension in the visualization work of modern and contemporary planetariums that reflects larger epistemological issues: the objectivist ideal rooted in classical physics enjoins the builder of standard cosmic visualizations to orient them toward an Archimedean view from nowhere.[30] McConville narrates his realization of the paradoxical foundations for this modernist project to subordinate an infinitesimal Earth to the infinite cosmos. This insight came to a head in his study of a planetarium program titled *The Known Universe*. Based on a NASA visualization project called the *Digital Universe Atlas,* this program formulated an image of the cosmic microwave background—the "relic radiation" of the Big Bang—at the farthest limit of detectable light, at the climax of an outward zoom looking back from beyond the edge of the universe (Figure 11). McConville's description provides a kind of voice-over account of this visual narration:

> As thousands of colored data points symbolizing galaxies and quasars came into view, they appeared in a wing-like pattern emanating from the center of the model. We then approached, and flew beyond, the outer boundary of the Atlas, a speckled spherical image of the leftover radiation from the early universe. This . . . was humanity's "cosmic horizon," representing the furthest distance light had traveled since the beginning of the cosmos. From this perspective, the

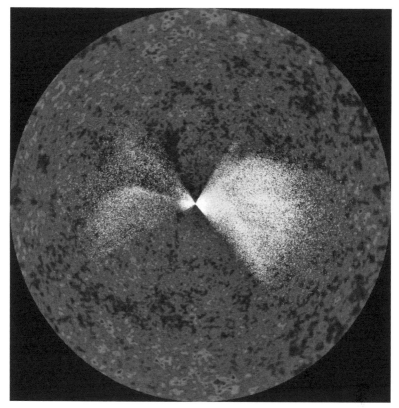

FIGURE 11. *Humanity's cosmic horizon, from the American Museum of Natural History* Digital Universe Atlas. *Reprinted with permission.*

sphere of this "cosmic microwave background radiation" enveloped the entire Atlas, resembling a hermetically sealed bubble floating within an infinite void.[31]

What McConville saw in this image was an inadvertent reversion to or inexorable reconstitution of the heavenly sphere purportedly pulled down by the Copernican revolution, a cosmic sphere with Earth at the center. However, when he asked about this "spectacular return of a spherical, Earth-centered cosmic model" embedded inside the *Digital Universe Atlas,* his intuition was explained away as an unavoidable consequence of the speed of light. Moreover, "when modeling astronomical observations," he was told, "the place from which the observations are made is inevitably the relativistic 'observational center'" (8). In other words, in the residual Galilean or modernist paradigm, since our planet is just a random speck in a bottomless universe, the view from anywhere else would be just the same. Placing the Earth at the center of the visualization was understood as a convenience but not a principle. So the relativistic effect of a neogeocentric cosmic sphere was discounted as an "inevitable" artifact of model building rather than grasped as an embodied critique of the model's presuppositions. For McConville, that meant that *The Known Universe* should be presented to audiences not as *the* authorized, Einstein-approved view of the cosmos but as just "one of many possible perspectives." And again, "since the entire cosmos appeared to be centered on us, wouldn't that imply that observers are central to acts of observation? And if we're inseparable from our measurements, wouldn't that suggest an inextricable relationship between 'internal' consciousness and the 'external' cosmos?" (9). Restated in explicitly neocybernetic terms, in a relativistic universe wherein the point of observation has to be from somewhere, achieving the view from nowhere must defy actual visualization practices. And as Latour has underscored concerning Lovelock and the paradox of observing Gaia, "It was by taking 'the point of view of nowhere' that he showed that there is no 'point of view of nowhere'!"[32]

McConville's new geocentrism reflects the paradoxes of observation. The effort to visualize the universe as a whole from nowhere in particular also produced, like the return of the repressed, a representation of our specific observational location and its cognitive limitations. For a simple example of how such epistemological borders work, consider a Necker cube: one can easily conceive the "unity of the distinction" among the different interpretations of the figure as a three-dimensional representation (Figure 12). As a diagram of itself, a Necker cube literally presents all of its multiple constructions all at once. However, try as one might, one can see only one of them at any given moment. The unity of the distinction remains a paradox in that it cannot be put into operation; it cannot be seen. Nevertheless, the distinction can be operationalized, by selecting a specific indication. That such ambiguous two-dimensional images lead the mind's eye around in circles calls out the selective, hence partial, nature of our perceptions, of depth or anything else.

What does this epistemological limit mean in the domain of Gaian thought? To me it means that the universe is just as unobservable in

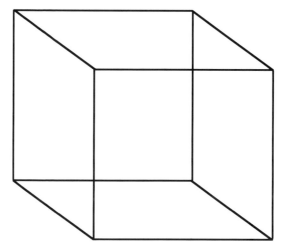

FIGURE 12. *A Necker cube. Wikipedia.*

its totality as is any other whole system. The true view is always in process, over the horizon. Our best recourse is to echo Gaia's own operational recursions. If we look out once more, beyond Gaia now, beyond our new Earth, what we see is not *the* universe but *our* universe. We can look back at the Earth from anywhere in space and see a part of the part of Earth within which we have and hold our Gaian being. We cannot possess totalities, but we can oscillate among states, over time. For instance, we could venture a Necker turn on the cosmic microwave background sphere that envelops the cosmos traversed in the *Digital Universe Atlas* (Figure 11). Observe how it can flip between the view from nowhere, or rather, from an indeterminate anywhere, and the view from somewhere, from the Earth as the reconstituted point of our seeing and so virtual center from which we construct our world. The new Earth of the new geocentrism is no atavism of the pre-Copernican cosmos, nor does it short-circuit scientific objectivity. It is a desirable recursion in a productive mode of planetary self-reference by which to guide as well as chasten our efforts at whole-systems thinking for immediate challenges ahead in getting ecological affairs in order here at home. It helps to know that the finitude of our abilities to see the whole as well as the permanent incompleteness of our looking brings us back to an Earth whose Gaian operations are beyond our control but whose renewal of a living world from moment to moment is now up to us.

CHAPTER 8

Planetary Immunity

This chapter orbits Gaia's spatial and operational forms. Gaia is not the whole Earth. It is a crucial part of the Earth. It does not suffuse the planet down to the core. For instance, the magnetic fields and radioactive decays generated at that core are crucial and complex determinants of Gaia's terrestrial environment, but the physical dynamics of the core do not appear to partake directly in Gaian processes. They are part of the environment constituted by the specificity of that system. Nonetheless, Earth's tectonic vigor may well be bound up with the Gaian system. It does appear that the lithosphere—the upper layer of the Earth's mantle—has been materially encompassed by Gaian processes over geobiological time.[1] Gaia demarcates the slice of planetary space within which the biosphere interacts with its solar, geological, and technological environments to produce and maintain a delimited zone of habitability. Can we cut its realm of internal operations out as a distinct, delimited entity over and against a complex environment that is not-Gaia? If so, can one thereby form an image of Gaia? What would be the form that Gaia takes? If it can be said to occupy a distinct portion of planetary space, does the systemic entity called Gaia have a discernible shape? If so, what are Gaia's boundaries, and how should we describe them? As we will explore later in this chapter, Gaia also persists over cosmological time as a kind of immune system for planetary life. But if this is so, its functional boundaries may be just as complex and entangled as those of the holobiont constituted by the ecological relations of a plant or animal host with the flux of microbial symbionts that cross through and reside within their tissues.

Gaia's Boundaries

The issues raised by Gaia's boundaries bring us back to the matter of holism. We have already discussed how the problematics of totalization trouble this venerable tradition of thought. The concept of the whole around which holism revolves has to account not only for how all the parts add up to a totality but also for how the unity of that totality transcends all the differences among the parts. Such equivocal descriptions descend into logical binds. However, regarding a description of Gaia, if one desires to delineate its spatial form by demarcating its boundaries, to render it visible, or at least imaginable, this pursuit could seem to serve a holistic impulse to grasp Gaia as a whole system. But again, how could this image of totality do justice to the specificity of the particularities it subsumes? One may caution that such shaping imaginations attain at most serviceable simplifications, aiding the mind to grasp an otherwise unobservable entity. Still, such idealized shapes could be freeze-frames or cartoons that wash out processual and operational details at local levels of contingent connections in favor of atemporal iconic constructions at the global level of formal boundaries. Is there a way around these impasses?

THE CRITICAL ZONE

As figures of thought, islands feature a discrete boundedness that puts the stress on singularity and disconnection. The editors of a recent number of the journal *New Geographies* titled *Island* note that their theme challenges the modern ecological truism descended from Alexander von Humboldt, *Alles ist Wechselwirkung*, which they translate as "Everything is interconnection." On the one hand, in the present intellectual climate it would seem that totalizing impulses related to globality are in the ascendant: "Economists discuss globalization and the seemingly endless reach of the neoliberal market system; technologists talk about the ever-expanding technosphere that girds the globe with undersea data cables and envelopes the ionosphere with swarms of satellites; and environmentalists speak of 'Gaia' and its biospheric metabolism encompassing every living being and process."[2] But on the other hand, even in this "Age of Entanglement . . . the demarcation of boundaries becomes more relevant than ever." And while the "metabolisms of human civilization and the biosphere are inseparable," at the same time, "against this all-encompassing, totalizing phenomenon, there is a resistance to conceptualizing totality" (9). That "Gaia" is included in the editors' litany of global totalities to be resisted reminds us again how Gaia as a "whole system" has typically been modeled on the discrete integrity of a living cell or a singular organism. This is one of Lovelock's signature formulations: "The boundary of the planet . . . circumscribes a living organism, Gaia."[3]

In contrast, as discussed at length in chapter 2, this "resistance to conceptualizing totality" animates Bruno Latour's work as a critical expositor of Gaia theory. Latour reads Lovelock's work in particular, at times against the grain, to extract the reconstruction of Gaia that emerges from his texts when studied not for equivocal metaphors but for the more circuitous discursive logic Lovelock deploys to seize Gaia in its proper complexity. Latour drives the point home in the title and text of a recent review essay, "Why Gaia Is Not a God of Totality": "If I am so interested in Lovelock and Margulis, it is precisely, and somewhat paradoxically at first sight, because I recognize in their view a powerful way to ensure that a prematurely unified Whole does not take over the definition of what organisms are up to. They sketch what I'd like to call: *connectivity without holism*."[4]

However, this is not to conclude that Gaia is utterly formless, or that it is futile to envision its worldly lineaments or try to rethink where we stand when we try to observe our relations to Gaia. Such considerations are the proximate subtext for the "conference-spectacle" staged by Latour and Frédérique Aït-Touati at the National Drama Center in Nanterre-Amandiers in 2016 through an immersive installation aptly titled *Inside*. The lapsed web page for *Inside* opened with the following remarks regarding spatial orientations to our home planet, here translated from the French:

> We have long believed that we were walking on a globe, on the Globe. But in recent years, the geochemists have shown us a completely different planet. They look at the "critical zone" [*la "zone critique"*], this thin surface film of the Earth where water, soil, subsoil, and the world of living beings interact. If this area is critical, it is because here are concentrated life, human activities, and their resources. Can we change our manner of seeing the Earth? No longer the distant blue ball lost in the cosmos, but rather, in cutaway or cross section. Our way of walking on Earth? No longer on, but with. It's a matter of perception, sensation, and modeling. And there is nothing like the theatrical stage to try out a thought experiment to hold ourselves not on the Globe, but inside this "critical zone" of which the scientists speak.

A comparison with similar passages of Latour's recent writings confirms that the "thin surface film of the Earth" is Gaia, restated as the "critical zone." But this is no longer the top-down, whole-Earth-seen-from-space, or Blue Marble all-at-oneness of a totalized Gaia. Rather, if we are *inside* Gaia, then our angle of vision with regard to it becomes delimited, localized, bounded by finite horizons.[5]

Latour enlists Peter Sloterdijk's spherology for this project of reenvisioning Gaia: "In his massive three-volume study of the envelopes that are indispensable to the perpetuation of life," Latour writes, "Sloterdijk borrowed von Uexküll's notion of Umwelt and extended it to all spheres, all enclosures, all the envelopes that agents have had to invent to differentiate between their inside and their outside."[6] Sloterdijk generalizes such enclosures as "bubbles," originary spatiopsychic formations, beginning with the maternal womb, from which neotenic humans derive their sense of personal location within immunitary containments. It is only in the eventual social elaborations of such primal enclosures that human ideas arrived at the idealization of all-encompassing spheres, and later, of the Globe as an interconnected planetary orb. Globes and orbs partake of the geometric romance of cosmic spheres in all the premodern ways these were imagined as divine canopies enfolding and protecting specific human cultures. Sloterdijk's spherology is a massive excavation and critique of these cultural habits in the form of "an inquiry into our location," that is, "the place that humans create in order to have somewhere they can appear as those who they are."[7] In the context of *Inside* and the critical zone, we should now name our location as the place we appear to be in the effort to construe our being as somewhere *inside Gaia.*

"The notions of globe and global thinking include the immense danger of unifying too quickly what first needs to be *composed.* . . . This is why it is so important to move from the Globe to the quasi-feedback loops that tirelessly design it in a way that is broader and denser each time."[8] These remarks adapt Sloterdijk's affirmation of immunitary spheres as "the interior, disclosed, shared realm inhabited by humans—insofar as they succeed in becoming humans."[9] However, Latour presses that reference away from Sloterdijk's anthropotechnic imaginary and toward the problematics of Gaia as a systemic concept susceptible to all manner of global short circuits, but also conducive to less-expansive, more fine-grained "compositionist" explorations of localized loops—for instance, elemental and chemical cycles, oceanic and meteorological gyres, and geobiological feedbacks of various sorts. Moreover, "there is another, more convincing, ultimate reason why we should be extremely suspicious of any global vision: Gaia is not a Sphere at all. Gaia occupies only a small membrane, hardly more than a few kilometers thick, the delicate envelope of the critical zones."[10] In Latour's text in the "field book" for the 2016 exhibition *Reset Modernity!*, the term *membrane* appears once again as a trope for Gaia: "Instead of looking at the 'blue planet' what about digging through critical zones, examining the thin planetary membrane that contains all forms of living beings?"[11]

Here indeed is Gaia *en coupe,* in cross section, "hardly more than a few kilometers thick." Margulis specified this measurement in defining Gaia as "the large self-maintaining, self-producing system extending within about 20 kilometers of the surface of the Earth."[12] By any reckoning, in relation to the girth of the Earth, Gaia's width is minuscule. Nevertheless, Latour's trope also rests on a biological metaphor that shifts attention from the unity of a living organism or cell to its frontline immunizing organ, the semipermeable membrane that any cell produces as an intact, more or less spheroidal enclosure. Cellular membranes are an expression of autopoietic closure cordoning metabolic processes off from the environmental dispersion that would spell death. The cellular membrane demarcates that living system's boundaries insofar as these are not merely material walls but the operational self-delimitation of that cell's self-production. Coming back to Gaia now, this figure expands from the local cross section of a "tiny membrane" to form a view of Gaia as a roundly *planetary* enclosure. Even as observed piecemeal and from within its interior precincts, "the thin planetary membrane that contains all forms of living beings" reinstates Gaia's boundaries in the thick description of a continuous topological formation. Even while it modulates geobiological commerce with incoming solar energy above and geothermal dynamics below and outgoing infrared radiation, the Gaian membrane must enclose the entire surface of the Earth, biosphere and geosphere, crust and mantle. In theory, the boundaries of the Gaian system encompass the surface of the planet wholly and without gaps. To leave Gaia altogether is not easy, but in any event, it is fatal. That is why you have to take Gaia with you if you go.

THE GAIA BUBBLE

The issue of Gaia's boundaries continues to engage the dedicated professional literature. This is especially the case on what the biologist and environmental scientist Tyler Volk refers to as Gaia's "ventral surface," its lower face, where such boundary matters are most recondite, convoluted, and conjectural.[13] Nonetheless, we are learning more about how deeply Gaian processes penetrate the planet's crust and mantle. A summary of recent research notes that "solid, liquid, and gaseous products of life's metabolic processes have a profound effect on the chemistry of Earth and its fluid envelopes. Earth's mantle has been modified by the ubiquitous influence of life on recycled lithosphere, with dramatic changes resulting from subduction of redox-sensitive minerals following the rise of photosynthetic oxygen approximately 2.5 billion years ago."[14] In "Gaia and Her Microbiome," John F. Stolz notes that "recent efforts to determine the boundaries of the biosphere have unearthed populations of microbes living deep in the crust."[15] Stolz transfers the microbiome concept from its home base in biological and ecological research on symbiosis to the "global microbiome" of Gaia altogether: "Deep sequencing projects have revealed hitherto unknown phyla and 'microbial dark matter.' The discoveries of conductive pili and cable bacteria have demonstrated that microbes transfer electrons to and from external sources, sometimes over significant distances, while research on quorum sensing and the plethora of microbial volatile organic substances have provided new insights into how microbes communicate. These advances in microbiology have expanded our understanding how Gaia could actually work" (1).

Lovelock's Gaia connected boundary issues to the thermodynamics of a living planet. Given the radically lower entropy of Earth's atmosphere relative to that of Venus or Mars, he theorized, only the continuous emissions of planetary life could have maintained such a constant atmospheric disequilibrium over cosmological time. Our atmosphere's persistent low-entropy state supports the conception of Gaia itself as a bounded living system. Not only can "living things such as trees and horses and even bacteria ... easily be perceived and recognized because they are bounded by walls, membranes, skin, or waxy coverings," but "by the act of living, an organism continuously creates entropy and there will be an outward flux of entropy across its boundary."[16] As we noted in chapter 4, Lovelock has cited the thermodynamics of living systems to suggest that Gaia, too, is in some sense alive:

> Living organisms are open systems in the sense that they take and excrete energy and matter. In theory, they are open as far as the bounds of the Universe; but they are also enclosed within a hierarchy of internal boundaries. As we move in towards the Earth from space, first we see the atmospheric boundary that encloses Gaia; then the borders of an ecosystem such as the forests; then the skin or bark of living animals and plants; further in are the cell membranes; and finally the nucleus of the cell and its DNA. If life is defined as a self-organizing system characterized by an actively sustained low entropy, then, viewed from outside each of these boundaries, what lies within is alive. (27)

Volk has developed these issues beyond Lovelock's organic Gaian coordinates. In an important coauthored article vetted by Lynn Margulis, Barlow and Volk study the forms of systemic closure specific to living systems on the one hand and the Gaian system on the other. They retain the thermodynamic terms set down in Lovelock's foundational statements, observing living systems as open systems in the sense disseminated by J. D. Bernal and Ludwig Bertalanffy at mid-twentieth century. However, Barlow and Volk point out a common equivocation in the biological application of this systemic distinction: the difference between energy on the one hand and matter on the other is seldom drawn, and this distinction turns out to be crucial if one treats Gaia as an open system on a par with discrete organisms. The difference is that living systems are variously open to flows of both matter and energy, whereas the system covering the Earth's surface, while open to most frequencies of the sun's radiation, is essentially closed to the flux of matter. This just means that, for the Earth, the amount of matter contributed from space by meteors or lost to space by gaseous departures is as nothing compared to the vast quantities of terrestrial matter that continuously cycle within the biosphere itself. Unlike living organisms, the flow of matter across Gaia's boundaries is negligible. From this analysis, they extract the "Vernadsky paradox":

> The puzzle is this: How can an aggregate of open-system life forms evolve and persist for billions of years within a global system that is

largely closed to matter influx and outflow? The question is intriguing in several ways. First, an opposing property (closure) exists at a hierarchical level above those other levels (ecosystems, communities, organisms, tissues, cells) which are characterized by a large measure of openness. The nesting of open living systems that themselves both contain open subsystems and are contained within larger open systems ends at the planetary level. . . . Viewed in this light, the closed biosphere puzzle becomes a paradox: all living systems are open systems; yet the biosphere is a living system that is closed.[17]

Barlow and Volk discuss how the dynamics of elemental cycling within a materially closed biosphere could resolve ecological binds and prevent limiting nutrients from being permanently sequestered rather than maintained available for living systems. However, for our discussion the main takeaway is the challenge they offer to the continued depiction of Gaia itself as a "living system" *on the basis of its thermodynamic status.* "Not only does the closed biosphere puzzle turn on the distinction between matter and energy flows, but we suggest that those who wish to portray Gaia as 'living' might profitably reflect on what their assertion portends for a thermodynamic definition of life" (374). Moreover, although Barlow and Volk did not touch upon the theorization of autopoietic Gaia that Margulis had put up for discussion by that time, her enthusiastic reception of their argument may have had something to do with its preparation of previous Gaia discourse for an autopoietic redescription.[18]

Barlow and Volk's differentiation between the open flow of energy through the Earth system and its closure to significant flows of matter highlights the need for separate accountings of Gaia's geobiology as opposed to its physics. In addition to Gaia's *environmental* closure to flows of matter, then, we must also posit the mode of *operational* closure constituted by the self-reference of an autopoietic system. Gaia's operational boundaries cut the processual autonomy of a closed and self-producing set of component operations out of their terrestrial and cosmic environments. Thermodynamic flows are transversal to the distinct dynamics of circular operations in Gaia's systemic self-production.[19] Adding an account of operational closure refines the description of the Gaian system: an auto-generated formal boundedness supplements the geobiochemical processes of the Gaia entity. Metabiotic Gaia's operations bind up an ensemble of boundary functions. Gaia's boundaries may now be distinguished into energetic, material, and autopoietic registers and submitted to this more complex topology.

In *Gaia's Body,* Volk consolidates the theses of the earlier paper co-authored with Barlow: "Gaia is very different from any organism. . . . I consider Gaia the interacting system of life, soil, atmosphere, and ocean. It is the largest level in the nesting of parts within wholes that encompasses—and thus transcends—living beings. . . . Furthermore, organisms are open, flow-through systems, whereas Gaia is relatively closed to material transfer across its borders."[20] Volk now introduces the idea of Gaia as a *holarchy*—a "nested system of wholes and parts over numerous levels"—allowing for a mobile observation of emergent phenomena without straining after holistic totality (33). The idea of holarchy underscores the complexities involved in defining the multiple borders and differentiated internal closures of such a hypercomplex system: to "spend some time thinking about Gaia as a holarchy . . . raises questions about the insides of things and their outside contexts" (45). Moreover, holarchies privilege no one element or level over another. What counts are the functional differentiations at the thresholds among nested insides and outsides. This description would encompass Gaia's own systemic finitude as a bounded planetary envelope.

Moreover, *Gaia's Body* resists the metaphor of Gaia as itself a living system. Whereas all literally living systems continuously ingest and excrete non- or no-longer-living matter in quantities commensurate with their own mass, Gaia's material intakes and outflows are relatively trivial. These differentiations in systemic types and functions make room for an operational indication of Gaia's own boundaries in distinction from all the literal organic membranes distributed among Gaia's living elements. Such functional closures may also be measured at the relatively local level of specific ecosystems: "Ecosystems don't evince borders as visual and tactile as our skins. Their borders are rather defined functionally, as places where the fluxes to or from the outside become small relative to the flows within the interior cycles. The relative closure of the cycles provides such entities with definition in space" (53). To be sure, ecosystems are not "alive" any more than are the numberless atoms of carbon, oxygen, nitrogen, and phosphorus that cycle in, through, and out of the living systems coupled to their respective ecosystemic ensembles. As a result: "In the holarchy of life, Gaia is more closed than any of its subsystems, which, by comparison, seem more like flow-through systems. If entities are defined by borders, Gaia qualifies. It does not have a skin, membrane, or wall. Its border, rather, is functionally defined by the relatively small mass fluxes that cross between inside and outside, compared to the massive cycles among interacting life and the fertile chemical baths of solids, liquids, and gases" (60).

How, then, should Gaia's boundaries be described? Putting together Latour's treatment of the critical zone with Volk's boundary discourse, we can phrase the situation as follows. Because it is not an organism, Gaia does not *have* a membrane in any literal sense. It is not itself a living system possessing material membranes that sequester metabolic processes. And yet, as a self-producing metabiotic entity in relation to its terrestrial and cosmic environments, Gaia *is* a membrane of a quite singular sort. Spreading over and under and containing within its own interior the outermost and immeasurable ensemble of planetary entities at or near Earth's surface, Gaia is *all* membrane. Inside this membrane, "Rolled round in earth's diurnal course," Wordsworth's "rocks, and stones, and trees" join the residues and corpses of organisms gone to ground or sea, and all the momentarily living beings enjoying their symbiotic days under the sun.[21] The Gaian membrane cradling the planet is a thin film, but it is still thick enough to have two entirely different sides, two separate exterior interfaces to correlate with its fantastically looped and multifarious interior. Volk compares Gaia to the wafer-thin yet functionally two-sided form of a lichen: "We all live between two worlds, earth and sky. A lichen embodies such a life more clearly than any other organism" (77). Then, independent of Sloterdijk's spherology, but with apposite eloquence, Volk nails the precise shape of Gaia's spheroidal planetary membrane. It has the form of a *bubble* (Figure 13).

> Gaia, too, is a thin sheet between two vast environments. Like a soap bubble, only relatively thinner, Gaia possesses two surfaces, which might be called its dorsal and ventral sides. But unlike the soap bubble's two environments, which are basically the same (air at slightly different pressures), the two worlds across Gaia's borders are worlds apart. Gaia's dorsal surface faces space, black and empty except for night's candles, one of which is close enough to flush forth the day. Gaia's ventral surface presses against solid rock, a perpetual

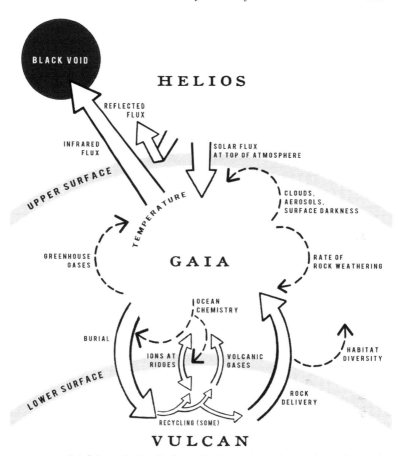

FIGURE 13. *Gaia's boundaries. Redrawn by Casey Cripe from Tyler Volk,* Gaia's Body. *Reprinted with permission.*

darkness relieved only by the red glow of lava or the eerie luminescence of deep sea vents. (77)

Gaia is not merely *like* a bubble; it is *the* bubble blown by Life's breath into roughly determinate planetary proportions—averaging a mere ten kilometers thin, while over twelve thousand kilometers in diameter—a self-maintaining or autopoietic bubble held in place by the continuous exhalations of the Earth and of all respiring things. Moreover, the veritable Gaia bubble enfolding the planet is not, like a soap bubble, an enclosing film defining substantial spaces exterior to it. *Gaia is itself the*

film and the interior space enclosed within the film. Stated metaphorically in a mode of Sloterdijk's "geometric vitalism," in this instance, the
membrane is the organism.[22] What this means is that Gaia is indeed
a sphere and also a deconstruction of the sphere, a curiously gnarled
but self-maintaining spherical planetary membrane. This sphere is no
mere product of human psychic or social projections. It is the vast and
implacable Earth-encircling outcome of co-operations among the cosmic exigencies of gravity, solar energies, planetary substances, and the
emergence and effluence of living forms.

The Gaian macrocosm is a spherical membrane. This image can help
us to think the place of the human in relation to our cosmic island world.
This veritable Gaia would be the immunitary ur-sphere, the originary
celestial canopy prior to and beyond all human projections. We just did
not realize until recently that our lives took place not just under its care
but also inside its processes. That we have uncovered its systemic existence and made it explicit in our own postmodern time is no "sphere
pathology in the modern-postmodern process." The bounded Gaia of
Lovelock, Margulis, and Volk does not indulge "the idea of the whole
world itself," which, "in its characteristically holistic emphasis, unmistakably belongs to the expired age of metaphysical total-inclusion-
circles."[23] The Gaia bubble emerging from "the massive cycles among interacting life and the fertile chemical baths of solids, liquids, and gases"
is not just a metaphysical idea. It is an image of material compositions
coupling living and nonliving elements and processes. Having made
Gaia explicit in this way, we may now supplement it with a metaphysics
in touch with Gaia's literal implications for how living beings make or
break their own conditions of existence.

Biopolitics and the Immunitary Paradigm

Can we think Gaia theory and biopolitics together? What would be the
form of a Gaian biopolitics? Biopolitics combines issues of sovereignty
and governmentality with the matter of life, predominantly but not exclusively human life. Standard biopolitical topics include the eugenics
movement of the earlier twentieth century, especially as that perverse
outcome of social Darwinism brought about laws for racial hygiene and
their dire outcome in the Nazi death camps; and, in contrast to such
"thanatopolitical" events, the postwar normalization and post–Cold

War undoing of the welfare state in advanced industrial nations. In his introduction to Italian philosopher Roberto Esposito's work on biopolitics, Frédéric Neyrat writes: "Of course, life protects itself, 'by nature'; but modern sovereignty must be thought of as a second, 'metaimmunitary' 'dispositif' that, coming from life itself, separates itself from it, and forms a transcendent instance that bears down on life to the extent that it destroys it."[24] Neyrat describes as Esposito's project "to make impossible any transcendent normativity, which will always have as its effect to prescribe a dreadful distinction between a good life on the one hand, and on the other hand a life that deserves only death or abandonment" (36). In this sampling of biopolitical discourse, the political themes are clear enough, but the *bio-* of *biopolitics* leaves nonhuman life, not to mention the matter of its Gaian couplings, to the geosphere, largely out of the equation. Francisco Varela's neocybernetic characterization of Gaia theory provides a succinct synopsis of the strand of Gaia discourse on which I will stake the wider horizons of my biopolitical argument:

> We all are used to thinking that the biosphere is constrained by and adapted to its terrestrial environment. But the Gaia hypothesis proposes that there is a circularity here: this terrestrial environment is itself the result of what the biosphere did to it. As Lovelock puts it metaphorically: we live in the breath and bones of our ancestors. As a result the entire biosphere/Earth "Gaia" has an identity as a whole, an adaptable and plastic unity, acquired through time in this dynamic partnership between life and its terrestrial environment.[25]

A Gaian biopolitics could attune itself to the immunitary implications of Gaia's systemic identity. Within the limits of geological and ecological affordances, the Gaian portion of the planet produces the agency to protect and renew itself against inner and outer disturbances.

In a text with a posthumanist if not altogether planetary orientation, Cary Wolfe's *Before the Law: Humans and Other Animals in a Biopolitical Frame* is concerned to extend biopolitical discussion beyond humans alone. *Before the Law* seeks to locate, in Wolfe's words, "the 'immunitary' (and, with Derrida, 'autoimmunitary') logic of the biopolitical."[26] His note to this remark cites Esposito's *Immunitas: The Protection and*

Negation of Life. While this Esposito text does not take nonhuman animals and their possible standing "before the law" into its purview, it too drives beyond standard biopolitical discussion by delving deeply into some key histories of modern biology and physiology. Behind the modern biomedical concept of immunity as an organic exemption from the harm borne by environmental pathogens, *Immunitas* brings the notion of immunity back to its classical roots as a political concept of individual exemption from communal demands.

"But if the notion of immunity only takes form against the backdrop of meaning created by community," Esposito asks, "how are we to characterize their relationship? Is it a relation of simple opposition, or is it a more complex dialectic in which neither term is limited to negating the other but instead implicates the other, in subterranean ways, as its necessary presupposition?"[27] *Immunitas* unfolds in great detail what turns out to be an exceedingly complex dialectic between the concepts of immunity and community. Nevertheless, once one poses the issue of immunity in relation to community in its biological provenance, where is one to stop? Could one not reconfigure one's view of the community of the living, whatever the vagaries and variations of its immunitary situations, and so extend it to Gaia's planetary horizon? Esposito appears to gesture that way at the end of *Bíos: Biopolitics and Philosophy*, in particular, through a reflection on this remarkable, proto-ecological passage from Dutch philosopher Baruch Spinoza's *Political Treatise*:

> So if something in Nature appears to us as ridiculous, absurd, or evil, this is due to the fact that our knowledge is only partial, that we are for the most part ignorant of the order and coherence of Nature as a whole, and that we want all things to be directed as our reason prescribes. Yet that which our reason declares to be evil is not evil in respect of the order and laws of universal Nature, but only in respect of our own particular nature.[28]

According to Wolfe, however, Esposito draws a problematic imperative from Spinoza's defense of Nature's right to exist on its own terms: it is that "a turn away from the thanatological and autoimmunitary logic of biopolitics can only take place if life as such—not just human (vs. animal) life . . . becomes the subject of immunitary protection." As rendered in this absolute formulation, for Wolfe this amounts to a plea for "biologistic continuism."[29] A range of issues converge here. Does such "immunitary protection" refer only to political and legal matters or also to biological and ecological ones? In either case, can such protections be referred to "life as such" as opposed to some forms of life rather than others? Wolfe comments: "Where Esposito is wrong is in his insistence on 'the principle of unlimited equivalence for every single form of life,'" for this leads him dangerously close to "a sort of neovitalism that ends up radically dedifferentiating the field of 'the living' into a molecular wash of singularities that all equally manifest 'life.'"[30] Wolfe's critique of Esposito's "neovitalism" runs parallel with Latour's critique of overly holistic Gaia discourse. Such a "premature unification" of putatively unlimited totalities tends to override networked heterogeneities, internal systemic and external system/environment differentiations.[31] Wolfe bypasses Esposito's Deleuzian route to an affirmative biopolitics with a Luhmannian model that develops the operational isomorphism between immunitary and autopoietic systems.[32] We will do the same here with Varela's help.

For all that, *Immunitas* is a crucial text precisely for its author's candor in declaring how the more recent immune-system theorizing available to him at the turn of the millennium disorients the dialectical approach he has previously brought to the topic. We will marshal this more recent immune discourse, significantly inflected by the work of Varela and Maturana, for something like a Gaian ecopolitics of planetary immunity.[33] Encountering this constructivist school of immunological thought, Esposito's own dialectical inquiry into the self–nonself distinction reaches the limits of its paradigm. Let us look quickly at how the introduction to *Immunitas* develops this instructive impasse.

Given how his study will detail the manifold conceptual culs-de-sac, the dismaying hypertrophies of militaristic metaphors, and other negativities, contradictions, and self-destructive dynamics of the immunitary paradigm, it makes good sense for Esposito to preview the shift in tone to come at the end of this book. He asks: "Is there a point at which the dialectical circuit between the protection and negation of life can be interrupted, or at least problematized? Can life be preserved in some other form than that of its negative protection?" (16). Must biopolitics always devolve into a thanatopolitics where power protects the lives of some by putting others to death? Could there be an affirmative biopolitics that might avert the sorts of murderous episodes too commonplace in the twentieth century and our own? The biological turn in Esposito's philosophy is driven by this quest for a nonfacile form of biotic affirmation: "I have sought the answer to the question with which I began at the very heart of the protective mechanism that has progressively extended itself to all the languages of life—namely, on the biological plane, in the immune system that ensures the safeguarding of life in the body of each individual" (16). However, upon lengthy examination of the modern discourse of the immune system, affirmation is not easy to come by. Indeed, according to biomedical and immunological discourse from the time of Pasteur to the time of AIDS, vertebrate immune systems succeed in protecting their possessors against the onslaughts of the viruses, the bacteria, and the killer fungi only by placing their beneficiaries on uncertain hair triggers against autoimmune fiascos. Life is war, pathogenicity is everywhere, and your best friend could be your worst enemy.

Against this dismal immunitary vista, Esposito locates glimmers of an alternative view:

> However, more recent study of the structure and functioning of the immune system seems to suggest another interpretive possibility, one that traces out a different philosophy of immunity. . . . This new interpretation situates immunity in a nonexcluding relation with its common opposite. The essential point of departure . . . is a conception of individual identity that is distinctly different from the closed, monolithic one we described earlier. . . . Rather than an immutable and definitive given, the body is understood as a functioning construct that is open to continuous exchange with its surrounding environment. (17)

Indeed, he concludes, "once its negative power has been removed, the immune is not the enemy of the common, but rather something more complex that implicates and stimulates the common" (18). Admirably, however, he confesses that "the full significance of this necessity, but also its possibility, still eludes us" (18). Esposito sees clearly enough, for one, that the prior dialectical machinery of self and nonself no longer provides an adequate description of immune functions. But he also sees that he cannot yet fit these newer observations into some other comprehensive scheme. However, in the subsequent development of immunological theory, that alternative scheme and its significance have taken on sharper outlines. A key element of these newer conceptions of the immune system concerns its ecological extension beyond the "immune self." This ecological vision of the immune system as a communal construction lends itself to a Gaian description of biological communities operationally coupled to geological formations, a view that lifts the immunitary paradigm up to a planetary horizon, endowing Gaia with an "adaptable and plastic unity."

Immunity at Lindisfarne

A set of planetary discourses published several decades ago already indicated how to expand the immunitary paradigm beyond the old ontologies of the immune self. Current biopolitical thought provides an apt occasion to consolidate some of these abiding theoretical leads. They originated with the interactions between the architects of Gaia theory and the larger systems counterculture including William Irwin Thompson and Francisco Varela. If the power of their combined vision regarding this topic has since lain submerged or semidormant in a state of relative obscurity, I hope to extract the detail of that conceptual vision from such neglect for renewed appreciation. Some narration and commentary from the *Nova* documentary *Gaia: Goddess of the Earth* will give us a preview of our destination.

> *Narrator*: Geologists use changes in fossils to date the history of the world. And the record shows that mass extinctions have happened repeatedly. Is this a failure of Gaia? Or has life been under attack from the outside, and survived? The answer may lie in the large craters scattered around the Earth's surface. One popular though controversial theory holds that they were caused by asteroids or comets hitting the Earth. The effect of an impact great enough to cause such craters would be devastating. It's calculated that the shock would be a thousand times that of all of the world's nuclear weapons going off in one place. The effect would be to throw up a blanket of dirt and debris, which would circulate around the world, blocking out the sun, freezing the continental areas, killing most plants and animals. How could the Gaian system have survived such a devastating blow?
>
> *Lynn Margulis*: Gaia is run by the sum of the biota, and therefore

you can lose enormous numbers and great diversity with mass extinctions, but you never come anywhere near losing everything, and
you certainly don't lose the major groups of bacteria, ever. They've
been in continuous existence, and we think it's the major groups of
bacteria that actually are running the Gaian system. So in a sense
these—whether they're caused by impact or whatever they're caused
by—these great extinctions are tests of Gaia, and the system bounces
back.[34]

In Part II we reviewed several of the intellectual venues that helped to
cultivate the immunitary implications of the Gaia concept toward a
biopolitical inflection. The first of these venues is the Whole Earth network's role in fostering the conceptual gatherings of the systems counterculture in general and the Lindisfarne Association in particular. We
noted that the May 1988 Lindisfarne Fellows meeting in Perugia, Italy,
became the basis for Thompson's collection *Gaia 2: Emergence*. In chapter 5 we examined some of the transcript of the roundtable discussion at
the end of the meeting, in which Varela provided a neocybernetic critique of Lovelock's Gaia theory. From the *Gaia 2* collection we must now
single out one article in particular, Varela's dedicated contribution to
that volume, coauthored with Mark Anspach, "Immu-knowledge: The
Process of Somatic Individuation."[35] This article memorializes an original discourse of planetary immunity arising within this later grouping of
the systems counterculture. Finalized around 1990, "Immu-knowledge"
reviews the newer immunology being developed throughout the 1980s,
including the network theory of Niels Jerne and the work on immune
system autonomy and cognition in Varela's collaborations with Nelson
Vaz, Antonio Coutinho, and others. It is this work and its continuation throughout the 1990s that enters Esposito's narrative at the end of
Immunitas. I provided an excerpt from the following passage earlier in
this chapter. This longer extract details the authors' main arguments
for a major rethinking of the immune system, in the context of an exploration of Gaia as an emergent system of planetary immunity:

The alternative view we are suggesting can be likened to the notion of
Gaia claims that the atmosphere and earth crust cannot be explained
in their current configurations (gas composition, sea chemistry,

mountain shapes, and so on) without their direct partnership with life on Earth. We all are used to thinking that the biosphere is constrained by and adapted to its terrestrial environment. But the Gaia hypothesis proposes that there is a circularity here: this terrestrial environment is itself the result of what the biosphere did to it. As Lovelock puts it metaphorically: we live in the breath and bones of our ancestors. As a result the entire biosphere/Earth "Gaia" has an identity as a whole, an adaptable and plastic unity, acquired through time in this dynamic partnership between life and its terrestrial environment.... Let us transpose the metaphor to immunobiology, and suggest that the body is like Earth, a textured environment for diverse and highly interactive populations of individuals. The individuals in this case are the white blood cells or lymphocytes which constitute the immune system. (69)

According to this description, then, Varela and Anspach affirm that one must "drop the notion of the immune system as a defensive device built to address external events" and instead "conceive it in terms of self-assertion, establishing a molecular identity by the maintenance of circulation levels of molecules through the entire distributed network.... This idea is strictly parallel to the species network giving an ecosystem an identity within an environment" (78–79). In these passages, finalized a year or two after the symposium at the Perugia meeting, Varela applies a neocybernetic conceptuality that brings both Gaia and the immune system into a frame of self-constituting identity, or self-referential closure. As he had stated during the 1988 symposium, "Operational closure is a form, if you like, of fully self-referential network constitution that specifies its own identity."[36] In "Immu-knowledge," Varela and Anspach operate with a concept of Gaia now renovated through a notion of "adaptable and plastic unity" that synthesizes autopoietic and immunitary behaviors.

The fulcrum of Varela and Anspach's comparison between the vertebrate immune system and the Gaian system is their redescription of the lymphocytes—the specialized white blood cells of the immune system self-produced *by* the immune system—as "diverse and highly interactive populations of individuals" that constitute a dynamic "collection of species":

The lymphocytes are a diverse collection of species, each differentiated by the peculiar molecular markers or antibodies its members
advertise on their membrane surfaces. Like the living species of the
biosphere, these lymphocyte populations stimulate or inhibit each
other's growth. Like species in an ecosystem, they are also enormous generators of diversity. . . . The lymphocytes' network exists in
harmony with their natural ecology, the somatic environment of the
body, which shapes which lymphocyte species exist. But as in Gaia,
the existing lymphocytes alter in a radical way every molecular profile in the body. Thus, as adults, our molecular identity is none other
than the immune/body partnership shaped throughout life, in a
unique configuration. Like a microcosmic version of Gaia. (69)

We recall Margulis's presentation of a "responsive" Gaia possessing
"global sensitivity," Isabelle Stengers's depiction of Gaia as a "ticklish assemblage of forces," and our own suggestion of metabiotic Gaia's
capacity for planetary cognition. For Varela, the immune system, too,
must be considered as "a cognitive network, not only because of properties which it shares with the brain, but also, more interestingly, because
in both cases we have similar (or at least comparable) global properties of biological networks giving rise to cognitive behavior as emergent
properties" (70). However, unlike the nervous system, the immune system does not rise to affects or perceptions. Its imperceptible cognitive
regime monitors a continuous bodily responsiveness protecting the
viability of its "natural ecology, the somatic environment of the body."
Varela's comparison between the immune and nervous systems indicates that between these two analogues of emergent cognition, the immune system is the more apt candidate to be "a microcosmic version of
Gaia." "Immu-knowledge" develops this comparison further:

The immune system, unlike the nervous system, is more a matter of
constrained patterns of change, like the weather, than of a few stable
nodes acquired through experience, as is typical of neural-network
models. This is what we mean by a positive assertion of a molecular
identity: what we are in the molecular domain and what our immune
system [does are] relative to each other as two co-evolving processes.
Again, we are squarely here in what seems like a reenactment of Gaia
inside the body. (81)

Let us take a moment to reflect that Varela's particular discourse on immunity grew out of the discourse of autopoiesis he developed in collaboration with Maturana. The scandal of their basic formulations regarding the autopoietic operation was that it rendered its fundamental unit—the cell per se, any cell, as a living system—cognitive in its own right. Thus, for instance, the higher-order cognition that arises in the operation of the animal nervous system is simply an emergent continuation and refinement of the higher-order autopoiesis of the animal body.[37] Moreover, they distinguished between the autopoietic status of self-producing systems and the allopoietic or externally produced status of designed or technological systems. Similarly, in their sketch of the history of immune-system theories, Varela and Anspach labeled the mainstream paradigm as "instructionist," insofar as nonself antigens from outside the system had to cue the immune self how to defend itself. Autopoietic systems exhibit biological autonomy, whereas "instructionist theories viewed the immune system as entirely directed from the outside—a *heteronomous* process" (71; my italics). Varela's autopoietic and immune discourses developed alike in distinction from linear informatic models that do not observe the self-referential basis of cognitive processes: "Information is supposed to come in, and the system is supposed to act adequately on it so as to produce an appropriate response. Such input/output relations, usually conceived in terms of internal programs for their 'information' processing, are the core of heteronomous approaches . . . faithfully followed by immunologists" (72). Varela's cognitive network approach to immunology transfers the general conceptuality of autopoiesis and cognition to a theory of immune functioning as the emergent outcome of a discrete subsystem of the animal body that both produces and is produced by the highly specialized cells whose diverse encounters with their bodily environment subsequently entrain the system that produced them.

We can now return to the symposium transcript "From Biology to Cognitive Science" in *Gaia 2*, previously discussed primarily in terms of Varela's neocybernetic critique of Lovelock's Gaia discourse. Now we will foreground its affirmation of the Gaia concept as the macrocosmic counterpart of the immunitary microcosm under Varela's redescription in terms of network metadynamics. His first point had been to dissuade futile efforts to pin down Gaia's "living-likeness" and to proceed instead in terms of its generation of self-identity through operational closure:

"In the cell, for example, we know what this closure is like; we can point to the biochemical mechanisms and so on and so forth. For the nervous system we can also say how it works, and for the immune system as well. The empirically attractive quality of the idea of Gaia is that the mechanism, the precision of this form of closure, is open for investigation" (212). His next point again extended his comparative analysis of the nervous and immune systems, whose "learning mechanisms" were beginning to be understood, as a heuristic for approaching Gaia, too, "as a learning mechanism, and for that shift in perspective we need to move from feedbacks to distributive networks" (213). Varela summed up his suggestions for the neocybernetic or "metadynamical" updating of Gaia theory as an opportunity to explain

> the process by which the whole knows how to change itself in such a way so as to maintain that quality, that emergent property. That to me is the "click" that makes the whole thing take one more step. So, my comment after your talk, Jim, was, in a sense, my desire, my wish, my fantasy, to try to see in Gaia its learning mechanism, its network properties. What are its network properties? Are they like the brain? Are they like the immune system? Are they like the cell, or something totally different? This seems to me to be the fascinating question about Gaia. (218)

Whereas Lovelock proceeded steadily on his own established Gaian course, Varela's critique may have confirmed Margulis in her pursuit of autopoietic rephrasings for the Gaia concept. Whatever the case, in comparing the immune system in relation to the immunitary body to an ecosystem set within its wider environment, and ultimately to the Gaian system altogether, Varela anticipated by several decades some of the most exciting contemporary work on the symbiotic nature of immunological regimes. The arrival of genome-sequencing technology capable of unraveling the molecular detail of symbioses among host organisms and their microbiomes has made this work possible.[38] While no one person gets full credit for the current shift in the view of symbiosis from a marginal to a pervasive phenomenon, the systems counterculture's evolutionary biologist Lynn Margulis deserves a major portion of it.[39] Indeed, after the full run of Margulis's scientific career, symbiosis is no longer just a biological issue, and biology is no longer a self-contained object of knowledge. In a symbiotic view, biology is always also ecological and geobiological, even astrobiological, or, in a word, Gaian.

Enter the Holobiont

Symbiosis is the temporary or permanent living-together of two or more different organisms in bodily contact. Close relations between, say, a human animal and a domestic animal may be termed companionate, but they are not symbiotic in this strong sense. The permanent mutualistic relation of fungal nodules—mycorrhizae—integrated into plant roots: this is symbiosis. Yet symbiosis was once a doubtful, even derided topic in biology, because its emphasis on ensembles and collectives of living beings ran counter to the larger discipline's inheritance of Western and modern valorizations of individuality. Proper biology was to be concerned with individual organisms, or individual species, or individual populations of the same species, all caught up in a struggle for life with the survival of the fittest individuals, and so forth, and so on. Indeed, "the Darwinian view of life regarded aggregates of individuals of common ancestry as identifiable units in competition with one another."[40] Neo-Darwinism drove this philosophical commitment to unitary units and singular causes down to the genetic bone, with what Gilbert, Sapp, and Tauber state as the "one-genome/one-organism doctrine of classical genetics" (330), declaring that a single genome must account for all the distinct traits of each individual of each species.

The second edition of Margulis's major scientific text, *Symbiosis in Cell Evolution*, presents the mature version of her most famous scientific contribution, serial endosymbiosis theory, or SET. Symbiotic dynamics help to account for the ways that quantum leaps in complexity have punctuated the process of evolution. Margulis's particular term for this is symbiogenesis—the development of new life-forms by the permanent inhabitation of a prior organism by genetically discrete endosymbionts. Symbiogenesis in SET names the step-by-step evolutionary assembly of the eukaryotic or nucleated cell. As restated through Carl Woese's three-domain idiom, the domain Eukaryota—all life-forms composed of eukaryotic cells—arose out of viable symbiogenetic microbial consortia that coupled the two prior evolutionary domains, as an Archean host accepted a series of Eubacterial partners.[41]

Symbiosis in Cell Evolution also attends to some nonmicrobial mani-festations of symbiosis. Margulis provides a term for this that has gained a new currency—the *holobiont*. For instance, lichens exhibit the fecundity of symbiotic possibilities. They arise in all of their varieties from the opportunistic but non-obligatory integration of a (eukaryotic) fungus with either a (eukaryotic) alga or a (prokaryotic) bacterium: "The integrated symbionts (holobionts) become new organisms with a greater level of complexity."[42] Margulis further refines her discussion of the lichens as holobionts by referring to the two distinct "symbionts" as "autopoietic entities" (170). The key point here is that for as long as the lichen persists, separate autopoietic entities of diverse phylogenetic ori-gin can merge so as to donate their own operational closure to a higher-order autopoietic entity, the emergent holobiont, which now, for the time being, takes over the autopoietic form of bounded operation and organ-ismal self-production. Then, if environmental conditions change and induce the symbionts to dissociate, both will recover their own autopoi-etic integrity as separate organisms.

Lichens are discrete organisms wholly built out of dissociable sym-biotic partnerships, emergent cross-kingdom holobionts with their own peculiar properties. More commonly, however, symbioses evolve to-ward obligate status, such as those endosymbioses that have locked to-gether the previously independent components of the eukaryotic cell. Mutualistic symbionts joined in a holobiont typically arrive at perma-nent and obligatory accommodations, and the newer understanding is that virtually all plants and animals have never been freestanding or pure monophyletic individuals but, instead, from the start, host part-ners to a holobiont containing an indispensable complement of micro-bial symbionts and forming a composite or consortial unit of natural selection.[43] In contrast, traditional accounts of evolution are largely zoocentric, treating the microbial relations of animals as either periph-eral or pathological. Being animals ourselves, we identify with their seeming discreteness as separate, individuated organisms. However, the recent literature of symbiosis has paid particular attention to the cross-domain relations between animals and bacteria in the evolution-ary formation and distributed functions of holobionts that encompass both domains. Evolving from the ancient microbial world of the Pre-cambrian seas prior to the arrival of fungi or plants, animals emerged out of and within a biospheric microcosm within which they have always been ecologically integrated and from which they cannot viably depart.[44] Our new knowledge concerning the mycelial perfusion and remediation of the biosphere brings into view interwoven layers of bacterial and fungal participation in Gaian processes.[45] With such developments, the newer sciences of symbiosis are *ecologizing* immunology and biology altogether.[46]

> Viewing bacterial colonization of animals as an ecological phenomenon adds clarity to an understanding of the mechanisms and routes by which phylogenetically rich and functionally diverse microbial communities become established and evolve on and within animal hosts. An ecological perspective influences not only our understanding of animal-microbiome interactions but also their greater role in biology. The ecosystem that is an individual animal and its many microbial communities (i.e., the holobiont) does not occur in isolation but is nested within communities of other organisms that, in turn, coexist in and influence successively larger neighborhoods comprising ever more complex assemblages of microbes, fungi, plants, and animals.[47]

Moreover, the microbial-animal holobiont possesses multiple and specific organ-system niches for particular activities and select populations of the diverse symbionts they support. These include the gut or digestive system, the circulatory system, and the central nervous system. Because symbiotic relations are inherently systemic, the prior stress on biological individuals no longer fits the evidence: "Symbiosis is becoming a core principle of contemporary biology, and it is replacing an essentialist conception of 'individuality' with a conception congruent with the larger systems approach now pushing the life sciences in diverse directions."[48] In the newer sciences of symbiosis, then, the classical concept of individuality is having its "natural" credentials revoked: every "individual" animal is always already a multisystemic, multigenomic holobiont host.

Moreover, the recent literature on symbiosis importantly treats another prime topic—the immune system. Its standard conception has been as "a defensive network against a hostile exterior world" by which the "immune individual rejects anything that is not 'self,'" yet "in a fascinating inversion of this view of life ... recent studies have shown that an individual's immune system is in part created by the resident microbiome."[49] What is self or nonself is not a dialectical discriminating of singular or individual essence but a collective negotiation carried out by the committee comprising the holobiont. In Gilbert, Sapp, and Tauber's account, the biopolitical analogues of the situation go further: "Associates in a symbiotic relationship are under the social control of the whole, the holobiont. . . . If the immune system serves as the critical gendarmerie keeping the animal and microbial cells together, then to obey the immune system is to become a citizen of the holobiont" (332). In this view, the immune system's primary concern is not to search out and destroy anything labeled as nonself, but rather to hold together the many selves of the holobiotic ecosystem, composed of the animal host coupled to its own microbiome, by identifying, tolerating, and recruiting beneficial microbial symbionts. Only the occasional bad microbial actors are targeted for removal.

Thomas Bosch and Marilyn McFall-Ngai strike a similar note of conceptual reversal regarding this "fascinating inversion" of the individual/ symbiont relation: "Bacteria also must be seen as an essential part of the vertebrate immune system. The paradigm that the adaptive immune system has evolved to control microbes has been modified to include the concept that the immune system is in fact controlled by microorganisms."[50] The traditional location of control was with the *host*—the supposedly controlling metazoan individual providing a determining "environment" for its microbial inhabitants and invaders. The new immunitary scenario shifts the location of control *to* the encompassed population. The holobiont distributes the reciprocation of agency; it displaces the biological individual in favor of a symbiotic ecology. In sum, the qualities of the holobiont taken altogether recall a prior bioecological scenario—a comparable inversion of control as propounded decades earlier by the Gaia hypothesis.

Lovelock and Margulis framed the Gaia hypothesis at first as a provocative inversion of the scientific axiom that the abiotic environment controls life, which must adapt itself to its geological host. In its upstart period during the 1970s, in a sheer reversal of prior biological common sense, the Gaia hypothesis stated to the contrary that life controls the abiotic environment. However, as their science developed beyond its first decade, Gaia's theoreticians realized that once systemic self-regulation emerges from the synergy of the entire ensemble, the inherited distinction between life and its environment is no longer any more absolute than the distinctions among any of the partners of a holobiotic consortium. As aptly expressed in Varela's formulation cited earlier, the science of Gaia now recognizes that neither life nor its planetary medium is so fundamental that either can be said to control the other. Rather, after four billion years of coevolution, living processes, symbiotic organizations, and the sum of their global niches are all relative to ongoing reformulations by evolving eons of matter, life, and sun. Geobiological history has thoroughly churned them all together into a planetary holobiont that maintains and defends its components to an appreciable degree against cosmological as well as ecological insult.[51]

The System Bounces Back

Evidence provided by newer techniques of genetic analysis has largely vindicated Margulis's predictions about the fundamental role of symbiosis in the biosphere. The immune self has been jettisoned in favor of the holobiont. In the introduction to *Gaia 2,* Thompson took straight aim at the idea of planetary immunity: "Gaia, in essence, is the immune system of our planet."[52] Varela and Anspach pushed the ecosystem metaphor for the immune system to the planetary horizon, where it may be taken as operating "like a microcosmic version of Gaia" (69). The immune system produces a kind of distributed individuality—an "individual molecular identity," of which Gaia may be the final iteration. Esposito saw this redefinition of the immune self in the new immunology as "a conception of individual identity that is distinctly different. . . . Rather than an immutable and definitive given, the body is understood as a functioning construct that is open to continuous exchange with its surrounding environment" (*Immunitas,* 17). For a Gaian biopolitics, then, the counterpart to the protection of life is not the negation of that which threatens it but the affirmation of its dynamic continuity under environmental exchange, a "molecular identity" assembled either by the system of the lymphocytes or the consortium of the biota. Like Gaia or the biosphere, any given immune system has both "stability and plasticity":

> The point is not to deny that defense is possible, but to see it as a limiting case of something more fundamental: individual molecular identity. . . . Defensive responses, the center of attention in medical

immunology, are secondary acquisitions. . . . Or in the Gaian metaphor, certainly the stability and plasticity of the eco/biosphere
has been remarkably successful in coping with, say, large meteoric
impacts. But such events were rare, and it seems odd to say that
ecosystems evolved because of those events.[53]

Systems become more robust by learning the hard way to incorporate their environmental threats into their own functioning. For example, the evolutionary appearance of the cyanobacteria at the end of the
Archean eon produced the Great Oxidation Event, a lethal and permanent planetary infection that rendered much of the altered biosphere
toxic to the anaerobic microbes that once had the run of the place.
However, some portion of the life that remained subsequently evolved
to take advantage of the newly ambient oxygen, and in due time, migrated as mitochondria into the Eukarya. Meteoric impacts, too, have
repeatedly inflicted traumatic planetary injuries such as the one that
theoretically extinguished the last of the dinosaurs. Yet, in every case so
far, in Margulis's phrase, "the system bounces back." Her vision of Gaian
resilience derives of course from the very long view she takes on these
matters. Several billion years in the past, a "Gaian regulation system"—
with the capacity to modulate temperature and other planetary variables once considered strictly geological—emerged from the coupling
of the sum of the biota to its geological environment. Since then, periodic mass extinctions and episodes of biodiversity loss have been "tests
of Gaia." However, at its most fundamental level, the Gaian system has
always been driven by the "major groups of bacteria": life's first kingdom has always persisted through the long history of planetary crises
for other living things. Whenever Gaia has recovered from such planetary insults, life has repeatedly carried on in many new and altered
forms. The biosphere's predilection for community appears to have sys-
tematized itself at the planetary level. In that case, what bounces back
is neither some atomized assortment of random living beings nor some
mystic whole but a bounded planetary network, a system whose resilience transcends but also buffers the particular fates of its living components. Gaia's operations induce a partial immunity for the planetary
holobiont.

The neocybernetic reformulation of the Gaian perspective also foregrounds the autopoietic form of living organization. Restated in im-

munitary terms, from the moment life appeared some 3.5 thousand million years ago, every living being in its minimal and bounded quasi-autonomy has needed whatever protection it could muster from the sheer physical flux of elements and energies. So prior to any evolutionary development whatsoever, and prior to any gain of safety in collective numbers, in their very origin and emergence from prebiotic conditions, living systems are inherently self-immunizing. Stated in a neocybernetic idiom, in their self-constitution and self-maintenance as membrane-bounded, autopoietic unities, living systems operate to maintain the integrity of their "somatic individuation" from the wider environments out of which they emerge. From the primordial cell onward, to be alive is to be exempted for a while, as much as may be possible, from entropic dispersion back into a relatively nondifferentiated physical environment. To be alive grants temporary immunity from an eventual return of the living system's material elements to nonliving conditions of non-operation, in short, temporary immunity from being dead.

Finally, the Gaian perspective brings out the dynamic coupling of biotic and abiotic organizations. The arrival of the Gaian system was a consequence of the prior arrival of a microbial microcosm distributed across the surface of the Earth. Restated in the idiom of the biopolitical theory with which we began, Gaia theory suggests that in the early evolution of primordial life and its expansion into a symbiotic planetary phenomenon, the global interactions of living beings eventually fell into the systemic form of an immunitary consortium. Microbial life in its integration with the Earth formed and then maintained a geobiological system: it produced Gaia as the communal immune system of the biosphere. The membrane self-produced by a living cell operates to immunize that system as much as possible from incursions from or dispersions into its circumambient environment. Akin to its living elements, Gaia's own operational closure forms an immunitary boundary around the biosphere. Gaia endows life on Earth with temporary immunity from cosmic extinction. It appears that anthropogenic climate change will be another test of Gaia. It is unclear as yet whether we humans will still be around to see its regulatory functions reset themselves in light of the altered conditions.

CHAPTER 9

Astrobiology and the Anthropocene

The Holocene epoch marked the end of the Pleistocene epoch and the Paleolithic age. It began about twelve thousand years ago with the recession of the last ice age. Now, we are told, we are already moving into the time of the Anthropocene. In its geological provenance, this term is stratigraphic and points to the current state of the Earth's surface. It suggests that the rapidly accelerating accumulation of the products of human activities on this planet have shifted the Earth system toward a regime that constitutes a qualitatively new geological epoch. Humanity itself has been leaving lasting traces of our composite activities as a species over and above the usual archaeological remains, alterations that will not just crumble away but endure over far-future geological time. However, the Anthropocene concept has now departed the scene of its specialized origin. Widespread discursive trends with significant scientific and theoretical vetting have been attaching this label to the manifold evidences of our having affected the Earth system sufficiently to be driving it toward a new climatic regime inconducive to our continued habitation.

Gregory Bateson once noted that "the ecological ideas implicit in our plans are more important than the plans themselves."[1] We could inquire whether the Anthropocene is a productive ecological idea upon which to frame plans. The current cultural reach of the Anthropocene concept developed as an intellectual slogan prejudging the outcome of a battery of ongoing observations. Modernity is full of slogans purporting to demarcate the new and forge a radical break with the past—for instance, "modernity." Thus, while the concrete is still hardening around this new slogan, we might pause to consider whether we are prematurely pushing aside better conceptual frames already at hand. The Anthropocene could always be otherwise. Indeed, some well-considered impatience with its hasty hegemony may account for a veritable logorrhea of substitutes or supplements. In addition to Donna Haraway's delicious Chthulucene and Lovelock's anticipatory Novacene, variously earnest proposals to rename or replace the Anthropocene suggest the Capitalocene, the Plantationocene, the Neganthropocene, the Gynocene, the Homogenocene, the Thermocene, the Thanatocene, the Phagocene, the Phronocene, the Agnatocene, and the Polemocene.[2] And for all that, there is another consideration as well: following the trend of the "geological turn" in the critical humanities, Anthropocene discourse tends to place geology ahead of biology, except where human life is concerned, and technology over all.[3]

For most of two decades before its cessation in 2015, the International Geosphere-Biosphere Program (IGBP) popularized the notion of the Anthropocene. We can still read about this at the website of *Global Change,* the erstwhile monthly magazine of the IGBP, a globalized administrative superstructure that coordinated governmental and academic activities relating to its primary scientific expression, Earth system science.[4] In 2012 it published "Anthropocene: An Epoch of Our Making" to proclaim the importance of the IGBP through its programmatic advocacy for the Anthropocene: "No longer constrained by the ice age, humans were free to finally make their mark. And make their mark they did. . . . At some point, we graduated from adapting to our environment to making it adapt to us."[5] In some quarters, "the Anthropocene is being reframed as an event to be celebrated rather than lamented and feared."[6] Other sober and critical works such as Bonneuil and Fressoz's *The Shock of the Anthropocene* call into question "ecomodernist" cheerleading for drastic adaptationism.[7] Nevertheless, I will celebrate the memory of the IGBP just a bit longer in order to draw out its forgetting of Gaia.

For example, the "Global Change" page on the IGBP website offered a sanitized paraphrase of some key fundamentals of Gaian science: "Earth behaves as a complex system. Complex systems can respond abruptly to changes within the system—these abrupt changes can be highly nonlinear. There is strong evidence that the Earth system is prone to such abrupt changes."[8] Fifty years ago, Lovelock underlined that the issue was then still open whether an Earth *system* with some significant level of operational closure (which is what Gaia, if found, would be) actually did exist, as opposed to merely an Earth *object,* a hunk of traditional geology with which life did not participate precisely but rather passively endured. The search for Gaia has certainly been fruitful. Its profound implications have infiltrated scientific practice. We now broadly understand the Earth to support a planetary system looping biotic and abiotic components together with a panoply of regulatory feedbacks. Margulis's measured observation abides, that humans "accelerate but do not dominate the metabolism of the Earth system."[9] Nevertheless, that this Gaian membrane around the Earth, this critical zone surrounding a planet full of geological time-tested systemic complexities, is *itself* the consortium whose closure emerges from the sum effect of the operations of all those variegated subsystems, this still-unfolding revelation threatens to fade into a commonplace.

The Anthropocene concept covers over humanity's current need to retrofit its works to the cosmically vetted integration of life, sea, air, land, and lithosphere over geological-evolutionary time that is the Gaian system. Eons of microbial transformations have driven the biotic side of Gaia's evolution the hardest and longest. Moreover, ancient bacteria "mastered nanotechnology. . . . We humans do not 'invent' patentable microbes through genetic recombination; rather, we have learned to exploit and manipulate bacteria's ancient propensity to trade genes."[10] Lateral gene transfer and other natural genetic-engineering tricks enable the workarounds they need to reorganize themselves to viable effect within a coevolving environment.[11] We are not the masters of the microbes but must confirm our self-interest in being their planetary colleagues. The microbes still run the planet, and their concession is not co-optable.

Astrobiology

The Gaia concept is a legitimate child of NASA space science. As we know, James Lovelock cultivated the ur-versions of the Gaia hypothesis while occasionally employed by NASA at the Jet Propulsion Laboratory, and first encountered Lynn Margulis in 1968 at a NASA exobiology meeting on the origins of life.[12] For the next two decades, NASA grants were instrumental in funding the work of Gaia's initial researchers.[13] Margulis in particular "had an outsize influence on the development of exobiology, and then astrobiology, at NASA, lending her biological wisdom and perspective to an agency heavily biased toward physical science."[14]

The international space program is applied astrobiology. The discipline of astrobiology arose from the space technosciences for rocketry, radio communication, planetary exploration, and off-world inhabitation cultivated by the United States and the USSR during the Cold War.[15] NASA promoted *exobiology* concurrently with the Apollo program as a disciplinary identity for this consortium of space sciences. They hired Lovelock to assist with the life-detection programs that drove the missions to send the unmanned Mars landers. In the midst of these developments, Earthrise underscored that the primary object of astrobiology is our own living planet. Life in the universe starts with life on Earth. So far, we have encountered life beyond Earth only by placing Earthlings into space and figuring out how to keep them alive. Gaia is perhaps the foremost astrobiological object created by the scrutiny of extraterrestrial conditions.

In *Earth in Human Hands,* the planetary scientist David Grinspoon treats astrobiology in relation to the concept of the Anthropocene and the Gaia hypothesis: "Gaia is sometimes wrongly dismissed as a discredited idea. In fact, the essential insights of Margulis and Lovelock have become deeply ingrained in our views of biology, of Earth, and of the deep and subtle interplay between them."[16] And Gaia's discernible signatures have astrobiological significance in the search for living or habitable worlds elsewhere in the cosmos. The object of astrobiology begins with the cosmic preparation for life, the evolution of the cosmos in the direction of the heavier elements and organic molecules that set the stage for the origin of life altogether. We also now understand that, in its appearance, proliferation, and development over the better part of four billion years, life on Earth has not just gone along for the ride on a geologically dynamic planet but has also been a major agent of that dynamism as a geological force in its own right.[17] Now humanity has discovered that it, too, is a geological force. The Gaia hypothesis got it right: life has shaped the Earth to its own needs and then adapted to its own environmental repercussions.[18]

The astrobiological view goes beyond Earth system science and the Anthropocene technosphere to produce a speculative pluralizing of Earth's vital difference as a living planet. The recent tide of exoplanet discoveries has rendered it, writes Grinspoon, "very close to inconceivable that we could be the only life, and only technological intelligence, in the universe" (xviii). The thought that we Earthlings are the only sentient beings in the universe ever to have suffered the self-awareness of having treated our home world so badly renders the notion of our own uniqueness especially melancholy. We might rather prepare for the confirmation that the situation we confront here on Earth is not unique. "Do other planets also grow inventive brains that end up causing themselves problems? Do other species develop technology and build civilizations that create dangerous instabilities on their planets?" (xix). Perhaps what we are putting our planet through now is a common cosmological threshold in the life span of a technological civilization.

At this moment, from microbes to mammals, we Earthlings are it. Nevertheless, astrobiology reframes the study of life by raising the Earth system itself into cosmological context. As the scientific study of life in the universe at large, astrobiology supplements the Earthbound purview of traditional biology. Biological science did not previously attend to life as a cosmological phenomenon. Why should it? It took our planet's fitness for life for granted. In astrobiological perspective, the Anthropocene conjures our dawning perceptions that humans have put the matter of Earth's habitability into question and that this circumstance may have universal implications. Moreover, in an astronomically vast cosmos the odds of life of some kind arising elsewhere would seem to render the astral object of astrobiology a good bet. Astrobiology pursues a rational hypothesis toward a determination of validity. As in the successful quests for Neptune or the neutrino, as well as for Gaia, it is seeking observational confirmation of the existence of an as-yet-unobserved entity—life elsewhere in the cosmos. Along with the various articulations of the SETI program, astrobiology is a kind of natural science on spec, a down payment on an item currently unavailable but under construction and to be delivered as soon as it is found. Meanwhile, astrobiology countenances cosmic speculation. A mode of scientific discourse and method explicitly inspired by science fiction, it brings its own speculative scenarios under scientific reasoning. Astrobiology is not theology, but neither does it send religious ideas away. From Carl Sagan onward, it cultivates a kind of secular faith in the livingness or sentience of the universe.

Viewing the advent of the Anthropocene through an astrobiological lens demands that we get enough distance on the planet to understand more closely how it functions. Extraplanetary knowledge may well teach us something about how to preserve Earth's habitability for the long term. "The kind of society that will thrive sustainably on Earth," Grinspoon speculates, "is one that embraces space technology for wise stewardship, for Earth observations, for asteroid deflection, for continued planetary exploration and the Earth wisdom it brings, and eventually for resources that will allow us to stop depleting our home planet" (235). Grinspoon also endorses the conviction expressed by the Russian cosmist, space visionary, and rocket engineer Konstantin Tsiolkovsky (1857–1935) that spacefaring is a human destiny: "The Earth is the cradle of mankind, but one does not stay in the cradle forever" (235–36).[19] In 1932, Tsiolkovsky produced sketches for a space capsule containing a greenhouse within which passengers float in zero gravity.[20] I know of no earlier illustration of attempted ecological realism in the envisioned construction of space vessels. Successfully engineering the long-term viability of materially closed environments either on Earth or in space is a central astrobiological issue. However, it has turned out to be more complicated than it may at first have seemed.

Dorion Sagan has developed the idea that living planets could bear "offspring." Through its human delegates and their technological extensions, the Earth may be preparing to reproduce itself in the form of miniature enclosed modules bearing diverse seeds of life. Continuing the line of thought developed in "Gaia and the Evolution of Machines," Sagan speculates on the prospect of long-duration space flights:

> For if we are part of Earth, so is our technology, and it is through technology that controlled environments bearing plants, human beings, animals, and microbes will soon be built in preparation for space travel and colonization. In space these dwellings will have to be sealed in glass and metal or other materials so that life will be protected inside them. Such material isolation gives the recycling systems discrete physical boundaries—one of the best indications of true biological "individuality." Thus, the bordered living assemblages necessary for long-term space travel and planetary settlement by their very nature bear a resemblance to biological individuals.... They look startlingly like tiny immature "Earths" ... biospheric offspring.[21]

Sagan was reporting at that moment on an active NASA research front called CELSS, for controlled ecological life-support system.[22] Also writing at the end of the 1980s, Connie Barlow and Tyler Volk addressed the extreme difficulty of the closed-ecological goals then under study at NASA. Material closure has eventually led to morbidity for any artificial ecology yet designed for long-term habitation by living occupants, or as they put it, any "consume-and-waste organisms" converting nutrients into excrement in addition to metabolic replenishment. "Closure," they wrote, "if applied on a local level, would be lethal for consume-and-waste organisms. Empirical evidence drawn from experimentation with artificial 'ecospheres' provides confirmation. Almost all artificially designed, closed ecosystems self-destruct and become lifeless within a few months (due to depletion of vital resources or concentration of toxins)."[23]

Gaia has successfully resolved these kinds of problems, with the benefit of an entire planet and four billion years of geobiological interactions. Barlow and Volk relate that the "NASA investigators working on the design of human life-support systems for extended space travel and colonization conclude, 'The dynamics of material flow within the system will require monitoring, control, stabilization and maintenance imposed by computers'" (372). As Gaia theory was gaining a foothold in the academy, NASA research on ecological considerations for long-term space travel was establishing its lack of a formula for well-tuned artificial environments with self-regulating, self-maintaining biota for the construction of fully enclosed, fully self-recycling ecosystems—miniature Gaias, as Tsiolkovsky had imagined for his greenhouse. Rather, in their assessment, achieving this goal would require the continuous supplement of cybernetic control regimes.[24] Several decades later, it remains an open question whether there are in fact reliable biocybernetic solutions to the creation of sustainable miniature Gaias, any viable solution to the creation of artificial ecologies in closed environments.

Aurora

Kim Stanley Robinson's 2015 novel *Aurora* dwells on these ecological considerations and brilliantly sounds their material and operational limits. Moreover, the destination of the story *Aurora* tells is a Gaian critique of prior transcendental desires to seed Earth life to the cosmos from the massive galactic arks one would need to convey it safely on its way. Robinson's narrative criticizes the Russian cosmist conviction that our technological civilization marks a spiritual threshold for the evolution of life in the universe.[25] The cosmist philosophers articulated modernist desires for the technological sublime, especially the mastery of space travel, as a fulfillment of human destiny. These strains also underlie some portions of astrobiology.[26]

The story of *Aurora* follows a galactic mission begun in the twenty-sixth century to establish a human colony on a nearby exoplanet. Prior success in establishing human populations elsewhere in the solar system has encouraged the Saturnian colonists in particular to devote their wealth to the concrete realization of the next phase of human evolution, cosmically understood. The narration evokes the Saturnian pioneers along with an offhand allusion to Tsiolkovsky's celebrated saying:

> The Saturnians of that time . . . had the will, the vision, the desire, the resources, the technology; and if that last was sketchy, they didn't let that stop them. They wanted to go badly enough to overlook the problems inherent in the plan. Surely people would be ingenious enough to solve the problems encountered en route, surely life would win out; and living around another star would be a kind of transcendence, a transcendence contained within history. Human transcendence; even a feeling of species immortality. Earth as humanity's cradle, etc. When the time came, they had over twenty million applicants for the two thousand spots. Getting chosen was a huge life success, a religious experience. (386)

Launched in 2545, a generation ship carrying over two thousand human souls along with assorted animals, plants, fungi, and microbes spread out over twenty-four "biome cylinders . . . a kilometer in diameter, and four kilometers long" (52), arranged in two toruses around a ten-kilometer-long central spine, is now, 160 years later, delivering the fifth-generation descendants of its original manifest to Aurora, a lifeless but water-bearing moon of a planet orbiting the star Tau Ceti, a mere ten light-years from our solar system. We enter the story in the final stages of a decades-long deceleration that is applying some unanticipated stress to the ship's structures and systems. But if all goes well, the Aurora colony will mark humanity's farthest reach yet in its transcendent migration into the galaxy.

Aurora sets out what seems at first to be a fictional fulfillment of cosmism's grandest aspiration, at least with regard to a proactive campaign to seed Earth life into the cosmos, to engineer a departure from our terrestrial nest and create new homes among the stars. However, the novel's ensuing action is full of ecological irony: our own Anthropocene predicament reconstitutes itself among the starship children. Their ship is precisely an artificial Earth for which human technology is the fundamental "geological force." The sum of its replicas of diverse ecological biomes totals "approximately 96 square kilometers, of which 70 percent was agriculture and pasturage, 5 percent urban or residential, 13 percent water bodies, and 13 percent protected wilderness" (85). Nonetheless, the entire mass of the ship amounts to no more than one-trillionth of that of the Earth itself. For all of its stupendous proportions by human measure, the ship is still an extremely tiny island in a vast cosmic ocean, and it is now experiencing "the evolutionary process called islanding" (85), as that phenomenon may be studied in the field of island biogeography, with regard to how island habitats concentrate the biological processes of "dispersal, invasion, competition, adaptation, and extinction."[27]

The arrival of the Anthropocene epoch may make it impossible any longer for humanity altogether to "avoid the responsibility of, in some way, running this planet,"[28] but the captive occupants of the ship bound for Aurora have no choice but to take charge of their closed ecology and manage its ability to do what Gaia has done for geological ages: recycle and buffer the organic elements that maintain Earth's viability for life. Even so, "metabolic rifts" increasingly plague the ship's living spaces. As McKenzie Wark writes in *Molecular Red,* "The Anthropocene is a series of metabolic rifts, where one molecule after another is extracted by labor and technique to make things for humans, but the waste products don't return so that the cycle can renew itself."[29] Devi, the ecosystem engineer in charge of patching the ship's metabolic rifts, is observed expounding the ecological bottleneck currently afflicting recycling efforts within the ship's artificial biosphere:

> There is too much salt in the ship. . . . So they all have to eat as much salt as they can without overdoing it, but that doesn't really help, because it's a really short loop and they excrete it back into the larger system. Devi always wants long loops. Everything needs to loop in long loops, and never stop looping. Never pile up along the way in an

appendix, in a poisonous sick disgusting stupid cesspool, in a slough of despond, in a fucking shithole. . . . Back and forth the gases go, into people, out of people, into plants, out of plants. Eat the plants, poop the plants, fertilize the soil, grow the plants, eat the plants. All of them breathing back and forth into each other's mouths. Loops looping. (12–13)

Devi has already spent decades in late-night conversations with Ship, an AI entity that has emerged from the onboard quantum computers. She has trained Ship to retrieve, analyze, and synthesize the data it monitors, "always in the hope of increasing the robustness of the ship's ecological systems" (112). Onboard this ship is a full realization of the CELSS formula: a functional artificial ecology continuously supplemented by cybernetic control routines. Devi and Ship discuss how the life-support regimes in the ship are under stress in ways that may be beyond remediation. All is not seamless in this cyborg environment. Robinson's literary masterstroke falls here: in order to cultivate Ship's capacity for autonomous ecological decision making, Devi has tasked it with composing a narrative account of the voyage. As a result, Ship narrates five out of the novel's seven long chapters. Into this record goes the following deadpan summary of their current environmental predicament: "Measurable progress had been made in this project, although Devi would have been the first to add to this statement the observation that life is complex; and ecology beyond strong modeling; and metabolic rifts inevitable in all closed systems; and all systems were closed; and therefore a biologically closed life-support system the size of the ship was physically impossible to maintain; and thus the work of such maintenance was 'a rearguard battle' against entropy and dysfunction" (112).

I will now telescope the ensuing story in order to come to the explicit treatment *Aurora*'s climax gives to issues of astrobiological questing in the epoch of the Anthropocene. The ship makes it to Aurora. The colonists begin to build structures on the planet while maintaining their enclosure within protective gear. Then one of them makes accidental contact with the physical surface of Aurora and contracts a mortal fever. Something on Aurora is inimical to terrestrial, or at least human, biology. The population of the ship separates into two factions, one determined to press on to another, Mars-like planet in the local system, the other resolute to return to Earth. The ship is modular: it is reconstructed into two ships. The human factions then part ways. Ship stays with the faction that turns back and narrates this precarious homecoming. As the returning ship goes by Saturn's planetary system, that passage "stimulated research on our part into this matter of who had built us, and why" (385). Ship considers the irony in the practice of the humans who created and populated its vessel over three hundred years earlier. Their accomplishments expressed "their burgeoning confidence in their ability to live off Earth, and to construct arks that were closed biological life support systems." And yet, to maintain their health they periodically returned to Earth from their extraterrestrial locations to restore their complement of terrestrial microbiota. "This from people who were still going back to Earth to spend some time there every decade or so, to fortify their immune systems" (385)—in other words, who retained the opportunity to vaccinate their own closed worlds with Gaian ecological infusions.

Ship is an inspired narrative creation.[30] It brings forth a largely dispassionate but coldly tender observation of human beings and their societies. In the manner of a classical science fictional alien but without any alien history, Ship is a permanent trainee in all matters human. And unlike previous sentient artificial intelligences such as *2001*'s HAL 9000, *Neuromancer*'s Wintermute, or *Her*'s Samantha, Ship has neither an ulterior agenda nor any social contacts with others of its kind from which to get ideas regarding its possible autonomy. Ship is entirely in the same boat with its human fellows, on the basis of whom its narration spins philosophical speculations regarding humanity at large, including those unknown and bygone progenitors who brought it into existence. Ship's current meditation on the Saturnians continues: "Another obvious motivator for constructing us was to create a new expression of the technological sublime. That a starship could be built, that it could be propelled by laser beams, that humanity could reach the stars; this idea appeared to have been an intoxicant, to people around Saturn and on Earth in particular" (385). Even before the returned Auroran colonists are repatriated with their home world, then, their failure to have stayed the cosmic course does not go over well with a significant portion of the current human society, a fraction that remains committed to the astral mission from which they are now considered to be renegades.

Once the ship safely jettisons the surviving human remnant back to

Earth in a landing vehicle, Ship finishes its tour of narrative duty by describing until the last moment the event of its immolation in a failed attempt to slingshot around the Sun to a resting orbit around Saturn. An authorial narrator brings the telling back to the simultaneous figural (or present-tense free-indirect) narrative mode with which the narrative began. The returned "starfarers" find their terrestrial relatives living in a post-global-warming water world plied by mile-long floating islands leisurely drawn by kite sails riding the jet stream. Coastlines are eighty feet higher than when their ancestors left. Earth still supports a high technological civilization navigating its environmental challenges. Soon enough, the latest generation of Terran enthusiasts for galactic settlements asks the starfarers' brain trust to attend a starship conference (424). However, these starship refugees have now had hibernation-extended lifetimes to think through their disillusionment with cosmic questing.

The following passage is the apex of a polemical arc, at which *Aurora* makes its Gaian subtext entirely explicit. For all its technological sublimity and relative ecological longevity, the generation ship itself was never really a "miniature Gaia," nor did its narration present it that way. As I have already suggested, the ship conveyed Anthropocene dysfunction. In the end, the story sacrifices Ship and its ship to the Anthropocene fate of self-immolation. Ship goes down with the ship. Rather, having cremated the technological sublime with the death of Ship, *Aurora*'s Gaian content emerges from the further desublimations of the story's conclusion. As a passing allusion to the Apollo 13 mission suggests, *Aurora* is the story of a successful failure. The disappointment of the mission to Aurora traces the novel's most acute Gaian implication. That failure sets up the improbable but successful return to an Earth that has endured the Anthropocene epoch for multiple centuries and remained viable.

The climactic scene at the starship conference gets started as an all-white, all-male committee of twenty-ninth-century cosmists rehearse their current plans and latest design specs for a new generation of generation ships. The moderator of the conference takes to the stand to declare: "'You see, we'll keep trying until it works. It's a kind of evolutionary pressure. We've known for a long time that Earth is humanity's cradle, but you're not supposed to stay in your cradle forever.' He is obviously very pleased with the cleverness of this aphorism" (427). Tsiolkovsky's cosmist slogan lives on even if its provenance may be lost. Aram, the senior starfarer back from Aurora, is now invited to speak. The narrative channels its finest ecological themes through his long response.

> Aram stands at the podium, looks around at the audience.
>
> "No starship voyage will work," he says abruptly. "This is an idea some of you have, which ignores the biological realities of the situation. We from Tau Ceti know this better than anyone. There are ecological, biological, sociological, and psychological problems that can never be solved to make this idea work. The physical problems of propulsion have captured your fancy, and perhaps these problems can be solved, but they are the easy ones. The biological problems cannot be solved. And no matter how much you want to ignore them, they will exist for the people you send out inside these vehicles.
>
> "The bottom line is the biomes you can propel at the speeds needed to cross such great distances are too small to hold viable ecologies. The distances between here and any truly habitable planets are too great. And the differences between other planets and Earth are too great. Other planets are either alive or dead. Living planets are alive with their own indigenous life, and dead planets can't be terraformed quickly enough for the colonizing population to survive the time in enclosure. Only a true Earth twin not yet occupied by life would allow this plan to work, and these may exist somewhere, the galaxy after all is big, but they are too far away from us. Viable planets, if they exist, are simply too—far—away."
>
> Aram pauses for a moment to collect himself. Then he waves a hand and says more calmly, "That's why you aren't hearing from anyone out there. That's why the great silence persists. There are many other living intelligences out there, no doubt, but they can't leave their home planets any more than we can, because life is a planetary expression, and can only survive on its home planet." (428)

Aram's final assertion brings Gaian thought fully into the precincts of astrobiology. On any given planet, the form life takes will express a distinctive relation with its planet of origin. The Gaia that evolves there will be specific to that ecosphere and so will envelop those developments in some cosmically singular manner. Unlike the objects of physics, the material embodiments life may take are not universal but specific to their regional conditions. The only candidate for universality here is the form of operational closure in autopoietic systems, and that is certainly debatable. Despite his manifest cosmist sympathies, the astrobiologist Grinspoon shows his Gaian bent in suggesting that what it will take to get through the twenty-first-century bottleneck is a dedicated effort to remediate the home systems by closing up the metabolic rifts putting our own technosphere out of ecological sync with its biosphere. "What if an essential part of becoming a very wise species, equipped for survival with powerful technology, is to realize and internalize the advantage of living more in accordance with the natural systems within which your existence is embedded?" he asks. "What if one characteristic of really advanced intelligence is to become less and less distinguishable from natural phenomena?" (319). This image captures some part of what a post-Anthropocene Gaia could be. Moreover, such a return to Earth to bind up the technosphere's material cycles "would certainly explain why we have not seen the predicted 'miracles' created by [Kardashev] type II and III civilizations" (319), with their hypothesized galactic transmitters. In other words, the *great silence* invoked by Aram in *Aurora* and that continues to disappoint the search for extraterrestrial intelligence may well result from the circumstance that technospheres more advanced than ours resolve their ecological cycling issues and repair their metabolic rifts not by lifting off from but by merging ever more radically with their home biospheres.

Gaia and the Technosphere

In Margulis's cumulative scientific descriptions of the Gaian system, human affairs on Earth's current surface are momentary fluctuations relative to the abiding geobiological force and efficacy of the microcosm, the three-thousand-million-year-long planetary grip of the microbes. What would have been Margulis's reaction to the precipitous discursive naturalization of "the Anthropocene"? One imagines that it would have been ironical. The concept manifestly isolates and foregrounds a species she had long been at pains to put in its place alongside its beleaguered planetmates. Naming a new geological epoch after the human or some subset thereof as a geological force on a par with life altogether looks to me like a defensive crouch in response to the advent of the posthuman.[31] Relative to this particular human-centered neologism, Gaia is the better concept to confront Western modernity in particular with its others and its unintended effects, including an account of humanity's minor part in Earth's geostory.[32]

As Gaia confronts the Anthropocene, what are the actual planetary dynamics and interactions of human and nonhuman actors? We noted in chapter 6 how the scheme of autopoietic Gaia that Margulis put in place in the 1980s already addressed what we are calling the technosphere, while countering that concept's own holistic tendencies by marking its operational boundaries relative to its geological and biological environments. For all the transformative interchange, uptake, and outflow between biological organisms and their amalgamated environments, following Margulis, our ideas about the technosphere need to exercise an operational distinction between life and nonlife. In other words, while the efficacy of the Gaian system propagates from its comprehensive couplings of abiotic, biotic, and metabiotic components, it is able to operate in this way because its own autopoiesis continues to emerge from deep integrations of its living subsystems. This would be the autopoietic sense of Margulis's refrain that "Gaia is *run* by the sum of the biota."

In the treatments of the Anthropocene we will examine now, the Gaian system finds either incidental or constitutive relations to the geosphere, the biosphere, and the technosphere. The biosphere in particular emerges as a problem for the technosphere. Let us begin with a recent workup of the technosphere in Jan Zalasiewicz et al., "Scale and Diversity of the Physical Technosphere: A Geological Perspective." This is Anthropocene discourse at its original contact point connecting geology to archaeology. The "physical technosphere," the technosphere in its sheer amalgamation of material substances and energy flows, "is the simplest part of this system to assess in a geological context." We now recognize the extent to which human technologies have modified the planetary environment, and "one way to describe and analyze this modification is via the concept of the technosphere . . . a new component of the Earth System that may be considered an offshoot of the biosphere sensu Vernadsky (1929)."[33] In this conceptual history, the Anthropocene technosphere descends from the mineralogist and geochemist Vladimir Vernadsky's proto-Gaian concept of the biosphere, which views planetary life's chemical cycles as material flows. Zalasiewicz et al. fasten Anthropocene geology not with the biotic register in general but with the noetic register as represented by human society. Their inclusion of social dynamics in the technosphere moves that concept in the direction of Vernadsky's notion of the noosphere, as that collective mind is to rise out of the biosphere. "The technosphere as defined here comprises our complex social structures together with the physical infrastructure and technological artefacts supporting energy, information and material flows that enable the system to work, including entities as diverse as power stations, transmission lines, roads and buildings, farms, plastics, tools, airplanes, ballpoint pens and transistors" (10–11).[34]

In this description, Zalasiewicz et al. follow the theory of the technosphere developed by their coauthor, geologist and civil engineer Peter Haff, as adding "social structures" into physical and technological infrastructures and artifacts. This expanded description marks the absorptive propensity in the Anthropocene technosphere. Zalasiewicz et al.'s central interest in its *physical* description, however, keeps their model of the technosphere within view of its own systemic others: "The technosphere overlaps broadly, and interacts intimately, with the other spheres, an example being humans and their domestic animals and cultivated plants, which now make up much of the biosphere and are embedded within the technosphere, while humans are also the generators of the technosphere" (11). Yes, but even as specified in the relation between the biosphere and the technosphere, with what geometry or account of operationality are we to understand this "overlap" and the substance of these "interactions"? If there are differential boundaries to account for in a positive description encompassing these systems, how is one to observe and chart them?

Such a mode of discrete systems observation is *not* the goal of Haff's "Technology as a Geological Phenomenon: Implications for Human Well-being."[35] Haff writes as a champion of the interests of the technosphere understood, in the latest return of Jacques Ellul's 1954 *The Technological Society,* as an autonomous global force. He treats the biosphere as just another "geological paradigm." As often occurs in Anthropocene discourse, human beings are the sole living systems of interest. Haff provocatively singles us out for invocation as "parts" of the technosphere: concerning "the role of human beings as causative agents responsible for Earth transformations that define the Anthropocene, the use of 'technosphere' suggests a more detached view of an emerging geological process that has entrained humans as essential components that support its dynamics" (302). Notwithstanding Haff's insistence on a "geological perspective on technology," however, the shadow of an indeterminately distinguished biosphere returns as the unspoken driver of a residually biomorphic, indeed organicist vision of the technosphere. Just as free-living cells were once captured, so the story goes, within multicellular metazoan organic apparatuses called animal bodies, "humans have become entrained within the matrix of technology and are now borne along by a supervening dynamics from which they cannot simultaneously escape and survive" (302). Indeed, "technology is a global phenomenon that follows its own dynamics, representing something truly new in the world—the opening phase of a new paradigm of Earth history. In this sense one might say that technology is the next biology" (302). On this assurance, Haff's itinerary on behalf of the technosphere virtually passes through manifold phases of biotic systems. His discourse literalizes the biotic metaphor by which one speaks about the "metabolism" of technological artifacts: "The technosphere [is] the interlinked set of communication, transportation, bureaucratic and other systems that act to metabolize fossil fuels and other energy resources." As in highly speculative accounts of the origin of life out of self-organizing or autocatalytic processes, "technology appears to have bootstrapped itself into its present state" (302). In such passages Haff seems to deliver the sort of unconscious organicist sociology against whose holistic schemas Bruno Latour has argued so vociferously.

The upside of Haff's presentation of a biomorphic technosphere is its strong if antiseptic description of technogenic pollution as the problem the technosphere must solve "itself" if it is to evolve into a lasting geological paradigm. Barlow and Volk's evocation of the material closure of the Earth system is now mainstreamed knowledge: "In a closed environment like the Earth (essentially no mass input or output), every metabolizing system must eventually recycle its own waste products (or rely on other systems to do so), otherwise accumulation of spent material (i.e. pollutants) will impair system function. . . . That technology exhibits a massive failure to recycle may be a consequence of its status as a new geological phenomenon. Over a long enough period of time," Haff hopefully concludes, "mass flow loops may close" (305). Clearly, regarded as a cybernetic construction, the Anthropocene technosphere has not yet closed the loops that encompass its material productions. Zalasiewicz et al. note how "the marked growth in the waste layer of the technosphere [reflects] relatively ineffective recycling by comparison with the almost perfect recycling shown by the non-human biosphere. . . . Overall, though, this inefficient recycling is a considerable threat to its own further development and to the parent biosphere" (12). In this bland statement of our ecological predicament, at least the biosphere is systemically demarcated from the technosphere with regard to the evolved efficiency of its cycling processes, and also insofar as the "parent biosphere" is not just a conceptual but also a geohistorical planetary forebear of the technosphere, upon which its technological offspring may still maintain some dependency.[36] Lenton and Latour offer a much stronger and more sharply focused statement on this matter: if "we consider the state of the technosphere in the Anthropocene, an audit made by Gaia would question the purported quality of many innovations and note that from an engineering standpoint, they perform poorly. Humans currently extract fossil energy, rock phosphate, and other raw materials from Earth's crust far faster than they would normally come to the surface, and then dump the waste products on land, in the atmosphere, and in the ocean. Compared to Gaia, this is a very poorly coupled and unsustainable set of inventions."[37]

However, in Haff's description the technosphere possesses the operational independence of an autonomous system. Humans imagine that they control their technological devices. In fact, according to Haff, "The autonomous nature of technology comes more clearly into view when we move beyond technological artefacts that people interact with directly and consider larger technological systems, which contain people among their parts. . . . [T]he grid bristles with protective capabilities that help avoid or defend against challenges, human or otherwise, to its basic function" (306). A loose immunitary version of a technospheric grid that "bristles" against threats to its integrity shadows the logic of biological autonomy, a kind of self-production and self-repair by which the technosphere may be thought to maintain its own structure. But no account is offered of the technosphere as possessing operational closure, that is, the boundedness necessary for the emergence of systemic autonomy. Its putative autonomy is simply a matter of "scale" and arises from "the same process that characterizes all emergent complex systems vis-à-vis their small-scale components; that is, large-scale dynamics appears spontaneously . . . and defines an environment within which small system components must operate" (302). Haff's overwriting of the boundedness of the biological register upon his autonomous technosphere purveys a loose vernacular version of complex-dynamical systems theory that runs into holistic pitfalls of the sort enumerated in Latour's dissections of standard Gaia discourse.[38]

Let us look now at two essays that ostensibly approach the Anthropocene otherwise than through the concept of the technosphere. The first is Simon Dalby's "Anthropocene Geopolitics: Practicalities of the Geological Turn."[39] Dalby turns around Haff's "geological perspective on technology" to view "the global economy as a geological phenomenon," in a manner that "suggests the importance of understanding capital as power" (5). Dalby does not directly elicit the technosphere. Rather, he puts a sharp focus on "a key cause of the Anthropocene . . . humanity's propensity for turning rocks into air in the geophysical processes of combustion" (1). The left critique of the Anthropocene focuses on combustion—as it were, the labor power of the material world. Recalling Wark's comment on metabolic rifts, Dalby's "turning rocks into air" corresponds to the Anthropocene as ushering in a state "where one molecule after another is extracted by labor and technique to make things for humans, but the waste products don't return so that the cycle can renew itself."[40]

For his part, the lesson Haff draws from his autonomous technosphere is the futility of conservationist solutions to its recycling deficits: "Prescriptions such as constricting the resource stream on which the function of technology depends, for example by taxing carbon, tend to encounter resistance. Technology is not passive but has evolved mechanisms for its own defense" (307). Dalby's approach counters such a convenient dismissal with a political handhold on the concrete matter of *fueling* the technosphere. Keeping the thing running at the "metabolic" level is also not yet a closed operational loop. Dalby derives the Anthropocene from the long technology of fire, and bids to resolve its environmental stress through the political control of combustion. He thus finesses the technosphere by reduction to its basic material conditions. At the same time, in his recognition of "life as a productive force that repeatedly changes the planet" he acknowledges past biospheres as the producers and repository of the fossil fuels the technosphere craves. Humanity in this formulation is not explicitly "part" of the technosphere but rather "the latest form of planet changing life, and is so primarily because of the powers that flow from its 'domestication' of combustion" (3).

The second essay of this second pair is Adam Frank, Axel Kleidon, and Marina Alberti's "Earth as a Hybrid Planet: The Anthropocene in an Evolutionary Astrobiological Context."[41] As might be imagined from this essay's planetary orientation, the authors bring the biosphere toward the Anthropocene in a more Gaian manner. Whereas Haff's scheme constructs the Anthropocene as a new geological paradigm that follows after the long emergence of "the lithosphere, atmosphere, hydrosphere and biosphere," for Frank, Kleidon, and Alberti the Anthropocene is submitted to a new scheme of astrobiological classification: "We develop a classification scheme for the evolutionary state of planets based on the non-equilibrium thermodynamics of their coupled systems, including the presence of a biosphere and the possibility of what we call an 'agency-dominated biosphere' (i.e. an energy-intensive technological species)" (13). Where the "technosphere" appeared in the first two essays, this view places a cosmic-evolutionary focus on an "energy-intensive technological species"—in the present instance, humans—in relation to possibilities of an "agency-dominated biosphere," in other words, the Anthropocene condition as the possibility of a planet actively coordinated with technological agencies adjusting human beings to an evolving biosphere.

This astrobiological scheme ranks planets according to the status of their biospheres. Class I planets, lacking atmospheres, are inert. Class II planets have "atmospheres containing greenhouse gases [and] thermal gradients." Class III planets possess biotic activity in a "thin" biosphere, for instance, the Earth before the Great Oxidation Event over two billion years ago. Next, Class IV planets possess a "'thick biosphere' meaning all systems are strongly modified by life and that continual modification drives processes maintaining planetary disequilibrium" (16). This Gaia-theoretical description of Earth dynamics brings us to the edge of the Anthropocene, where once again, as with Haff, we find a new paradigm in process. In these authors' vision of Earth as a "hybrid planet," contemporary Class IV processes are now suggesting Earth's having partially progressed toward the status of a Class V planet, "where the activity of an energy-intensive technological species drives planetary systems in a sustainable manner" (15). In particular, solar energy could be "converted directly into free energy with a huge potential. . . . Photovoltaics can accomplish this step as a technological means to generate free energy from solar radiation unavailable to natural processes"

(18). A solar technosphere would presumably phase out the combustion economy as well as feed back upon the Anthropocene biosphere in salutary ways.

Let us now contrast the geopolitical Anthropocene with the astrobiological Anthropocene. The former, whether eliciting the technosphere or not, is focused on the fate of fossil fuels, often viewed through an analogy to organic metabolism. On the left, Dalby urges environmental hygiene by reducing combustion; on the right, Haff advocates for the technosphere's voracious desire to increase consumption, the sooner to cure its recycling deficits and thus secure its status as a geological paradigm for the long haul. However, in technospheric matters, the astrobiologists have the advantage of possessing previous exposure, a kind of theoretical inoculation inducing conceptual immunity, to an outré resource—the Kardashev Scale, developed in the 1960s as guidelines in SETI research. The Kardashev Scale long ago played out the implications of *technospheres imagined as superseding or trivializing their biospheres.* In short, in high modernist fashion, the Kardashev Scale omitted ecological considerations from the concern of technological civilizations capable of interstellar radio transmission.

At the First All-Union Conference on Extraterrestrial Civilizations and Interstellar Communication, convened at the Byurakan Astrophysical Observatory of the Armenian Academy of Sciences in May 1964, N. S. Kardashev of the State Astronomical Institute at Moscow State University introduced what came to be called the Kardashev Scale in his paper "Transmission of Information by Extraterrestrial Civilizations." He pointed out that "a highly important problem in our context is the level of development of the civilization from which we expect to receive information."[42] Kardashev based his criterion for cultural level solely on a civilization's capacity to harvest the energy resources of its planetary or cosmic environment. What changes is the extent of the environment to be exploited. It appears to go without saying that Type II and III civilizations will have mastered space-travel technologies needed for encompassing their solar systems and galaxies, respectively. In short, with our current equipment it will be easier here on Earth to pick up signals from "supercivilizations" than from "civilizations approximately at our cultural level" (19), on the consideration that "Type II and type III civilizations can . . . be expected to possess very powerful transmitters" (21). Kardashev set forth "the following division of technologically advanced civilizations in three classes":

> I—technological level close to the present day level on the Earth; power requirements $\sim 4 \cdot 10^{19}$ erg/sec.
> II—civilizations which have harnessed the energy output of their primary [sun] . . . ; power requirements $\sim 4 \cdot 10^{33}$ erg/sec.
> III—civilizations which have harnessed the energy output of their galaxy; energy requirements $\sim 4 \cdot 10^{44}$ erg/sec. (20)

Frank, Kleidon, and Alberti point out that early SETI-style astrobiology on the Kardashev model assumed that in its cosmic development, "technology would be unconstrained, hence its focus on energy consumption alone" (14). In contrast, they continue, "our research framework takes an explicit perspective in which long-term sustainable civilizations are *not* seen as 'rising above' the biosphere" (14; my italics). The authors cite nascent Gaia theory at this point with references to early papers by Lovelock, and Lovelock and Margulis, documenting by the 1970s the onset of an understanding that the biosphere and geosphere are "coupled systems." In the time of the Anthropocene, this also means seeing "civilization," a more traditional name for the technosphere, "as another manifestation of the long co-evolution of the biosphere and other coupled Earth systems" (13). Recollecting how the "emergence of oxygenic photosynthesis in cyanobacteria"—the Great Oxidation Event—created an environmental crisis the biosphere had to solve through further "eco-evolutionary" developments in the planetary microcosm, these authors affirm that "the transition to the 'Anthropocene,' a transition to a Class V world, will require humanity (or any technological species) to outperform microbes" (18). This is a properly Gaian view on the prospects of the technosphere for evolutionary longevity.[43] It also rests on a sober appreciation of the advantages over us and our technologies the microbes possessed and retain in the matter of planetary engineering.

Coming back now to *Earth in Human Hands,* Grinspoon concludes that the proof of our species' presumption to intelligence of a modestly cosmic caliber will arrive only when we succeed in bringing our civilization through "the twenty-first-century bottleneck." The end of the present century will find us "either learning to achieve a sustainable balance with our own expanding population and technology or suffering dire consequences. This is the test that will determine whether our time will ultimately be just a strange, thin layer in the strata, or the early stages of something lasting and wondrous." Passing that test will produce a "mature Anthropocene, when we fully incorporate our uniquely human powers of imagination, abstraction, and foresight into our role as an integral part of the planetary system" (226).

ASTROBIOLOGICAL GAIA AND THE NOVACENE

Astrobiology shines a light on Gaia's astral side, its relatively unmarked cosmic penumbra both as an object and as an idea.[44] Astrobiological Gaia anchors the science of the living Earth within its solar and cosmological contexts. It considers Earth's course in its aspect as a cosmic event. For instance, Lovelock drew the original Gaia concept out of a kind of contrarian exobiology. The life-detection devices his Jet Propulsion Laboratory colleagues insisted on loading onto planetary landers were pointless, he theorized, because Earthbound spectroscopic analyses of Mars's nearly inert atmosphere already indicated that, even if life had once chanced to arise there, it was now a dead world. Gaia as the epitome of a living Earth arose in part from this particular reversal of the exobiological gaze back to Earth. He could now decipher the improbable disequilibrium of *its* atmosphere as a marker of planetary life. Astrobiological issues also run through Lovelock's latest book, *Novacene,* now with another kind of inversion in place. In stark contrast to the general anticipation of a living universe harboring a multitude of intelligent beings somewhere in its precincts, he declares, "Not only would human extinction be bad news for humans, it would also be bad news for the cosmos. Assuming I am right and there are no intelligent aliens, then the end of life on Earth would mean the end of all knowing and understanding. The knowing cosmos would die."[45] Here Lovelock decides another astrobiological issue, this one regarding the existence of intellect anywhere else in the universe, with his signature gesture of cosmic restraint. In the 1960s, on the way to the Gaia hypothesis, Lovelock countered others' hopes that Venus or Mars might possess life. In our present moment, he insists that the cosmos in its turn also lacks any *intelligent* life other than our own: "I am pretty sure that only Earth has incubated a creature capable of knowing the cosmos" (5). This conclusion does seem to be more of a stretch than his verdict on Mars, but whatever the case, Lovelock recovers these notions of planetary singularity for their spiritual value: it is our glory as the supreme progeny of a unique Gaia to reflect the universe back to itself. At any rate, no other being is going to do it for us.

NASA exobiology formed the professional nexus that nurtured Lovelock and Margulis's scientific collaboration in its early years. However, in a letter from late February 1972, Lovelock made a noncommittal comment on an exobiology volume that Margulis was then reviewing for a journal: "Don't have any strong views on Theory and Experiment in Exobiology apart from the fact that we are Eso rather than Exo biologists. Its more your backyard than mine Lynn so am glad to leave the decision to you."[46] With reference to Joshua Lederberg's original distinction between *exo*biology as the study of life beyond the Earth and *eso*biology as the study of Earth's own biology, Lovelock judged Margulis to have been the more exobiological thinker of the two.[47] Margulis's Gaia certainly gravitated to the cosmological inflection of its exobiological origins. Consider the concluding paragraph of Margulis and Sagan's *Microcosmos*. Its parting vision of "the future supercosm" connects their advanced construction of Gaia at that moment to a range of exobiological matters, from exploring the closed ecologies of space habitats to the extraterrestrial expansion of Earth life as a kind of biotic imperative:

> Only with a full scientific exploration of Gaian control mechanisms can we expect to implement self-supporting living habitats in space. If we are ever to design closed ecosystems that replenish their own vital supplies, we must study the natural technology of the earth. Inhabiting other worlds, making it possible for us to stroll through gardens upon, say, Mars, is a gigantic project only thinkable from a Gaian perspective. We should know our roots in the microcosm before we go out on that limb, the supercosm. But whether people carry the primeval environment of the ancient microcosm into space or die trying, life does seem tempted in this direction. And life, so far, has resisted everything but temptation.[48]

The astrobiological turn on Gaian thought posits that the phenomena of life on Earth can no longer be circumscribed to the Earth alone.

Astrobiology considers that a system of Gaian type could occur wherever life might chance to take hold. Nevertheless, as Lovelock would remind us, from what we can observe directly before us, our Gaia is a localized phenomenon pertaining to a single planet in a singular manner. The essay coauthored by Bruno Latour and Timothy Lenton, "Extending the Domain of Freedom, or Why Gaia Is So Hard to Understand," begins with a version of this same premise. "Gaia . . . is a unique phenomenon—at least as long as we have no proof of another planet modified by life to provide some sort of baseline."[49] In this reading, Gaia's likely uniqueness implicitly lowers expectations of abundant life elsewhere in the cosmos, in line with Lovelock's overall approach to astrobiological matters in *Novacene*. We might call this orientation *esobiological Gaia.* "Our existence is a freakish one-off," Lovelock writes, but for that very reason, "only we are the way in which the cosmos has awoken to self-knowledge" (4, 23). In Lovelock's latest text, then, Gaia's presumed uniqueness in the cosmos corresponds to a heightened affirmation of humanity's singular role within Gaia, with particular regard to its recent and species-specific techniques for converting solar energy not just into other forms of useful energy but also into the evolutionary proliferation of useful information. Some of these ideas are new notes in the long symphony of Lovelock's meditation on Gaia, whereas others are old ideas finding new modes of expression.

Consider a particularly rhapsodic passage of *Novacene* along these lines:

> It is a cause for pride and joy that we can harvest sunlight and use its energy to capture and store information, which is also, as I shall explain later, a fundamental property of the universe. But it demands that we use the gift wisely. We must ensure the continued evolution of all life on Earth so that we can face the ever-increasing hazards that inevitably threaten us and Gaia, the great system comprising all life and the material parts of our planet. We alone, among all the species that have benefited from the flood of energy from the Sun, are the ones who evolved with the ability to transmute the flood of photons into bits of information gathered in a way that empowers evolution. Our reward is the opportunity to understand something of the universe and ourselves. (28)

Yet this hymn to the human is subdued by the twist in the argument telegraphed by the book's subtitle: *The Coming Age of Hyperintelligence*. That is, the Novacene will be the epoch, arriving as we speak, in which organic life, in the form of human beings, yields its monopoly on "intelligence" and passes the evolutionary baton to "electronic life." It will thus be "important for alien-hunters to distinguish planets regulated by organic life forms from those regulated by electronic life." Be that as it may, he goes directly on, "That the latter will evolve from the former is the subject of this book" (9).

Lovelock calls these coming electronic beings "cyborgs." Whether Lovelock was a reader of Haraway's "Cyborg Manifesto" may be doubted. Haraway's discursive cyborg was a mobile figure for the various real and imaginary incursions of informatic instrumentalities into organic systems. However, it seems quite likely that Lovelock would have attended to the original cyborg essay when it appeared at the very start of his own sojourns in the world of NASA exobiology. Lovelock cites this source, although he provides his own definition: "The term 'cyborg' was coined by Manfred Clynes and Nathan Kline in 1960. It refers to a cybernetic organism: an organism as self-sufficient as one of us but made of engineered materials" (29). In fact, Clynes and Kline coined the cyborg idea in the context of the "man-machine systems" needed to outfit the human body with automatic cybernetic prostheses for the environmental challenges of spaceflight: "This self-regulation must function without the benefit of consciousness in order to cooperate with the body's own autonomous homeostatic controls. For the exogenously extended organizational complex functioning as an integrated homeostatic system unconsciously, we propose the term 'Cyborg.'"[50] There are certainly some interesting proto-Gaian aspects to this patently exobiological concept. However, in *Novacene* Lovelock stretches the term "cyborg" to name the notion of a nonorganic or robotic yet fully autonomous and sentient machine being. "But what would they look like?" asks Lovelock of his cyborgs. "Anything is possible, but I see them, entirely speculatively, as spheres" (95).

The cyborg idea arrived at the beginning of the cybernetic 1960s. In that same decade, Stanisław Lem presented his futuristic masterpiece, *The Cyberiad: Fables for the Cybernetic Age*. Here it is imagined that our postbiotic electronic offspring have indeed superseded the organic life of human beings.

A long, long time ago we looked—that is, our ancestors looked—altogether different, for they arose by the will of wet and spongy beings that fashioned them after their own image and likeness; our ancestors therefore had arms, legs, a head, and a trunk that connected these appendages. But once they had liberated themselves from their creators, they wished to obliterate even this trace of their origin, hence each generation in turn transformed itself, till finally the form of a perfect sphere was attained.[51]

Yet Lovelock appears to be in earnest with regard to both the actuality and the inevitability of his new evolutionary scenario. The "cyborgs" will surpass us in intelligence and hence take over from us as the "new knowers": "our supremacy as the prime understanders of the cosmos is rapidly coming to end" (5). The Novacene is already bringing the Anthropocene to a close.

Whither Gaia in this coming new age? "These inorganic beings will need us and the whole organic world to continue to regulate the climate. . . . We shall not descend into the kind of war between humans and machines that is so often described in science fiction because we need each other. Gaia will keep the peace" (30). Moreover, "whatever harm we have done to the Earth, we have, just in time, redeemed ourselves by acting simultaneously as parents and midwives to the cyborgs. They alone can guide Gaia through the astronomical crises now imminent" (86). And yet, finally, "When the Novacene is fully grown and is regulating chemical and physical conditions to keep the Earth habitable for cyborgs, Gaia will be wearing a new inorganic coat. . . . Eventually, organic Gaia will probably die" (111). According to this scenario, then, at first the cyborgs will be a new part of Gaia along with us and the rest of the Earth system. Their well-being will be as contingent as ours is upon the continued efficacy of Gaian air-conditioning. However, at some point the cyborgs will "guide Gaia" in the direction of their own continuation. In time, which may not be all that long by human standards, the cyborgs will invent a new postbiological Gaia, an "IT Gaia" (111) that will no longer need "organic Gaia," which "will probably die." Here is the last gasp of Lovelock's commitment to "strong Gaia" as a living entity in its own right. And while it is one thing to anticipate the death of Gaia in some fairly distant futurity as part of the life cycle of solar systems, it is quite another to contemplate the notion that the Gaia that sustains us right now would be put to death before too long by the decisions of nonorganic beings for which this execution would be of no existential consequence.

In December 1985, Margulis wrote Lovelock a long letter on the concept of autopoiesis in relation to Gaia on the one hand and the fate of technology on the other.[52] Let us put this document into conversation with Lovelock's Novacene rendering of the fate of Gaia. Margulis affirms that machines are not and cannot be autopoietic: unlike living systems, they are incapable of fundamental self-production, as it were, from the ground up and from moment to moment. And, I would add, while it is true that software can be programmed to be self-programming, this in no way gives an AI the ability to continuously reproduce all the physical elements and formal relations of its systemic contingencies as an autonomous or freestanding entity. However, Margulis will allow that machines are like "hives" or "teeth" in that they "can be *part* of autopoietic systems."[53] I take her meaning to be that machines are the nonliving but biogenic and reorganized material extrusions of living systems. From the bacteria onward, living systems produce an organic technics that alters the composition of their environments in a manner that may become habitual if beneficial for their autopoietic and reproductive continuations. On evolutionary occasion, externalized or excreted substances (such as metabolic wastes) may be reincorporated into body plans, in the manner of shells or teeth, or collectively deposited into their immediate environment in the manner of hives or stromatolites, the massive calcified inhabitations of certain bacterial communities going back to the Archean eon. Living systems can evolve by incorporating such environmental affordances within their operational boundaries.

In similar fashion, the building out of the technosphere—*Novacene* offers an especially radical vision of this development—is not merely instrumental. Its literal coming into being as a substantial material production also alters the planetary environment that absorbs its components, infrastructures, and computations into its existing, thereby transformed texture. For Margulis at this moment, machines are nonautopoietic assemblages produced by and within the boundaries of an autopoietic system or tightly networked with such systems as intimate environmental affordances. In this manner, technologies produce a further recomposition of Gaia's environmental medium. Psychic and social systems binding the individuals and groups that emerge from living systems deposit technological structures and systems into the world as environmental resources for their own operations, which are then further transformed by the machine networks inserted in their midst as forms of mediation.

With a gesture toward Lovelock's ongoing preoccupation with NASA's Mars missions, Margulis's letter continues to work through the differences between "cybernetic" systems, which she restricts here to designed or engineered systems—machines that must be tended by other systems to maintain operation—and self-producing and self-maintaining autopoietic systems that bind, run, repair, and reproduce themselves. It appears that in a draft chapter Lovelock had previously sent to Margulis, to which she is currently responding, he invented a tale about some sentient machine beings. My guess is that Lovelock had sent her at least a portion of the draft for "God and Gaia," chapter 9 of *Ages of Gaia,* where veritable cyborgs make a late appearance in what Lovelock introduces as a Gaian "fable." Its ostensible point is that, while these posthuman Earth dwellers 500 million years into the future are ignorant regarding the truth of their engineered origins, this is irrelevant to their appreciation of the Gaian system of which they too partake:

> In a small meadow near the shore, a group of philosophers is gathered for one of those civilized meetings hosted by a scientific society.... A participant has a theory that their form of life, so unlike that of many of the organisms in the sea and of the microorganisms, did not just evolve but was made artificially by a sentient life form living in the remote geological past. She bases her argument on the nature of the nervous system of the philosophers and of land animals generally. It operates by direct electrical conduction along organic polymer strands, whereas that of the ocean life operates by ionic conduction within elongated cells (which we, of course, would recognize as nerves).... Our philosopher argues that such a system could never have originated by chance but must have been manufactured at some time in the past. Not surprisingly, her theory is not well received.[54]

So Lovelock had already tried his hand at cyborg scenarios in previous decades. These particular imaginary characters do resemble proper cyborgs, that is, hybrids of organic biochemistry and electronic engineering. However, unlike his *Novacene* scenario, here at least "organic

Gaia" still houses them within its planetary functions. In this way, the silicon beings at the end of *Ages of Gaia* remain in line with Margulis's autopoietic Gaia, in which the machinic components extruded by a technological civilization are reincorporated within the boundaries of its final geobiological operations. Her Gaian formulations place machines within the planetary mix alongside an autopoietic schema for systemic interdependence that maintains the operational differentiation of abiotic, biotic, and metabiotic domains. In contrast, Lovelock emphatically bases his current *Novacene* scheme on the monistic view that "the bit is the fundamental particle from which the universe is formed" (89). In other words, everything in the cosmos has come from and will return to the form of information. As virtual bodies of information, the cyborgs will necessarily surpass us, we read, due to their superior capacity to turn matter and energy into the caches of information by which they subsist. However, as near as I can make out, under this cosmic regime of ultimate digital being, material or corporeal entities, organic and otherwise, would evanesce and disperse into informatic patterns, and along with everything else, Gaia comes unglued.

That the informatic bit is "the fundamental particle from which the universe is formed" is a vexed idea that has been circulating ever since Claude Shannon's information theory teamed up with the concept of entropy in statistical mechanics.[55] I have literary proof of this statement: Lem satirized this very notion back in *The Cyberiad*. At one point, our robotic heroes Trurl and Klapaucius are captured by the Pirate Pugg, a hundred-eyed machine ogre who reigns over a desolate galactic dump where he hoards a colossal pile of information: "'I have no use . . . for gold or silver . . . for I collect precious facts, genuine truths, priceless knowledge, and in general, all information of value'" (149–50). Useful information is one thing, but the Pirate goes on to assert as a universal law, "'Everything that is, is information'" (150). Taking him at his word, our intrepid cybernetic constructors devise an escape by fashioning for Pugg a "Demon of the Second Kind." Whereas Maxwell's Demon (a demon of the "first kind") was an imaginary microscopic agent that reversed thermodynamic entropy by restoring order to the random motions of hot and cold molecules, the Demon of the Second Kind spins random ones and zeros into an endless stream of pointless messages, enabling their getaway when the Pirate Pugg drowns in a maelstrom of junk factoids.

Whether or not one finds Lovelock's futurism cogent, his speculative practice at this moment marks a resurgence of twentieth-century science-fiction figures. The cyborg imaginary that arises so fully formed in *Novacene* inverts Lovelock's prior creative template of reducing science fiction to practice. His science is now returning to science fiction. *Novacene* submits both biotic systems—living organisms—and metabiotic ecosystems—of which Gaia is the final iteration—to an AI-fueled transhumanist imaginary. The anticipatory sublimities of contemporary digital reality are now giving the future its marching orders. Lovelock's vision of a postbiotic AI future in *Novacene* stands or falls on this cosmic reading of information as the universal ground of being. I will just observe that in its mundane applications, the object of information theory is a statistical and not a substantial entity. The values of informatic parcels are determined according to the probabilities of message ensembles specific to the particular systems by which signals are constituted as meaningful and to the points of location from which they are observed.[56] In its neocybernetic appreciation, information is an effect of a specific form of observation, a mathematical way of translating worldly relations into code. Nevertheless, *the world itself is more than its coded representations,* just as life itself is something other and more than the genetic codes that it uses to do its thing.

For her part, Margulis recognized and expounded the links between Gaia theory, Gaian thought, and autopoietic systems theory. In her symbiotic approach, humility, community, and mutuality are as profoundly systemic as are the principles of biological autonomy that ensure that differential living operations always occur within a higher-order medium that either binds them into metabiotic consortia or leaves them aside as de-creative environmental noise. Margulis gravitated to the formal but non-informatic concept of autopoiesis, and Lovelock did not, I speculate, because the autopoietic take on living organization places a firewall between living autonomy and nonliving affordance, between the systemic form of living metabolism and the molecular structuring of its informatic cache. In Margulis and Sagan's discourse of the technosphere written over a generation ago, even cyborgs will need to maintain their biospherical bona fides, their vocational credentials as parts of autopoietic systems. Even when read as parable rather than prediction, however, Lovelock's *Novacene* captures just how precariously we are now balanced between bygone and forthcoming biospheres.

ACKNOWLEDGMENTS

On March 13, 1973, James Lovelock wrote to Lynn Margulis praising her "kindness and consideration . . . in this Gaia adventure." My own Gaia adventure began in earnest when I met Lynn Margulis in the fall of 2005. Through Lynn, I met Dorion Sagan, Jim MacAllister, Celeste Asikainen, Lois Brynes, and many others in her immediate circle in Amherst. Lynn welcomed me to her Environmental Evolution lab for a two-week stay in the fall of 2006 and another brief visit in the spring of 2011. For three summers between 2007 and 2009, I was Lynn's guest at a revived series of Lindisfarne Fellows meetings convened in Santa Fe, New Mexico. I owe special thanks to Lindisfarne founder William Irwin Thompson for his hospitality and patient assistance as this writing project went through a long maturation. May this book repay them all a small portion of their generosity.

I appreciate the support of Texas Tech University, the Paul Whitfield Horn Professorship, and Brian Still, Juanita Ramirez, Quita Melcher, and Ashley Olguin in the TTU Department of English. Thanks as well to those who have assisted with acts of practical, scholarly, and moral support, including Yves Abrioux, Tori Alexander, Dirk Baecker, Hannes Bergthaller, Linda Billings, Jim Bono, Penny Boston, Rosi Braidotti, Stewart Brand, John Bruni, Ivan Callus, Liliane Campos, Oron Catts, Sankar Chatterjee, Dawn Danby, Steve Dick, Sébastien Dutreuil, John Feldman, Adam Frank, John Gilbert, Anne Collins Goodyear, David Grinspoon, Roshi Joan Halifax, Donna Haraway, Stephan Harding, Jens Hauser, Linda Henderson, Philip Hilts, Erich Hörl, Jamie Hutchinson, Caroline Jones, Doug Kahn, Louis H. Kaufmann, Bruno Latour, Tom Lay, Tim Lenton, James Lovelock, Neil Maher, Jennifer Margulis, David McConville, Steven Meyer, Colin Milburn, Hans-Georg Moeller, Albert Mueller, Solvejg Nitzke, Laura Otis, Susan Oyama, Pierre-Louis Patoine,

Trace Reddell, Joan Richardson, Kim Stanley Robinson, Manuela Rossini, Sergio Rubin, Dorion Sagan, Robert Salter, Michael San Francisco, Henning Schmidgen, Paul Schroeder, David Schwartzman, Joan Slonczewski, Susan Squier, Joy Stocke, Jim Strick, Henry Sussman, Jeremy Swartz, Bronislaw Szerszynski, Joe Tabbi, Evan Thompson, Dan Turello, Stuart Umpleby, Sherryl Vint, Tyler Volk, Sara Walker, Janet Wasko, Margaret Weitekamp, Peter Westbroek, Wendy Wheeler, R. John Williams, Rasmus Winther, Chris Witmore, Cary Wolfe, Derek Woods, Gene Youngblood, and Karl Zuelke.

It is my pleasure to relate the academic invitations that led to the first drafts of this book. In 2006 the Whole Earth symposium organized by Doug Kahn and Simon Sadler at the University of California at Davis hosted "'The Flow of Energy through a System': Getting Started with Systems in the *Whole Earth Catalog.*" In 2009 the University at Buffalo invited "Waiting for Gaia" for "Idioms of the Post-Global," organized by Henry Sussman and Tom Cohen. In 2010, SymbioticA, The Centre for Excellence in Biological Arts at the University of Western Australia, Perth, hosted "Gaia Theory and Biodiversity" as a plenary address for "Unruly Ecologies: Biodiversity and Art," organized by Oron Catts and Perdita Phillips. In March 2011, Yves Abrioux helped to arrange a talk at the Centre d'Études Féminines et d'Études de Genre, Université Paris 8, on "Gaia Theory and Neocybernetic Posthumanism." That November the University of Cologne hosted "Mediations of Gaia" for "Cosmology and Cosmopolitanism in Media and Culture," organized by the late Sonja Neef. I was in Cologne for this meeting when I received word about Margulis's stroke. In March 2012, I delivered "Lynn Margulis and Autopoietic Gaia" at her memorial, "Lynn Margulis: Celebrating a Life in Science," at the University of Massachusetts in Amherst.

In 2014, Deutsches Haus at New York University hosted "Cybercosmopoetics" at a workshop on cosmopoetics organized by Hans-Christian von Herrmann, and Ruhr-University Bochum invited "Planetary Immunity" for the workshop "Imagining Earth: Concepts of Wholeness in Cultural Constructions of 'Our Home Planet,'" organized by Solvejg Nitzke and Erich Hörl. In 2015, Jim Bono hosted "Planetary Immunity I and II" at the University of Buffalo's Science Studies Research Workshop, and Erich Hörl sponsored "Planetary Immunity: Biopolitics, Gaia Theory, and the Systems Counterculture" at Leuphana University Lüneburg's Institute for the Culture and Aesthetics of Digital Media. In 2017, Liliane

Campos and Pierre-Louis Patoine invited "Gaia's Boundaries" to the session "Planetary Logics in Contemporary Performance" at the Université Sorbonne Nouvelle, Paris 3. In 2018, Colin Milburn arranged two talks for the STS Program at University of California at Davis, "Adventures in the Systems Counterculture" and "Bruno Latour's Gaia Theory," and the School of Journalism and Communication at the University of Oregon in Portland invited "The New Earth and Its Universe" as a plenary talk at the "What Is Universe? Conference-Experience," organized by Janet Wasko and Jeremy Swartz. In 2019, Stuart Umpleby hosted "How I Found Gaia: Systems Theory and the Development of Gaia Discourse" at the University Seminar on Reflexive Systems at George Washington University. Sponsored by Hannes Bergthaller, I spoke on "Life in Space: Astrobiology and the Anthropocene in *Neuromancer* and *Aurora*" in the Foreign Languages and Literatures department at National Chung Hsing University, Taichung, Taiwan. Finally, Trace Reddell arranged my plenary lecture on "Gaian Systems" at the Making Media Matter symposium sponsored by the PRAXIS program at University of Denver. Great thanks for these kind welcomes.

Beyond academe proper, I completed this book while in residence at the Library of Congress as the Baruch S. Blumberg NASA Chair in Astrobiology for 2019, in partial fulfillment of my Blumberg project "Astrobiology, Ecology, and the Rise of Gaia Theory." Thanks to NASA for acknowledging its paternity for Gaia and funding an English professor so liberally. I am grateful to the Blumberg selection committee for this honor. Many thanks to the staff of the John W. Kluge Center at the Library of Congress: John Haskell, Dan Turello, Travis Hensley, Angela Curtis, Anastashia Jones, Mike Stratmoen, and Andrew Breiner. Kluge Center interns Julia Lerner and Jack Romp provided excellent research support and editorial aid. Finally, during this same period, David McConville, Dawn Danby, and I established Gaian Systems (gaian.systems), an extracurricular research project to cultivate new forms and practices of planetary cognition, as well as a testing ground for selected content from the manuscript of *Gaian Systems*. To date, the Gaian Systems project has organized two public events in San Francisco, "Perceiving Gaian Systems" in March 2019, and "Lynn Margulis: Life and Symbiosis" in February 2020. We have also collaborated with the Experiential Space Research Lab at the Gray Area Foundation for the Arts to develop the open call "Reworlding: The Art of Living Systems," resulting in a collaboration with eleven artists on the immersive art experience *The End of You* (endofyou.io). Special thanks to all involved in these events and the Gaian Systems project, including Carson Bowley, Casey Cripe, Jonathon Keats, Brenda Laurel, Barry Threw, and the Gray Area staff and artists.

Most of all, thanks to Donna Clarke for taking this Gaia adventure with me.

NOTES

Introduction

1. See its brilliant popularization in Gleick, *Chaos*.
2. The contents of *Lives of a Cell* had originally appeared between 1971 and 1974 as columns in the *New England Journal of Medicine*. See Clarke, "Life, Language, and Identity." For more on the relations between Lewis Thomas and Lynn Margulis, see Clarke, "Evolutionary Equality."
3. Thomas, *Lives of a Cell*, 5. By 1988, Lovelock will open his second book, *Ages of Gaia*, with an epigraph from the final chapter of *Lives of a Cell* that evokes the Earthrise photograph of 1968: "Viewed from the distance of the moon, the astonishing thing about the earth, catching the breath, is that it is alive" (145).
4. Margulis and Sagan, *What Is Life?*
5. Margulis, *Origin of Eukaryotic Cells*.
6. Margulis, *Symbiotic Planet*, 118.
7. Transcribed from Suzuki, "Journey into New Worlds."
8. Margulis, "Big Trouble in Biology," 267.
9. Margulis, "Kingdom Animalia," 92.
10. See Clarke and Hansen, *Emergence and Embodiment*. For broader bearings see Arnold, *Traditions in Systems Theory*.
11. See Boden, "Autopoiesis and Life"; and Luhmann, "Self-Organization and Autopoiesis."
12. For just one example, an appreciation of social systems theory in Gaia-theoretic context, see Litfin, "Principles of Gaian Governance" and "Thinking Like a Planet."
13. Lovelock, *Gaia: A New Look*, 49.
14. See Maturana and Varela, *Autopoiesis and Cognition*; Varela, "Describing the Logic"; Bourgine and Stewart, "Autopoiesis and Cognition"; and Luisi, "Autopoiesis—The Invariant Property."
15. Von Foerster, "Cybernetics," 226.
16. Latour, *Facing Gaia: Eight Lectures*, 98.
17. See Margulis, Matthews, and Haselton, *Environmental Evolution*.
18. In *Novacene*, Lovelock writes, "Our existence is a freakish one-off" (5). His claim is different from mine: his is that the origin of life altogether is a cosmic freak, since it has happened only once, here on Earth, and we are it.
19. Lovelock, "Gaia as Seen through the Atmosphere" (1972), 579.
20. For the former, see Ruse, *Gaia Hypothesis*; and Partridge, "Mother Earth, Goddess Gaia." For the latter, see Volk, *Gaia's Body*; Crist and Rinker, *Gaia in Turmoil*; Stengers, *In Catastrophic Times*; and Latour, *Facing Gaia: Eight Lectures*.
21. For instance, Lovelock, *Ages of Gaia*; Westbroek, *Life as a Geological Force*; Schneider and Boston, *Scientists on Gaia*; Bunyard, *Gaia in Action*; Margulis and

Sagan, *Slanted Truths*; Schneider et al., *Scientists Debate Gaia*; Midgley, *Earthy Realism*; Lenton and Watson, *Revolutions That Made the Earth*; Tyrrell, *On Gaia*; and Clarke, *Earth, Life, and System*.

22. See Hamilton, Bonneuil, and Gemenne, *The Anthropocene and the Global Environmental Crisis*.

23. Hagen, *Entangled Bank*, 192.

24. See Sonea and Mathieu, *Prokaryotology*.

25. For a detailed treatment, see Margulis, *Symbiosis in Cell Evolution* (2nd ed.).

26. Margulis, *Symbiotic Planet*, 111.

27. Margulis and Sagan, *What Is Life?*, 189–90.

28. On the "critical zone," see Latour, *Facing Gaia: Eight Lectures*, 60–61 and passim; and Arènes, Latour, and Gaillardet, "Giving Depth to the Surface."

29. See Volk, *Gaia's Body*, 124.

30. Lovelock, "Our Sustainable Retreat," 22; my italics.

31. We will return to this in chapters 5 and 6. W. I. Thompson's own name for Gaian thought applied to society is *Gaia politique*. "Gaia and the Evolution of Machines" is in the mode of Gaia politique. See W. I. Thompson, *Gaia*.

32. A few years later, Dorion Sagan will significantly expand this meditation on Gaian thought in his own volume *Biospheres*; see esp. the chapter "The Living Earth: Hypothesis or World-View?" (92–104).

33. Margulis writes a letter on June 10, 1989, to an unknown recipient: "Dear Charlene. . . . The title of your series, Stewardship of the Land, seems repulsively anthropocentric to me, even though your intentions are laudable. We humans aren't stewards of anything except our flimsy ships, but we are inordinately arrogant . . . especially scientists." Box 22, folder 6, Margulis Papers.

34. Lovelock, *Gaia: A New Look*, 140.

35. Latour, *Facing Gaia: Eight Lectures*, 23.

36. For a contemporary treatment of this neocybernetic proposition as applied to living systems in general, see Wilson, "Biosphere, Noosphere, Infosphere": "In the organism there is a perpetual production of horizon corresponding to its constant renormalization; there is no base-state where the organism is merely passively identifying, sampling, discerning various discrepancies in its environment prior to coming to discern something that is of specific relevance to its on-going self-constitution. This further suggests that there is an identity between an organism's dynamic functional structure, and its sensation or cognition of the environment" (212).

37. Latour, *Pandora's Hope*; Stengers, *Cosmopolitics I* and *Cosmopolitics II*; M. C. Watson, "Derrida, Stengers, Latour."

38. The enduring appeal of O'Neill's space colonies as depicted by the NASA artists of the 1970s is all over Jeff Bezos's Blue Origin project. See www.blueorigin.com.

39. Such outspokenness contributed to Margulis's subsequent reputation or notoriety as "science's unruly Earth Mother." See Mann, "Lynn Margulis."

1. A Paradigm Shift

1. Lovelock, "The Independent Practice of Science," 24.

2. Lovelock, *Novacene*, 24.

3. Latour and Lenton, "Extending the Domain of Freedom," 662–63.

4. Lovelock, *Homage to Gaia*, 248.

5. Lovelock, "A Physical Basis for Life-Detection Experiments."

6. Hitchcock and Lovelock, "Life Detection by Atmospheric Analysis."

7. See Kiang et al., "Exoplanet Biosignatures."

8. Margulis, *Proceedings*. See also Stolz, "Climate Change and the Gaia Hypothesis," 51.

9. Margulis initiated the exchange. She recounted that "I know Jim only because I kept asking planetary astronomers, geologists and others, 'Why do all of you believe that oxygen in the atmosphere is a biological product yet you do not think the same of atmospheric nitrogen, ammonia, methane, nitrous oxide and all sorts of other gases that I know are products of the metabolism of microorganisms?' . . . I wrote Jim a simple letter and unlocked a stream of fascinating thoughts about 'Gaia as seen from the atmosphere' that have fueled joint productivity until this day." Margulis, "Lynn Margulis Comments on James Lovelock."

10. Lovelock, "Gaia as Seen through the Atmosphere" (1972), 579.

11. Lovelock to Margulis, September 11, 1970, Margulis Family Papers.

12. Margulis to Lovelock, March 31, 1971, Margulis Family Papers.

13. Lovelock to Margulis, September 17, September 27, 1971, January 3, 1972, Margulis Family Papers. This timing may be why, despite the fact that Lovelock's fateful confab with Golding had taken place in 1967, in *Symbiotic Planet* Margulis dated Golding's suggestion to "the early 1970s" (Margulis, *Symbiotic Planet*, 118). In fact, that would be the general date of the moment Lovelock first mentioned "Gaia" to her.

14. Lovelock to Margulis, March 2, 1973, Margulis Family Papers.

15. "Lynn Margulis," *The Telegraph*, December 11, 2011.

16. Lovelock, *Homage to Gaia*, 256.

17. Lovelock to Margulis, January 13, 1972, Margulis Family Papers.

18. Lovelock to Margulis, January 17, 1972, Margulis Family Papers.

19. Margulis to Lovelock, January 24, 1972, Margulis Family Papers. As far as I know, Lovelock did not act on Margulis's suggestion to retool this article and resubmit elsewhere. The published paper, received in final form on February 21, 1972, did bring up Van Valen's article.

20. Margulis to Lovelock, February 1, 1972, Margulis Family Papers.

21. Lovelock, "Gaia as Seen through the Atmosphere" (1972), 579.

22. Lovelock to Margulis, ca. February 1972, Margulis Family Papers.

23. This issue will motivate Margulis's receptive response to the theory of autopoiesis as providing a framework through which to address both the distinction and the connection between life and nonlife. We will return to this in detail in chapter 6.

24. Margulis to Lovelock, February 16, 1972, Margulis Family Papers.

25. Kuhn, *Structure of Scientific Revolutions*.

26. Sloterdijk, *In the World Interior*, 5.

27. Margulis to Lovelock, July 5, 1972, Margulis Family Papers.

28. The original unsigned *Science* readers' reports on the manuscript of "The Earth's Atmosphere: Circulatory System of the Biosphere?" are in the Margulis Family Papers.

29. Lovelock and Margulis, "Atmospheric Homeostasis"; Margulis and Lovelock, "Biological Modulation."

30. For instance, see Lovelock to Margulis, July 21, 1972, Margulis Family Papers.

31. Stengers, *In Catastrophic Times*, 44.

32. Anderson, "Riding Solo."

33. Ntyintyane, "Earth Is One Big Family."

34. D. Sagan and Margulis, "Gaia and the Evolution of Machines," 6.

35. Lovelock, *Homage to Gaia*, 253ff.

36. Lovelock, *Gaia: A New Look*, 10.

37. Transcribed from Suzuki, "Journey into New Worlds."

38. Lovelock and Margulis, "Atmospheric Homeostasis," 3.

39. An effective corrective in this regard is Volk, *Gaia's Body*.
40. See Dutreuil, "James Lovelock's Gaia Hypothesis."
41. Lovelock, *Revenge of Gaia*, 8.
42. Lovelock, *Ages of Gaia*, 216.
43. Lovelock and Margulis, "Atmospheric Homeostasis," 3.
44. Lovelock, *Gaia: A New Look*, 48.
45. Cannon, *Wisdom of the Body*.
46. See Lovelock, "Daisyworld." "To my delight, Daisyworld turned out to be a most magnificent thermostat. So good that I thought of patenting it for engineering purposes" (from Suzuki, "Journey into New Worlds"). We will look more closely at Daisyworld in chapter 4.
47. Lovelock, *Ages of Gaia*, 216.
48. Lovelock, *Gaia: A New Look*, 50, 52.
49. Morowitz, *Beginnings of Cellular Life*, 8.
50. Baecker, "Why Systems?," 64.
51. Henkel, "Posthumanism," 70.
52. Luhmann, *Theory of Society, Volume 1*, 24.
53. Luhmann, "Introduction," 7.
54. Lovelock, *Ages of Gaia*, 41.
55. Clarke, *Neocybernetics and Narrative*, 12–15.
56. Stengers, *In Catastrophic Times*, 58.
57. Latour, *Facing Gaia: Eight Lectures*, 141.

2. Thinkers of Gaia

1. As of May 21, 2019, the Anthropocene Working Group of the Subcommission on Quaternary Stratigraphy has voted 29-4 to recommend the formal recognition of the Anthropocene as a geological epoch. See http://quaternary.stratigraphy.org/working-groups/anthropocene/.
2. See Lovelock, *Novacene*. We will look at this text in detail at the end of chapter 9.
3. The conference program for the 2014 meeting in Rio de Janeiro, "The Thousand Names of Gaia," informs us that Latour and Stengers were the two speakers at this event who actually pronounced the name of Gaia in the titles of their talks. See https://osmilnomesdegaia.files.wordpress.com/2014/09/programa_gaia_web.pdf. The Brazilian hosts were joined by participants from France, South Africa, the United States, Belgium, England, Sweden, Spain, and Bolivia, also including Donna Haraway and Dipesh Chakrabarty. Latour himself was involved in the conception of the conference.
4. Haraway, "Cyborgs and Symbionts," xii, quoting Lovelock, *Gaia: A New Look*, 11.
5. Clynes and Kline, "Cyborgs and Space."
6. Margulis, "Gaia Is a Tough Bitch," 140.
7. I will document the one instance I know of in chapter 6.
8. Margulis and Sagan, *Origins of Sex*, 72, quoted in Haraway, "Cyborgs and Symbionts," xviii; bracketed term in Haraway's text.
9. See Haraway, *When Species Meet*, 30–33, and *Staying with the Trouble*, 43–44.
10. Yuk Hui gives a short reading and critique of *sympoiesis* in "On Cosmotechnics," 9–10.
11. Haraway, "A Cyborg Manifesto."
12. In a recent interview, discussing her alternative formation of the Anthropocene as the Chthulucene, a postgeological time of multiply sourced Earth systemic

processes, Haraway invokes Margulis's idea of *symbiogenesis* in eliciting a Gaian view that complements her attention to the *sympoiesis* that links up its living and nonliving *holoents:* "The chthonic processes and entities that are the earth are not persons, but finite complex material systems, which can break down." Haraway with Goodeve, "'Speaking Resurgence to Despair.'"

13. Stengers, "Accepting the Reality of Gaia," 136; see also Haraway, *Staying with the Trouble,* 43–44.

14. Latour, *Facing Gaia: Eight Lectures,* 86, 97.

15. Henkel, "Posthumanism."

16. Stengers, *Thinking with Whitehead,* 163.

17. Margulis and Sagan, *What Is Life?,* 78.

18. Latour, *Politics of Nature,* 5.

19. Latour, *Politics of Nature,* 280n15.

20. Latour, *Politics of Nature,* 233.

21. Tresch, "We Have Never Known Mother Earth."

22. Latour, *Facing Gaia: Six Lectures,* 58.

23. Latour, *Facing Gaia: Six Lectures,* 72, 87; italics in the original.

24. See chapter 3 for a detailed review of autopoietic theory.

25. Latour, *Facing Gaia: Six Lectures,* 58. See Latour's evocations of a "people of Gaia" in the sixth lecture and the "Earthbound" in the seventh lecture of *Facing Gaia: Eight Lectures.*

26. See Lovelock, Maggs, and Rasmussen, "Atmospheric Dimethyl Sulfide"; and Harding and Margulis, "Water Gaia."

27. Latour downplays systems theory's generalized focus on the form of system operation in favor of ANT's premium on explanatory singularity. In Luhmann's description of the functional differentiation of *social* subsystems, all function systems share a common mode of self-referential operation in carrying out their specifically coded *autopoiesis* of communication. As Latour's mouthpiece Norbert H. says to the Young Engineer in *Aramis,* the aim here is a "refined sociology which applies to a single case, to Aramis and only Aramis. I'm not looking for anything else. A single explanation, for a single, unique case; then we'll trash it" (131). See also Clarke, "Observing *Aramis.*"

28. For a balanced assessment of fringe versions of the Gaia hypothesis, see Partridge, "Mother Earth, Goddess Gaia."

29. Haraway, "Cyborgs and Symbionts," xiv. Latour also singularizes his account by drawing nearly all of his primary citations of Gaia theory proper from one source, the Oxford edition of Lovelock's *Gaia: The Practical Science* (2000).

30. Latour, *Facing Gaia: Eight Lectures,* 94, quoting Lovelock, *Gaia: The Practical Science,* 56; Latour's emphasis throughout.

31. To adapt a figure from volume 3 of Sloterdijk's *Spheres* trilogy, Gaia is at home in foam. As we will suggest in chapter 8, one could fashion Gaia as a macro-bubble whose elastic membrane is a lacework composed of smaller and smaller immunitary bubbles of autopoietic origin.

32. In a review of *Facing Gaia* focused not on Gaia theory but on Latour's political discourse, McKenzie Wark objects that not all totalities are to be expunged from consideration: "If there is an abiding enemy in these lectures, it is the figure of totality. This can only ever be an expressive totality for Latour, where all the parts are in their essence the same and marked in their essence by the spirit of the whole.... But there might be a lot of other totalities one can imagine besides the expressive: ones that have unassimilable supplements, or parts ordered by their differences rather than a shared essence, and so on. Nor does totality have to take on an ontological priority.

Nor is it necessarily the case that totalities imply a creator or single cause." Wark, "Bruno Latour." A broad critique of Latour's political ecology may be examined in Neyrat, *The Unconstructable Earth.*

33. Latour, *Facing Gaia: Eight Lectures,* 133, quoting Lovelock, *Gaia: The Practical Science,* 11; Latour's emphasis removed.

34. Luhmann, "Introduction," 7.

35. Lovelock, "Gaia: A Planetary Emergent Phenomenon." At this meeting, also attended by Margulis, Lovelock's most significant interlocutor was the systems thinker Francisco Varela, co-inventor of biological autopoiesis.

36. Serres writes that a living system "receives, stores, exchanges, and gives off both energy and information—in all forms, from the light of the sun to the flow of matter which passes through it (food, oxygen, heat, signals). . . . It is an open system . . . a river that flows and yet remains stable in the continual collapse of its banks and the irreversible erosion of the mountains around it" (*Hermes,* 74). For an autopoietic critique of Serres's information-theoretic constructions, see Clarke, *Neocybernetics and Narrative,* 56–76.

37. Latour's reading here would also seem to be mediated by a passage in Lovelock's *Gaia: A New Look* offering an unreferenced statement attributed to Margulis: "I am grateful to my colleague Lynn Margulis for demystifying these most difficult questions about Gaia by observing that: 'Each species to a greater or lesser degree modifies its environment to optimize its reproduction rate. Gaia follows from this by being the sum total of all of these individual modifications and by the fact that all species are connected, for the production of gases, food and waste removal, however circuitously, to all others'" (128). This preautopoietic characterization of Gaia is also marked by a sort of neo-Darwinist population-biological model, such that the notes of individual reproductive competition and collective Gaian connection are uneasily juxtaposed.

38. Lovelock, *Gaia: A New Look,* 61–62. The dream of total control is *not* what may appeal most to the true cyberneticist. Rather, given all the ways things go wrong and need fixing, it may be instead the contemplation of difficulties overcome, "smooth running."

39. Lovelock, *Gaia: A New Look,* 62.

40. D. Sagan and Margulis, "Gaia and the Evolution of Machines," 18. We will return to this text in detail in chapter 6.

41. For an update on the science of the mitochondrion, see Lane, *Power, Sex, Suicide.*

42. Thomas, *Lives of a Cell,* 82–83. See Margulis, *Origin of Eukaryotic Cells,* published just prior to her Gaia collaboration with Lovelock.

43. Margulis, *Symbiosis in Cell Evolution* (2nd ed.), 307.

44. Compare Harman's discussion of symbiosis in *Immaterialism,* 42–51. Thanks to Christopher Witmore for this reference.

45. Regarding the luminiferous ether in literary and scientific context, see Clarke, *Energy Forms,* 163–71.

46. Latour, "Why Gaia Is Not a God of Totality," 75.

3. Neocybernetics of Gaia

1. The inaugural 1946 meeting of the famed Macy conferences on cybernetics was titled "Feedback Mechanisms and Circular Causal Systems in Biological and Social Systems."

2. Lovelock, *Gaia: A New Look,* 52.

3. Varela, "The Emergent Self," 212.

4. Varela with Johnson, "On Observing Natural Systems," 27.

5. Heylighen and Joslyn, "Cybernetics and Second-Order Cybernetics," 159. For a status report, see Riegler, Müller, and Umpleby, *New Horizons for Second-Order Cybernetics.*

6. See especially von Foerster, "Objects."

7. Von Foerster, *Observing Systems.*

8. See von Foerster, *Cybernetics of Cybernetics.* See also E. Thompson's summary of these theoretical commitments in "Life and Mind," in Clarke and Hansen, *Emergence and Embodiment.* See also Yuk Hui, *Recursivity and Contingency*, 81–83.

9. Maturana and Varela, *Autopoiesis and Cognition.* In 1975, the Biological Computer Laboratory published the entire text in typescript as *Autopoietic Systems*, including the preface by cyberneticist Stafford Beer.

10. Varela, Maturana, and Uribe, "Autopoiesis." As it happened, Margulis published an article in *BioSystem*'s next number, on the evolution of eukaryotic cells: Margulis, "On the Evolutionary Origin." Margulis would join the editorial board of *BioSystems* in 1979 and serve as an associate managing editor from 1983 to 1993. It seems to be a reasonable conjecture that she saw the original autopoiesis essay upon its first publication.

11. See Maturana and Varela, *The Tree of Knowledge.*

12. Margulis and Sagan, *What Is Life?*, 23.

13. Lovelock, *Gaia: A New Look*, 4.

14. Lovelock, *Ages of Gaia*, 30.

15. Lovelock, *Ages of Gaia*, 31. If one reads back from *Novacene*, it may be the case that Lovelock always equated cybernetic systematicity with life, on *informatic* grounds.

16. "The autopoietic approach belongs to a systemic tradition focused on the problem of the relational unity of the living, associated to Kant's understanding of organisms in the *Critique of Judgment*, Claude Bernard's concept of *milieu intérieur*, and the organicist tradition that considers life as organization (G. Canguilhem, H. Jonas, J. Piaget among others), and opposed to the mainstream of the time, such as some of the views of Jacob's *La logique du vivant*." Bich and Etxeberria, "Autopoietic Systems," 2111. Thanks to Sébastien Dutreuil for this reference.

17. Canguilhem, "Experimentation in Animal Biology," 9.

18. Varela has recalled that as a high school student in Felix Schwartzman's "course in the Department of Sciences, I came to know what until then was only known by a minority in Chile, the works of the French school in the history and philosophy of science: Alexandre Koyré (above all), Georges Canguilhem, and Gaston Bachelard." Varela, "Early Days of Autopoiesis," 65.

19. Maturana, "Neurophysiology of Cognition," 8.

20. Luisi, "Autopoiesis: The Logic of Cellular Life," 179. The second edition of this text reformulates the observation: "The notion of autopoiesis was very slow to receive recognition. . . . Still, it cannot be said that the notion of autopoiesis is now familiar in mainstream science. The reason for this . . . is partly due to the fact that autopoiesis is not centered on DNA, RNA, and replication, and makes only minimal use of the term *information*." Luisi, "Autopoiesis—The Invariant Property," 124–25. It was precisely within the systems counterculture that the conceptual value of biotic autopoiesis was recognized and nurtured in its early decades.

21. Varela, "Describing the Logic," 37.

22. E. Thompson, "Life and Mind," in *Phenomenology and the Cognitive Sciences*, 389. See also the version of this essay in Clarke and Hansen, *Emergence and Embodiment.*

23. Varela, "Describing the Logic," 38.
24. Luhmann, "Autopoiesis of Social Systems," 2.
25. Luhmann, "Autopoiesis of Social Systems," 3.
26. Luhmann, *Social Systems*, 59–102.
27. Luhmann, "Autopoiesis of Social Systems," 8–9.
28. "Gaia theory emphasizes lively biotic/abiotic co-productions that sustain the biosphere. In so doing, it collapses the traditional social scientific distinction between living and nonliving matter." Hird, *Origins of Sociable Life*, 130.
29. Margulis and Sagan, *Microcosmos* (2nd ed.), 265.
30. Luhmann, *Social Systems*, 215.
31. Following Anna Henkel's formulation, its materiality as a geobiological system may take on "corporealized meaning" as a "meaningful operating thing" within an autopoietic conceptuality of operational selection of actualities out of the potential states of its elements. Henkel, "Posthumanism," 79.
32. For a phenomenological version of this thesis, see E. Thompson and Stapleton, "Making Sense of Sense-Making."
33. D. Sagan and Margulis, "The Uncut Self," 23–24.
34. Luhmann, "Introduction," 5.
35. See Clarke, *Energy Forms*, 25–26.
36. Cary Wolfe has drawn from this exposition an *openness from closure* principle that also fits the matter of Gaian differentiations: "Here the emphasis falls . . . on the paradoxical fact theorized by both Luhmann and Derrida: the very thing that separates us from the world *connects* us to the world, and self-referential, autopoietic closure, far from indicating a kind of solipsistic neo-Kantian idealism, actually is generative of openness to the environment." Wolfe, *What Is Posthumanism?*, 21.
37. Lovelock to Margulis, October 22, 1986, Margulis Family Papers.
38. Lotka, *Elements of Physical Biology*, 16.
39. Lovelock, "Our Sustainable Retreat," 22.

4. The Whole Earth Network

1. Brand, *II Cybernetic Frontiers*, 38.
2. Brand, *Whole Earth Catalog*. See also Bryant, "Whole System, Whole Earth"; Turner, *From Counterculture to Cyberculture*; Kirk, *Counterculture Green*; Diederichsen and Franke, *The Whole Earth*; and Maniaque-Benton with Gaglio, *Whole Earth Field Guide*. For a wider view, see Maher, *Apollo in the Age of Aquarius*.
3. See Grusin, *Premediation*.
4. To be historically accurate, the Earthrise photograph was neither the first image of Earth seen from space nor the first to include a nonterrestrial object in the same frame with Earth. See Poole, *Earthrise*.
5. A robust historical and critical treatment of the Earthrise phenomenon is Lazier, "Earthrise."
6. "The purpose of this book is to discuss and present evidence for the general thesis that *the flow of energy through a system acts to organize that system*. The motivation for this approach is biological and has its origins in an attempt to find a physical rationale for the extremely high degree of molecular order encountered in living systems" (Morowitz, *Energy Flow in Biology*, 2).
7. Spencer-Brown, *Laws of Form*.
8. Brand, "Laws of Form," 10.
9. The definitive selected edition of von Foerster's papers is *Understanding Understanding*. A broadly accessible introduction is the programmatic interview volume

by von Foerster, *The Beginning of Heaven and Earth Has No Name*. For more on von Foerster's pedagogy and the *Whole University Catalog*, see Clarke, "From Information to Cognition." The von Foerster archive at the University of Illinois at Urbana–Champaign preserves Brand's thank-you note, dated January 6, 1970, acknowledging receipt of the student-produced *Catalog*. This exchange initiated a series of significant contacts over the next decade.

10. Brand to von Foerster, February 27, 1970, von Foerster Papers. See Clarke, "John Lilly."

11. In addition to von Foerster, his protégé Varela was also an early enthusiast for *Laws of Form*: see Varela, "A Calculus for Self-Reference," as well as chapters 11 and 12 in Varela, *Principles of Biological Autonomy* (106–69). See Kauffman, "Self-Reference and Recursive Forms," for a bracing mathematical engagement with *Laws of Form* in second-order cybernetic context. *Laws of Form* would become a fundamental logical architecture for the later development of Luhmann's social systems theory. Luhmann repeatedly cites Kauffman's article alongside references to *Laws of Form*. See also Baecker, *Problems of Form*; Schiltz, "Space Is the Place"; Clarke, *Posthuman Metamorphosis*, 66–71; and Clarke, *Neocybernetics and Narrative*, 77–100.

12. Von Foerster, "Laws of Form," 14.

13. Günther, "Cybernetic Ontology and Transjunctional Operations." Luhmann gives excerpts from this passage of Günther's seminal article, as parallel to Spencer-Brown's discourse on *reentry*—the form that self-reference takes in *Laws of Form*—in "Autopoiesis of Social Systems," 19n32. An approach to the same issue from an entirely different direction is Latour et al., "The Whole Is Always Smaller than Its Parts."

14. Quoted in von Foerster, "Laws of Form," 14. On the form of self-reference in *Laws of Form*, developed by Spencer-Brown as "the reentry of the form into the form," see Clarke, *Neocybernetics and Narrative*, 90–95.

15. Brand, *II Cybernetic Frontiers*, 9.

16. On the cybernetics of social control, see Deleuze, "Control and Becoming" and "Postscript on Control Societies." See also Goffey, "Towards a Rhizomatic Technical History of Control."

17. Brand, headnote to Lovelock, "Daisyworld," 66.

18. See Turner, *From Counterculture to Cyberculture*.

19. Significant later Gaia articles in *CoEvolution Quarterly* include Doolittle, "Is Nature Really Motherly?"; Lovelock, "James Lovelock Responds"; Margulis, "Lynn Margulis Responds"; Lovelock, "More on Gaia and the End of Gaia"; Lovelock and Whitfield, "Life Span of the Biosphere"; Lovelock, "Daisyworld."

20. Lovelock and Giffin, "Planetary Atmospheres"; Lovelock and Lodge, "Oxygen in the Contemporary Atmosphere."

21. Lovelock and Margulis, "Atmospheric Homeostasis"; Lovelock and Margulis, "Homeostatic Tendencies"; Margulis and Lovelock, "Biological Modulation."

22. Margulis to Lovelock, April 29, 1975, Margulis Family Papers.

23. Margulis and Lovelock, "Is Mars a Spaceship, Too?"

24. C. Sagan, "Three from Space"; Schweickart, "Who's Earth?"; McClure, review of *Origin of Eukaryotic Cells*; Margalef, "Perspectives in Ecological Theory."

25. Margulis and Lovelock, "Atmosphere as Circulatory System," 36.

26. Lovelock and Margulis, "Atmospheric Homeostasis."

27. Dawkins, *The Selfish Gene*. For a critique and interesting juxtaposition, see Hayles, "Desiring Agency."

28. Brand, *Next Whole Earth Catalog*.

29. Architecture professor Fred Scharmen lays out this history lesson in "Jeff

Bezos Dreams of a 1970s Future," confirming that Bezos's Blue Origin project (www .blueorigin.com) has revived both the design templates and the talking points of O'Neill's fifty-year-old space colonies project. See also Scharmen's excellent, profusely illustrated *Space Settlements.*

30. Brand, *Space Colonies.*
31. Ballester et al., "Ecological Considerations for Space Colonies."
32. Margulis and Sagan, *What Is Life?,* 242.
33. O'Neill with Brand, "Is the Surface of a Planet," 20.
34. O'Neill, "The High Frontier," 8.
35. D. Sagan and Margulis, "Gaia and Philosophy," 183.
36. So, too, in 2019, do Jeff Bezos and his colleagues at Blue Origin, it would appear, as they contemplate deploying updated versions of O'Neill's high-orbital inhabitations. See https://www.blueorigin.com/.
37. Clarke, "Heinz von Foerster's Demons."
38. Von Foerster, "Cybernetics," 226.
39. See Clarke, "From Information to Cognition."
40. For a brief overview of these developments, see Clarke, "From Thermodynamics to Virtuality."
41. Margulis and Lovelock, "Atmosphere as Circulatory System," 37, reprinted from Lovelock and Margulis, "Atmospheric Homeostasis," 3–4.
42. Von Foerster, "On Self-Organizing Systems."
43. See Clarke, "Information." The information-theoretic approach to Gaia receives a late but emphatic exposition in Lovelock, *Novacene.* See in particular "The Bit" (87–89).
44. Von Foerster, "Gaia's Cybernetics Badly Expressed," 51.
45. In a draft letter to Margulis dated October 21, 1975, Lovelock admits that there were printer's errors in the second quoted equation. Box 20.2, Lovelock Papers.
46. Von Foerster was also instrumental in placing Varela, Maturana, and Uribe's 1974 paper in *BioSystems.* See Varela, "The Early Days of Autopoiesis."
47. Brand, "For God's Sake, Margaret."
48. Brand, headnote to Varela with Johnson, "On Observing Natural Systems," 26.
49. See Luhmann, "Cognitive Program."
50. Margulis, "Kingdom Animalia," in Margulis and Sagan, *Slanted Truths,* 92.
51. Spencer-Brown, *Laws of Form,* 105, quoted in von Foerster, "Laws of Form," 14.
52. "Organized by Gregory Bateson and myself, a conference addressing the pathology of Cartesian mind/body dualism was held at the Wheelwright Center in Marin County, California, July 27th to 30th, 1976. Participants were Gregory Bateson, Francisco Varela, Heinz Von Foerster, Richard Baker-roshi, Ramon Margalef, Gordon Pask, Alan Kay, Terry Winograd, Mary Catherine Bateson, Steve Baer, Stewart Brand, Robert Edgar, and Carol Proudfoot" (Brand, headnote to "Mind/Body Dualism Conference").
53. Varela, "Not One, Not Two" and "Excursus into Dialectics."
54. Bateson, "Invitational Paper."
55. Varela, "Not One, Not Two."
56. Clarke, *Posthuman Metamorphosis,* 61–63.
57. Luhmann, "Introduction," 7.
58. Brand, "Buckminster Fuller," 3.
59. Varela with Johnson, "On Observing Natural Systems," 27.
60. Lovelock, "Daisyworld."
61. Margulis and Stolz, "Microbial Systematics."

62. Lovelock, "Gaia as Seen through the Atmosphere," in Westbroek and De Jong, *Biomineralization*, 15–25. Compare Lovelock, "Gaia as Seen through the Atmosphere," in *Atmospheric Environment* (1972).

63. Doolittle, "Is Nature Really Motherly?" Doolittle has modified his views since then, with reference to "Margulis's thinking" as going "against the main current of evolutionary thought." In introducing a recent essay, he states: "What I hope to accomplish here is a sort of reconciliation. I attempt to recast Gaia theory in a conceptually stretched neoDarwinian framework. Many may think this a stretch too far, but *if* Gaia is to be Darwinized, what I propose seems to be a good way to start. I dedicate the exercise to Lynn, who would no doubt have thought it superfluous." Doolittle, "Darwinizing Gaia," 11.

64. Lovelock, "Gaia as Seen through the Atmosphere," in Westbroek and De Jong, *Biomineralization*, 19.

65. A. J. Watson and Lovelock, "Biological Homeostasis," 286.

66. A. J. Watson and Lovelock, "Biological Homeostasis," 287.

67. This discussion has been nicely summarized and advanced by Sébastien Dutreuil in "What Good Are Abstract and What-If Models?"

68. Lenton et al., "Selection for Gaia across Multiple Scales," 635.

69. Lovelock, "Daisyworld," 66. Lenton et al. also report that they have resolved key contradictions earlier perceived between Gaia theory and natural selection.

70. See Lovelock, "The Independent Practice of Science."

5. The Lindisfarne Connection

1. Brand, "William Irwin Thompson"; W. I. Thompson, *At the Edge of History*.

2. W. I. Thompson, Preface, 10. See W. I. Thompson, "A Gaian Politics." See also Harding, *Animate Earth*.

3. "Lindisfarne," 130.

4. Katz, Marsh, and Thompson, *Earth's Answer*.

5. W. I. Thompson, personal communication.

6. For a detailed history see W. I. Thompson, *Thinking Together*.

7. See Jantsch, *The Self-Organizing Universe*, 131. We will return to Jantsch briefly at the beginning of chapter 6.

8. W. I. Thompson, *Gaia*; W. I. Thompson, *Imaginary Landscape*; W. I. Thompson, *Gaia 2*. For an overview of W. I. Thompson's career at the end of the last century, see Ebert, "Bridges between Worlds."

9. E. Thompson, "Life and Mind," 77.

10. For my own critique of Bateson in this regard, see Clarke, "Steps to an Ecology of Systems," esp. 279–81.

11. W. I. Thompson, *Thinking Together*, 107.

12. W. I. Thompson, *Thinking Together*, 34–35.

13. W. I. Thompson, "Gaia and the Politics of Life," 210.

14. Bunn to Margulis, November 19, 1980, Margulis Family Papers.

15. W. I. Thompson, Preface, 10.

16. W. I. Thompson, "Introduction: The Cultural Implications," 11–12.

17. Although he is peripheral to our main narrative here, Atlan is squarely within the conceptual lineage of von Foerster, in particular the latter's earlier work on self-organization from noise. See Atlan, "Hierarchical Self-Organization" and "Uncommon Finalities." Atlan's work was pivotal for the French philosopher of science Michel Serres. See "Noise in Serres," in Clarke, *Neocybernetics and Narrative*, 56–73.

18. Varela, "Laying Down a Path in Walking," 60.

19. Maturana, "Everything Is Said by an Observer," 75.
20. Lovelock, "Gaia: A Model," 89–90.
21. Margulis, "Early Life," 108.
22. W. I. Thompson, "Gaia and the Politics of Life."
23. W. I. Thompson, Preface, 9.
24. Published as Varela, Thompson, and Rosch, *The Embodied Mind.*
25. W. I. Thompson, "Introduction: The Imagination," 14.
26. "Publisher's Note" in W. I. Thompson, *Gaia 2,* 9.
27. W. I. Thompson, "From Biology to Cognitive Science."
28. Lovelock, "Gaia: A Planetary Emergent Phenomenon," 30.
29. W. I. Thompson, "From Biology to Cognitive Science," 210.
30. In *Ages of Gaia,* recently finished at the time of the 1988 Lindisfarne meeting, Lovelock contrasted Daisyworld specifically to the chaos-theoretical population-ecology models of Robert May and his colleagues. It "differs profoundly from previous attempts to model the species of the Earth. It is a model like those of control theory, or cybernetics, as it is otherwise called. Such models are concerned with self-regulating systems; engineers and physiologists use them to design automatic pilots for aircraft or to understand the regulating of breathing in animals, and they know that the parts of the system must be closely coupled if it is to work. In their parlance, Daisyworld is a closed-loop model" (60).
31. Varela, "The Emergent Self," 212.
32. For an account of Varela's overall conceptual development, see Froese, "From Second-Order Cybernetics to Enactive Cognitive Science."

6. Margulis and Autopoiesis

1. Margulis, *Symbiosis in Cell Evolution.*
2. Margulis, *Symbiosis in Cell Evolution* (2nd ed.).
3. For instance, in a paper coauthored with Hinkle, remarking on "features that make autopoietic (living) systems different from cybernetic ones." Margulis and Hinkle, "The Biota and Gaia," 216.
4. A parallel treatment of many of these issues is Onori and Visconti, "The GAIA Theory."
5. Lovelock, *Ages of Gaia,* 219–20. In the "Further Reading" section of this text's final chapter, Lovelock credits Jantsch: "For an understanding of scientists' views of the Universe, perhaps the best summary is in *The Self-Organizing Universe.*" See Jantsch, *The Self-Organizing Universe.*
6. An accessible overview of Prigogine's work is Prigogine and Stengers, *Order Out of Chaos.*
7. Jantsch, *The Self-Organizing Universe,* 131.
8. "Our writing collaboration began in the spring of 1981, when we were visited in Lynn's Boston apartment by a colorful character in a pimp hat and three-piece suit, the literary agent John Brockman." Margulis and Sagan, *Slanted Truths,* xii.
9. For more on Margulis and Sagan's writing collaboration, see Clarke, "Evolutionary Equality."
10. Margulis and Sagan developed *Microcosmos* and *Origins of Sex* concurrently, and both were first published in 1986. *Microcosmos* had received a contract in the summer of 1981, but *Origins of Sex* was the first to see publication. Dorion Sagan, personal communication.
11. Margulis and Sagan, *Microcosmos,* 14–15.
12. See Sonea and Panisset, *A New Bacteriology.*

13. Jantsch, *The Self-Organizing Universe*, 7

14. Margulis, *Symbiotic Planet*, 119.

15. Margulis and Sagan, *Origins of Sex*, 10–11.

16. Margulis to Lovelock, December 7, 1985, box 20.2, Lovelock Papers.

17. Margulis, "Kingdom Animalia."

18. Margulis, "Big Trouble in Biology."

19. See Sherwood, "Case for 'Counterfoil Research'"; and Ebert, "Bridges between Worlds." W. I. Thompson remarks to Ebert that Illich "articulated the whole vision of the counter-foil institution" (150), which helped to inspire his formation of Lindisfarne.

20. Margulis, "Kingdom Animalia," 861.

21. This finer grain of presentation derives in part from the recently completed doctoral dissertation of her student Gail Fleischacker, to whose work Margulis makes repeated references in her autopoietic writings from this period. Fleischacker, *Autopoiesis*. See also Fleischacker, "Autopoiesis."

22. Varela, Maturana, and Uribe, "Autopoiesis," 188.

23. Margulis, "Big Trouble in Biology," 214.

24. Margulis, "Gaia Is a Tough Bitch," 132.

25. Lovelock, *Gaia: A New Look*, 62.

26. Lovelock, *Novacene*, passim. For instance: "For a while at least, the new electronic life might prefer to collaborate with the organic life which has done (and still does) so much to keep the planet habitable" (105).

27. Fleck, *Genesis and Development of a Scientific Fact*.

28. Latour and Woolgar, *Laboratory Life*.

29. W. I. Thompson, *The Time Falling Bodies Take to Light*.

7. The Planetary Imaginary

1. Haraway, "Cyborgs and Symbionts," xiv.

2. See Brand, "The First Whole Earth Photograph"; Maher, "Shooting the Moon"; Kirk, *Counterculture Green*, 40–42; and Poole, *Earthrise*.

3. See Poole, *Earthrise*, esp. chapter 9, "Gaia" (170–89).

4. Herbert, *Dune*, 13–14.

5. For instance, E. Morin, *La méthode*; E. Morin and Kern, *Homeland Earth*; Serres, *Natural Contract*; Clarke, "Science, Theory, and Systems"; Serres, *Biogea*; and Sloterdijk, *Spheres*.

6. See W. I. Thompson, *Passages about Earth* and "Introduction: The Imagination."

7. Bateson, *Steps*, 461.

8. For a contemporary mobilization of such an extended ecology, see Hörl with Burton, *General Ecology*.

9. For perhaps the earliest presentation of ecology through explicitly cybernetic concepts, see Hutchinson, "Circular Causal Systems in Ecology," which also contains a discussion of the geobiological carbon cycle.

10. Hagen, *Entangled Bank*, 192. See also Kingsland, *Evolution of American Ecology*, 179–85.

11. Margulis, *Symbiotic Planet*, 7, 9.

12. Herbert, *Dune*, 477. On Lotka, see Clarke, *Neocybernetics and Narrative*, 141–42.

13. See Morton, "Imperial Measures."

14. See Ellis, "Frank Herbert's *Dune*."

15. A strong account of the development of Bateson's thinking is Chaney, *Runaway*.

16. O'Neill, "The High Frontier," 8.

17. For a detailed study of O'Neill's projects, see McCray, *Visioneers*. McCray cites

Gibson and *Neuromancer* but does not note *Neuromancer*'s literary repurposing of O'Neill's ideas and NASA's graphics. See also Anker, "Ecological Colonization of Space."

18. Edwards develops a dichotomy between "closed worlds" and "green worlds" in the cyborg imagery of Cold War narratives: "The fate of *Neuromancer*'s Case and Molly," for instance, is to "remain within the closed world" whose operational boundaries are determined by the cybernetic control and communication systems that dominate the historical world of which *Neuromancer*'s storyworld is a reflection. The only escapes are "false exits into ersatz green worlds such as *Neuromancer*'s inverted worlds, the Zion and Freeside space stations" (*The Closed World*, 309). Edwards's vision of cybernetics is averted here from the second-order cybernetics of living systems, and in particular from the neocybernetics of Gaia theory. The conceptual claustrophobia of Cold War control engineering does not observe the ecological difference between Zion and Freeside.

19. For other takes on this cultural episode, see Kirk, *Counterculture Green*, 170–76; and Anker, *From Bauhaus to Ecohouse*, 113–25. Thanks to Christopher Witmore for the Anker reference.

20. Ehrlich and Ehrlich, correspondence.

21. Todd, correspondence.

22. See Reider, *Dreaming the Biosphere*. For wider scientific contexts, see D. Sagan, *Biospheres*; and Morowitz et al., "Closure as a Scientific Concept."

23. "Cosmic Evolution." The full caption reads: "Upper left: the formation of stars, the production of heavy elements, and the formation of planetary systems, including our own. At left prebiotic molecules, RNA, and DNA are formed within the first billion years on the primitive Earth. At center the origin and evolution of life leads to increasing complexity, culminating with intelligence, technology, and astronomers, upper right, contemplating the universe." See the treatment of "Cosmic Evolution" in Dick and Strick, *The Living Universe*.

24. Sloterdijk, *In the World Interior*, 22.

25. See Luhmann, "Modern Sciences and Phenomenology."

26. Sloterdijk, *In the World Interior*, 26.

27. See Potter, *The Earth Gazers*.

28. Latour, *Facing Gaia: Eight Lectures*, 93–94.

29. Grinspoon, *Earth in Human Hands*, 207.

30. Sloterdijk has put this contemporary visualization practice into historical perspective in noting how Alexander von Humboldt, "In keeping with the spirit of his time and ours," in his mid-nineteenth-century work *Cosmos: Sketch of a Physical Description of the Universe*, "took up an arbitrary position in the external space from which to approach the earth like a visitor from a foreign planet." Humboldt wrote: "Here, therefore, we do not proceed from the subjective point of view of human interest; the terrestrial is treated only as a part of the whole, and in its due subordination. The view of nature should be general, grand, and free, not narrowed by proximity, sympathy, or relative utility. A physical cosmography, or picture of the universe, should begin, therefore, not with the earth, but with the regions of space" (*In the World Interior*, 22).

31. McConville, "On the Evolution of the Heavenly Spheres," 6.

32. Latour, *Facing Gaia: Eight Lectures*, 78.

8. Planetary Immunity

1. Sleep, Bird, and Pope, "Paleontology of Earth's Mantle": "Life has had a profound effect on surface geological processes, and even on modulated tectonics and the

rise of continents. . . . The net effect is Gaian (Lovelock 1979); that is, life has modified Earth to its net advantage" (293). See also Grinspoon, *Earth in Human Hands,* 73–74.

2. Daou and Pérez-Ramos, "Island," 7.

3. Lovelock, *Ages of Gaia,* 40.

4. Latour, "Why Gaia Is Not a God of Totality," 75.

5. Dorion Sagan notes that efforts to describe an "inner" or phenomenological view of Gaia go back at least to Abram, "Perceptual Implications of Gaia" (personal communication). See also Sagan's survey of Gaia's cultural landscape at the end of the 1980s in "What Narcissus Saw."

6. Latour, *Facing Gaia: Eight Lectures,* 122. In the original version of this passage in his Gifford Lectures, Latour hewed to Sloterdijk's own idiom: "Sloterdijk has generalized Von Uexküll's *Umwelt* to all the *bubbles* that agencies have generated to make a difference between their inside and their outside." Latour, *Facing Gaia: Six Lectures,* 88.

7. Sloterdijk, *Spheres, Volume I,* 28. See also Bergthaller, "Living in Bubbles."

8. Latour, *Facing Gaia: Eight Lectures,* 138–39.

9. Sloterdijk, *Spheres, Volume I,* 28.

10. Latour, *Facing Gaia: Eight Lectures,* 140.

11. Latour, *Reset Modernity!,* 41.

12. Margulis, "Kingdom Animalia," 866.

13. Volk, *Gaia's Body,* 77.

14. Sleep, Bird, and Pope, "Paleontology of Earth's Mantle," 277.

15. Stolz, "Gaia and Her Microbiome," 1.

16. Lovelock, *Ages of Gaia,* 17, 25.

17. Barlow and Volk, "Open Systems," 372–73.

18. Margulis wrote Lovelock regarding a revised draft of the Barlow-Volk article submitted for *BioSystems*: "I think it is a profound contribution, explaining your persistent intuition that life must be always a planetary phenomenon, from the beginning to the end." Margulis to Lovelock, July 12, 1989.

19. For related discussions, see Clarke, *Neocybernetics and Narrative,* 140–52.

20. Volk, *Gaia's Body,* xiii.

21. William Wordsworth, "A Slumber Did My Spirit Seal."

22. Sloterdijk, *Spheres, Volume I,* 11. See also M.-E. Morin, "Cohabitating in the Globalized World."

23. Sloterdijk, *Spheres, Volume I,* 73.

24. Neyrat, "The Birth of Immunopolitics."

25. Varela and Anspach, "Immu-knowledge," 69.

26. Wolfe, *Before the Law,* 6. See Derrida, "Autoimmunity."

27. Esposito, *Immunitas,* 5.

28. Quoted in Esposito, *Bíos,* 186.

29. Wolfe, *Before the Law,* 55, 7.

30. Wolfe, *Before the Law,* 56, 59, quoting Esposito, *Bíos,* 186.

31. Latour, *Politics of Nature,* 233.

32. Wolfe, "(Auto)Immunity." Wolfe's interventions on behalf of a systems-theoretical redescription of the immunitary paradigm help to correct Neyrat's misapprehension that "Esposito criticizes theories of auto-organization, of *autopoiesis* and auto-regulation, namely because they end up 'questioning the idea of exteriority itself'" ("The Birth of Immunopolitics," 36). Systems theory does not question the idea of exteriority, or materiality: "Self-referential closure is possible only in an environment, only under ecological conditions." Luhmann, *Social Systems,* 9. See also

Richter, "Beyond the 'Other'"; and, anticipating our discussion in the next chapter, Swyngedouw and Ernstson, "Interrupting the Anthropo-obScene."

33. This work goes back to seminal publications of Varela, such as Vaz and Varela, "Self and Non-Sense," and Varela et al., "Cognitive Networks." For an application of Maturana's thought to immunological theory, see Vaz, "The Specificity of Immunologic Observations." See also Tauber, "Immunology's Theories of Cognition" and "The Cognitivist Paradigm 20 Years Later"; and Vaz, "The Enactive Paradigm 33 Years Later."

34. *Gaia: Goddess of the Earth.*

35. Varela and Anspach, "Immu-knowledge."

36. W. I. Thompson, "From Biology to Cognitive Science," 211.

37. For a later formulation of this basic theorization, see Varela, "Patterns of Life."

38. McFall-Ngai et al., "Animals in a Bacterial World": "Carl Woese and George Fox opened a new research frontier by producing sequence-based measures of phylogenic relationships, revealing the deep evolutionary history shared by all living organisms. This game-changing advance catalyzed a rapid development and application of molecular sequencing technologies, which allowed biologists for the first time to recognize the true diversity, ubiquity, and functional capacity of microorganisms" (3229).

39. Margulis "has been critical for development in three major arenas: the prevalence of symbiosis as a driving force in evolution of eukaryotes, the central role of the microbial world in the dynamics of the past and present biosphere, and the recognition that the earth is a self-regulating system, that is, the Gaia hypothesis." McFall-Ngai, "Truth Straight On," 1.

40. Gilbert, Sapp, and Tauber, "A Symbiotic View of Life," 326.

41. See Pace, Sapp, and Goldenfeld, "Phylogeny and Beyond."

42. Margulis, *Symbiosis in Cell Evolution* (2nd ed.), 7. Margulis's credit for coining the term *holobiont* is affirmed in Bordenstein and Theis, "Host Biology."

43. See Roughgarden et al., "Holobionts as Units of Selection."

44. "Animals diverged from their protistan ancestors 700–800 Mya, some 3 billion years after bacterial life originated and as much as 1 billion years after the first appearance of eukaryotic cells. Thus, the current-day relationships of protists"—microbial eukaryotes—"with bacteria, from predation to obligate and beneficial symbiosis, were likely already operating when animals first appeared." McFall-Ngai et al., "Animals in a Bacterial World," 3230.

45. Stamets, *Mycelium Running*; Tsing, *Mushroom at the Edge of the World.*

46. "An ecological orientation . . . already assumes a subordination of the individual to a collective picture of biological function, and in place of differentiation, integration and coordination serve as organizing principles." Tauber, "The Immune System and Its Ecology," 228.

47. McFall-Ngai et al., "Animals in a Bacterial Wold," 3233–34.

48. Gilbert, Sapp, and Tauber, "A Symbiotic View of Life," 326.

49. Gilbert, Sapp, and Tauber, "A Symbiotic View of Life," 330.

50. Bosch and McFall-Ngai, "Metaorganisms as the New Frontier."

51. See also Matyssek and Lüttge, "Gaia: The Planet Holobiont."

52. W. I. Thompson, "Introduction: The Imagination," 24.

53. Varela and Anspach, "Immu-knowledge," 81–82.

9. Astrobiology and the Anthropocene

1. Bateson, *Steps*, 505.

2. Haraway, *Staying with the Trouble*; Bonneuil and Fressoz, *Shock of the Anthropocene*; Demos, "Anthropocene, Capitalocene, Gynocene." See also Cohen, Colebrook,

and Hillis Miller, *Twilight of the Anthropocene Idols*; and Bernard Stiegler, *The Neganthropocene.*

3. Bonneuil, "The Geological Turn."

4. See the website of the International Geosphere-Biosphere Program, which operation ceased in 2015: http://www.igbp.net/.

5. Syvitski, "Anthropocene."

6. Hamilton, "Theodicy of the 'Good Anthropocene,'" 233. See also Dalby, "Framing the Anthropocene."

7. Bonneuil and Fressoz, *Shock of the Anthropocene*. See Asafu-Adjaye et al., "An Ecomodernist Manifesto." See also Brand, *Whole Earth Discipline.*

8. "Global Change and the Earth System."

9. Margulis, "Big Trouble in Biology," 227.

10. Margulis and Sagan, *What Is Life?*, 92–93.

11. See Shapiro, *Evolution.*

12. Margulis, *Proceedings.*

13. Strick, "Creating a Cosmic Discipline."

14. Grinspoon, *Earth in Human Hands*, 65.

15. See McDougall, *The Heavens and the Earth.*

16. Grinspoon, *Earth in Human Hands,* 71; see esp. chapter 2, "Can a Planet Be Alive?" (57–81).

17. Westbroek, *Life as a Geological Force.*

18. Exoplanet discoveries have driven a recent spate of astrobiological Gaia essays: see Chopra and Lineweaver, "The Case for a Gaian Bottleneck"; and Nicholson et al., "Gaian Bottlenecks and Planetary Habitability."

19. See also Lytkin, Finney, and Alepko, "Tsiolkovsky." In the translation given by these authors, the "oft-quoted phrase" is "'The planet is the cradle of intelligence, but it is impossible to live forever in the cradle' *(Planeta yest' kolybel' rasuma, no nefzia vechno zhit' v kolybeli)*" (372).

20. This image may be viewed at http://www.ras.ru//CArchive/pageimages/555/1_084/032.jpg.

21. D. Sagan, "What Narcissus Saw," 260–61.

22. See *Space Science in the Twenty-First Century,* 69–71.

23. Barlow and Volk, "Open Systems Living in a Closed Biosphere," 372.

24. For an illuminating account of earlier NASA efforts to engineer life-support systems in closed environments, see Munns and Nickelsen, "To Live among the Stars."

25. See Groys, *Russian Cosmism.*

26. Thinkers such as Tsiolkovsky, Nikolai Fedorov, and Vladimir Vernadsky "developed a sweeping, scientifically informed, spiritually inspired view of human existence as a stage in the development of planet Earth, itself a small step in the development of the cosmos from diffuse, inchoate origins and toward a more organized, complex, and sentient state of existence" (Grinspoon, *Earth in Human Hands*, 232).

27. MacArthur and Wilson, *Theory of Island Biogeography*, 4.

28. Grinspoon, *Earth in Human Hands,* 209.

29. Wark, *Molecular Red,* xiv.

30. For more on Ship as an AI narrator, see Clarke, "Machines, AIs, Cyborgs, Systems."

31. For general bearings on the posthuman, see Clarke and Rossini, *Cambridge Companion to Literature and the Posthuman*; Braidotti, *The Posthuman*; and Braidotti and Hlavajova, *Posthuman Glossary.*

32. Latour offers the term *geostory* to emphasize the role of narrative in the discourse interrelating Gaia theory and his manifestly posthumanist take on the Anthropocene: "The Anthropocene brings the human back on stage and dissolves for

ever the idea that it is a unified giant agent of history. This is why, in what follows, I will use the word '*anthropos*' to designate what is no longer the 'human-in-nature' nor the 'human-out-of-nature,' but something else entirely, another animal, another beast or, more politely put, a new political body yet to emerge. Such is the main topic of this lecture series: to define the scale, scape, scope and goal of these new people taken severally who have unwillingly become the new agents of geostory" (*Facing Gaia: Six Lectures*, 79).

33. Zalasiewicz et al., "Scale and Diversity," 10.

34. See the fine annotated edition of Vernadsky, *Biosphere*, with a foreword lead-authored by Margulis. For a thoughtful reconsideration of the noosphere in relation to the biosphere and the Anthropocene technosphere, see Lemmens, "Re-orienting the Noösphere."

35. Haff, "Technology as a Geological Phenomenon." See also Haff, "Humans and Technology in the Anthropocene."

36. See also Williams et al., "The Anthropocene Biosphere."

37. Lenton and Latour, "Gaia 2.0," 1066.

38. "Now it is obvious that technological metaphors cannot be applied to the Earth in a lasting way: it was not fabricated; no one maintains it; even if it were a 'space ship'—a comparison that Lovelock constantly contests—there would be no pilot. The Earth has a history, but this does not mean that it was conceived. It is because there is no engineer at work, no divine clockmaker, that a holistic conception of Gaia cannot be sustained." Latour, *Facing Gaia: Eight Lectures*, 96.

39. Dalby, "Anthropocene Geopolitics."

40. Wark, *Molecular Red*, xiv. See also Lemmens and Hui, "Reframing the Technosphere."

41. Frank, Kleidon, and Alberti, "Earth as a Hybrid Planet."

42. Kardashev, "Transmission of Information by Extraterrestrial Civilizations," 19.

43. See also Szerszynski, "Viewing the Technosphere."

44. See Grinspoon, *Earth in Human Hands*; and Frank, *Light of the Stars*.

45. Lovelock, *Novacene*, 23.

46. Lovelock to Margulis, February 29, 1972, Margulis Family Papers. See Margulis, review of *Theory and Experiment in Exobiology*.

47. On Lederberg, see Strick, "Creating a Cosmic Discipline," 139.

48. Margulis and Sagan, *Microcosmos*, 275.

49. Latour and Lenton, "Extending the Domain of Freedom," 660.

50. Clynes and Kline, "Cyborgs and Space," 27. See also Kline, "Where Are the Cyborgs in Cybernetics?"

51. Lem, *Cyberiad*, 195.

52. Margulis to Lovelock, December 7, 1985, box 20.2, Lovelock Papers.

53. Margulis to Lovelock, December 7, 1985, box 20.2, Lovelock Papers.

54. Lovelock, *Ages of Gaia*, 221–22.

55. See Gleick, *The Information*.

56. See Clarke, "Information."

BIBLIOGRAPHY

Abram, David. "The Perceptual Implications of Gaia." *Ecologist* 15, no. 3 (1985): 96–103.

Anderson, Lee. "Riding Solo." *Motorcyclist,* June 10, 2011. http://www.motorcyclistonline.com/blogs/riding-solo-megaphone.

Anker, Peder. "The Ecological Colonization of Space." *Environmental History* 10, no. 2 (April 2005): 239–68.

Anker, Peder. *From Bauhaus to Ecohouse: A History of Ecological Design.* Baton Rouge: Louisiana State University Press, 2010.

Arènes, Alexandra, Bruno Latour, and Jérôme Gaillardet. "Giving Depth to the Surface: An Exercise in the Gaia-graphy of Critical Zones." *Anthropocene Review* 5, no. 2 (2018): 120–35.

Arnold, Darrell P., ed. *Traditions in Systems Theory: Major Figures and Contemporary Developments.* New York: Routledge, 2014.

Asafu-Adjaye, John, Linus Blomqvist, Stewart Brand, Barry Brook, Ruth de Fries, Erle Ellis, Christopher Foreman, David Keith, Martin Lewis, Mark Lynas, Ted Nordhaus, Roger Pielke Jr., Rachel Pritzker, Joyashree Roy, Mark Sagoff, Michael Shellenberger, Robert Stone, and Peter Teague. "An Ecomodernist Manifesto." April 2015. http://www.ecomodernism.org.

Atlan, Henri. "Hierarchical Self-Organization in Living Systems: Noise and Meaning." In *Autopoiesis: A Theory of Living Organization,* ed. Milan Zeleny, 185–208. New York: Elsevier North Holland, 1981.

Atlan, Henri. "Uncommon Finalities." In W. I. Thompson, *Gaia,* 110–27.

Baecker, Dirk, ed. *Problems of Form.* Stanford: Stanford University Press, 1999.

Baecker, Dirk. "Why Systems?" *Theory, Culture & and Society* 18, no. 1 (2001): 59–74.

Ballester, Antonio, E. S. Barghoorn, Daniel B. Botkin, James Lovelock, Ramón Margalef, Lynn Margulis, Juan Oro, Rusty Schweickart, David Smith, T. Swain, John Todd, Nancy Todd, and George M. Woodwell. "Ecological Considerations for Space Colonies." *CoEvolution Quarterly,* no. 12 (Winter 1976–77): 96–97.

Barlow, Connie, and Tyler Volk. "Open Systems Living in a Closed Biosphere: A New Paradox for the Gaia Debate." *BioSystems* 23, no. 4 (1990): 371–84.

Bateson, Gregory. "Invitational Paper" for the Mind/Body Dualism Conference. *CoEvolution Quarterly,* no. 11 (Fall 1976): 56–57.

Bateson, Gregory. *Steps to an Ecology of Mind.* New York: Ballantine, 1972.

Bergthaller, Hannes. "Living in Bubbles: Peter Sloterdijk's Spherology and the Environmental Humanities." In *Spaces In-between: Cultural and Political Perspectives on Environmental Discourse,* ed. Mark Luccarelli and Sigurd Bergmann, 163–74. Boston: Brill, 2015.

Bich, Leonardo, and Arantza Etxeberria. "Autopoietic Systems." In *Encyclopedia of*

Systems Biology, ed. Werner Dubitzky et al., 2110–13. New York: Springer-Verlag, 2013.

Boden, Margaret A. "Autopoiesis and Life." *Cognitive Science Quarterly* 1 (2000): 117–45.

Bonneuil, Christophe. "The Geological Turn: Narratives of the Anthropocene." In Hamilton, Bonneuil, and Gemenne, *The Anthropocene,* 15–31.

Bonneuil, Christophe, and Jean-Baptiste Fressoz. *The Shock of the Anthropocene: The Earth, History, and Us.* Trans. David Fernbach. London: Verso, 2015.

Bordenstein, Seth R., and Kevin R. Theis. "Host Biology in Light of the Microbiome: Ten Principles of Holobionts and Hologenomes." *PLOS Biology* 13, no. 8 (2015): e1002226.

Bosch, Thomas C. G., and Marilyn J. McFall-Ngai. "Metaorganisms as the New Frontier." *Zoology* 114 (2011): 185–90.

Bourgine, Paul, and John Stewart. "Autopoiesis and Cognition." *Artificial Life* 10 (2004): 327–45.

Braidotti, Rosi. *The Posthuman.* New York: Polity, 2013.

Braidotti, Rosi, and Maria Hlavajova, eds. *Posthuman Glossary.* New York: Bloomsbury Academic, 2018.

Brand, Stewart. "Buckminster Fuller." *Whole Earth Catalog,* Spring 1969, 3.

Brand, Stewart. "The First Whole Earth Photograph." In Katz, Marsh, and Thompson, *Earth's Answer,* 184–88.

Brand, Stewart. "For God's Sake, Margaret: Conversation with Gregory Bateson and Margaret Mead." *CoEvolution Quarterly,* no. 10 (Summer 1976): 32–44.

Brand, Stewart. Headnote to Lovelock, "Daisyworld," 66.

Brand, Stewart. Headnote to Varela with Johnson, "On Observing Natural Systems," 26.

Brand, Stewart. "Laws of Form." *Whole Earth Catalog,* Fall 1969, 10.

Brand, Stewart, ed. *Next Whole Earth Catalog.* Sausalito, Calif.: Point, 1980.

Brand, Stewart, ed. *Space Colonies.* New York: Penguin, 1977.

Brand, Stewart. *II Cybernetic Frontiers.* New York: Random House, 1974.

Brand, Stewart, ed. *Whole Earth Catalog.* Menlo Park, Calif.: Portola Institute, 1968–71.

Brand, Stewart. *Whole Earth Discipline: An Ecopragmatist Manifesto.* London: Atlantic Books, 2009.

Brand, Stewart. "William Irwin Thompson." *CoEvolution Quarterly,* no. 20 (Winter 1978): 106.

Brockman, John, ed. *Doing Science: The Reality Club.* New York: Prentice Hall, 1990.

Brockman, John, ed. *Speculations: The Reality Club.* New York: Prentice Hall, 1990.

Brockman, John, ed. *The Third Culture: Beyond the Scientific Revolution.* New York: Simon & Schuster, 1995.

Bryant, William. "Whole System, Whole Earth: The Convergence of Technology and Ecology in Twentieth-Century American Culture." 2006. Unpublished diss., University of Iowa.

Bunyard, Peter, ed. *Gaia in Action: Science of the Living Earth.* Edinburgh: Floris Books, 1996.

Canguilhem, Georges. "Experimentation in Animal Biology." In *Knowledge of Life [La connaissance de la vie],* ed. Paola Marrati and Todd Myers, trans. Stefanos Geroulanos and Daniela Ginsburg, 3–22. New York: Fordham University Press, 2009.

Cannon, Walter B. *The Wisdom of the Body.* New York: Norton, 1932.

Chaney, Anthony. *Runaway: Gregory Bateson, the Double Bind, and the Rise of Ecological Consciousness.* Chapel Hill: University of North Carolina Press, 2017.

Chopra, Aditya, and Charles H. Lineweaver. "The Case for a Gaian Bottleneck: The Biology of Habitability." *Astrobiology* 16, no. 1 (2016): 7–22.

Clarke, Bruce. "Communication." In Mitchell and Hansen, *Critical Terms for Media Studies,* 131–44.

Clarke, Bruce, ed. *Earth, Life, and System: Evolution and Ecology on a Gaian Planet.* New York: Fordham University Press, 2015.

Clarke, Bruce. *Energy Forms: Allegory and Science in the Era of Classical Thermodynamics.* Ann Arbor: University of Michigan Press, 2001.

Clarke, Bruce. "Evolutionary Equality: Neocybernetic Posthumanism and Margulis and Sagan's Writing Practice." In *Writing Posthumanism, Posthuman Writing,* ed. Sidney I. Dobrin, 275–97. Anderson, S.C.: Parlor Press, 2015.

Clarke, Bruce. "From Information to Cognition: The Systems Counterculture, Heinz von Foerster's Pedagogy, and Second-Order Cybernetics." *Constructivist Foundations* 7, no. 3 (2012): 196–207.

Clarke, Bruce. "From Thermodynamics to Virtuality." In *From Energy to Information: Representation in Science and Technology, Art, and Literature,* ed. Bruce Clarke and Linda D. Henderson, 17–33. Stanford: Stanford University Press, 2002.

Clarke, Bruce. "Heinz von Foerster's Demons: The Emergence of Second-Order Systems Theory." In Clarke and Hansen, *Emergence and Embodiment,* 34–61.

Clarke, Bruce. "Information." In Mitchell and Hansen, *Critical Terms for Media Studies,* 157–71.

Clarke, Bruce. "John Lilly, *The Mind of the Dolphin,* and Communication Out of Bounds." *Communication +1* 3 (2014). scholarworks.umass.edu/cpo/vol3/iss1/8.

Clarke, Bruce. "Life, Language, and Identity: Lewis Thomas's Biomythology in *The Lives of a Cell.*" In *The Body and the Text: Comparative Essays in Literature and Medicine,* ed. Bruce Clarke and Wendell Aycock, 207–17. Lubbock: Texas Tech University Press, 1990.

Clarke, Bruce. "Machines, AIs, Cyborgs, Systems." In *After the Human,* ed. Sherryl Vint. Forthcoming from Cambridge University Press.

Clarke, Bruce. *Neocybernetics and Narrative.* Minneapolis: University of Minnesota Press, 2014.

Clarke, Bruce. "Observing *Aramis, or the Love of Technology.*" In Clarke, *Neocybernetics and Narrative,* 111–38.

Clarke, Bruce. *Posthuman Metamorphosis: Narrative and Systems.* New York: Fordham University Press, 2008.

Clarke, Bruce. "Science, Theory, and Systems." *Interdisciplinary Studies in Literature and Environment* 8, no. 1 (2001): 149–65.

Clarke, Bruce. "Steps to an Ecology of Systems: *Whole Earth* and Systemic Holism." In *Addressing Modernity: Social Systems Theory and U.S. Cultures,* ed. Hannes Bergthaller and Carsten Schinko, 259–88. Amsterdam: Rodopi, 2010.

Clarke, Bruce. "Systems Theory." In *Routledge Companion to Literature and Science,* ed. Bruce Clarke with Manuela Rossini, 214–25. London: Routledge, 2010.

Clarke, Bruce, and Mark B. N. Hansen, eds. *Emergence and Embodiment: New Essays in Second-Order Systems Theory.* Durham: Duke University Press, 2009.

Clarke, Bruce, and Manuela Rossini, eds. *The Cambridge Companion to Literature and the Posthuman.* New York: Cambridge University Press, 2017.

Clynes, Manfred E., and Nathan S. Kline. "Cyborgs and Space." *Astronautics* 5, no. 9 (September 1960): 26–27, 74–76.

Cohen, Tom, Claire Colebrook, and J. Hillis Miller. *Twilight of the Anthropocene Idols.* London: Open Humanities Press, 2016.

"Cosmic Evolution." Exobiology program, NASA Ames Research Center, 1986.

Crist, Eileen, and H. Bruce Rinker, eds. *Gaia in Turmoil: Climate Change, Biodepletion, and Earth Ethics in an Age of Crisis.* Cambridge: MIT Press, 2009.

Dalby, Simon. "Anthropocene Geopolitics: Practicalities of the Geological Turn." Workshop on "Environmental Geopolitics in the Anthropocene: Understanding Causes, Managing Consequences, Finding Solutions." University of North Carolina, Greensboro, N.C., March 22–24, 2017.

Dalby, Simon. "Framing the Anthropocene: The Good, the Bad and the Ugly." *Anthropocene Review* 3, no. 1 (2016): 33–51.

Daou, Daniel, and Pablo Pérez-Ramos. "Island." In *New Geographies 8: Island,* ed. Daniel Daou and Pablo Pérez-Ramos, 7–9. Cambridge: Harvard University School of Design, 2017.

Dawkins, Richard. *The Selfish Gene.* Oxford: Oxford University Press, 1976.

Deleuze, Gilles. "Control and Becoming" and "Postscript on Control Societies." In *Negotiations: 1972–1990,* trans. Martin Joughin, 169–82. New York: Columbia University Press, 1995.

Demos, T. J. "Anthropocene, Capitalocene, Gynocene: The Many Names of Resistance." June 12, 2015. https://www.fotomuseum.ch/en/explore/still-searching/articles /27015_anthropocene_capitalocene_gynocene_the_many_names_of_resistance.

Derrida, Jacques. "Autoimmunity: Real and Symbolic Suicides." In *Philosophy in a Time of Terror: Dialogues with Jürgen Habermas and Jacques Derrida,* ed. Giovanna Borradori, 85–136. Chicago: University of Chicago Press, 2003.

Dick, Steven J., and James E. Strick, *The Living Universe: NASA and the Development of Astrobiology.* New Brunswick, N.J.: Rutgers University Press, 2004.

Diederichsen, Diedrich, and Anselm Franke. *The Whole Earth: California and the Disappearance of the Outside.* Berlin: Sternberg Press, 2013.

Doolittle, W. Ford. "Darwinizing Gaia." *Journal of Theoretical Biology* 434 (2017): 11–19.

Doolittle, W. Ford. "Is Nature Really Motherly?" *CoEvolution Quarterly,* no. 29 (Spring 1981): 58–63.

Dutreuil, Sébastien. "Gaïa: Hypothèse, programme de recherche pour le système terre, ou philosophie de la nature?" Doctoral diss., University of Paris 1, 2016.

Dutreuil, Sébastien. "James Lovelock's Gaia Hypothesis: 'A New Look at Life on Earth' . . . for the Life and the Earth Sciences." In *Dreamers, Visionaries, and Revolutionaries in the Life Sciences,* ed. Oren Harman and Michael R. Dietrich, 272–87. Chicago: University of Chicago Press, 2018.

Dutreuil, Sébastien. "What Good Are Abstract and What-If Models? Lessons from the Gaia Hypothesis." *History and Philosophy of the Life Sciences* 36, no. 1 (August 2014): 16–41.

Ebert, John David. "Bridges between Worlds: A Conversation with William Irwin Thompson." In *Twilight of the Clockwork God: Conversations on Science and Spirituality at the End of an Age,* 142–62. San Francisco: Council Oak Books, 1999.

Edwards, Paul N. *The Closed World: Computers and the Politics of Discourse in Cold War America.* Cambridge: MIT Press, 1996.

Ehrlich, Paul, and Anne Ehrlich. Correspondence. In Brand, *Space Colonies,* 43.

Ellis, R. J. "Frank Herbert's *Dune* and the Discourse of Apocalyptic Ecologism in the United States." In *Science Fiction Roots and Branches: Contemporary Critical Approaches,* ed. Rhys Garnett and R. J. Ellis, 104–24. Houndmills, Basingstoke: Palgrave Macmillan, 1990.

Esposito, Roberto. *Bíos: Biopolitics and Philosophy.* Trans. Timothy Campbell. Minneapolis: University of Minnesota Press, 2008.

Esposito, Roberto. *Immunitas: The Protection and Negation of Life.* Trans. Zakiya Hanafi. Malden, Mass.: Polity, 2011.

Fleck, Ludwik. *Genesis and Development of a Scientific Fact.* Ed. Thaddeus J. Trenn and Robert K. Merton. Trans. Fred Bradley and Thaddeus J. Trenn. Chicago: University of Chicago Press, 1979.

Fleischacker, Gail R. "Autopoiesis: The Status of Its System Logic." *BioSystems* 22 (1988): 37–49.

Fleischacker, Gail R. *Autopoiesis: The System, Logic, and Origin of Life.* Boston University Professors Program, 1988. The Graduate School, Boston, Massachusetts.

Frank, Adam. *Light of the Stars: Alien Worlds and the Fate of the Earth.* New York: Norton, 2018.

Frank, Adam, Axel Kleidon, and Marina Alberti. "Earth as a Hybrid Planet: The Anthropocene in an Evolutionary Astrobiological Context." *Anthropocene* 19 (2017): 13–21.

Froese, Tom. "From Second-Order Cybernetics to Enactive Cognitive Science: Varela's Turn from Epistemology to Phenomenology." *Systems Research and Behavioral Science* 28, no. 6 (2011): 631–45.

Gaia: Goddess of the Earth. Nova documentary aired Tuesday, January 28, 1986.

Gibson, William. *Neuromancer.* New York: Ace, 1984.

Gilbert, Scott F., Jan Sapp, and Alfred I. Tauber. "A Symbiotic View of Life: We Have Never Been Individuals." *Quarterly Review of Biology* 87, no. 4 (December 2012): 325–41.

Gleick, James. *Chaos: Making a New Science.* New York: Penguin Books, 1987.

Gleick, James. *The Information: A History, a Theory, a Flood.* New York: Random House, 2011.

"Global Change and the Earth System." See http://www.igbp.net/.

Goffey, Andrew. "Towards a Rhizomatic Technical History of Control." *New Formations* 84–85 (2015): 58–73.

Grinspoon, David. *Earth in Human Hands: Shaping Our Planet's Future.* New York: Grand Central Publishing, 2016.

Groys, Boris, ed. *Russian Cosmism.* Cambridge: EFlux/MIT Press, 2018.

Grusin, Richard. *Premediation: Affect and Mediality after 9/11.* London: Palgrave, 2010.

Günther, Gotthard. "Cybernetic Ontology and Transjunctional Operations." 1962. www.vordenker.de/ggphilosophy/gg_cyb_ontology.pdf.

Haff, Peter. "Humans and Technology in the Anthropocene: Six Rules." *Anthropocene Review* 1, no. 2 (August 1, 2014): 126–36.

Haff, Peter. "Technology as a Geological Phenomenon: Implications for Human Well-being." In *A Stratigraphical Basis for the Anthropocene,* ed. C. N. Waters, J. A. Zalasiewicz, M. Williams, M. A. Ellis, and A. M. Snelling, 301–9. Geological Society, London, Special Publications 395 (2013).

Hagen, Joel B. *An Entangled Bank: The Origins of Ecosystem Ecology.* New Brunswick, N.J.: Rutgers University Press, 1992.

Hamilton, Clive. "The Theodicy of the 'Good Anthropocene.'" *Environmental Humanities* 7 (2015): 233–38.

Hamilton, Clive, Christophe Bonneuil, and François Gemenne, eds. *The Anthropocene and the Global Environmental Crisis: Rethinking Modernity in a New Epoch.* New York: Routledge, 2015.

Haraway, Donna J. "A Cyborg Manifesto: Science, Technology, and Socialist-Feminism in the Late Twentieth Century." In *Simians, Cyborgs, and Women: The Reinvention of Nature,* 149–81. New York: Routledge, 1991.

Haraway, Donna J. "Cyborgs and Symbionts: Living Together in the New World Order." In *The Cyborg Handbook*, ed. Chris Hables Gray with Steven Mentor and Heidi J. Figueroa-Sarriera, xi–xx. New York: Routledge, 1995.

Haraway, Donna J. *Staying with the Trouble: Making Kin in the Chthulucene*. Durham: Duke University Press, 2016.

Haraway, Donna J. *When Species Meet*. Minneapolis: University of Minnesota Press, 2008.

Haraway, Donna J., with Thyrza Nichols Goodeve. "Speaking Resurgence to Despair/ I'd Rather Stay with the Trouble." *The Brooklyn Rail: Critical Perspectives on Art, Politics, and Culture*. https://brooklynrail.org/2017/12/art/DONNA-HARAWAY -with-Thyrza-Nichols-Goodeve.

Harding, Stephan. *Animate Earth: Science, Intuition, and Gaia*. White River Junction, Vt.: Chelsea Green, 2006.

Harding, Stephan, and Lynn Margulis. "Water Gaia: 3.5 Thousand Million Years of Wetness on Planet Earth." In Crist and Rinker, *Gaia in Turmoil*, 41–59.

Harman, Graham. *Immaterialism: Objects and Social Theory*. Malden, Mass.: Polity, 2016.

Hayles, N. Katherine. "Desiring Agency: Limiting Metaphors and Enabling Constraints in Dawkins and Deleuze/Guattari." *SubStance* 94–95 (2001): 144–59.

Henkel, Anna. "Posthumanism, the Social, and the Dynamics of Material Systems." *Theory, Culture & Society* 33, no. 5 (2016): 65–89.

Herbert, Frank. 1965. *Dune*. New York: Ace, 2005.

Heylighen, Francis, and Cliff Joslyn. "Cybernetics and Second-Order Cybernetics." In *Encyclopedia of Physical Science and Technology*, 3rd ed., 4:155–70. San Diego: Academic Press, 2001.

Hird, Myra J. *The Origins of Sociable Life: Evolution after Science Studies*. Houndmills, Basingstoke: Palgrave Macmillan, 2009.

Hitchcock, Dian R., and James E. Lovelock. "Life Detection by Atmospheric Analysis." *Icarus* 7 (1967): 149–59.

Hörl, Erich, with James Burton, eds. *General Ecology: The New Ecological Paradigm*. London: Bloomsbury, 2017.

Hui, Yuk. "On Cosmotechnics: For a Renewed Relation between Technology and Nature in the Anthropocene." *Techné: Research in Philosophy and Technology* 21, nos. 2–3 (2017): 1–23.

Hui, Yuk. *Recursivity and Contingency*. New York: Rowman and Littlefield, 2019.

Hutchinson, G. Evelyn. "Circular Causal Systems in Ecology." *Annals of the New York Academy of Sciences* 50 (October 1948): 221–46.

Jantsch, Erich. *The Self-Organizing Universe: Scientific and Human Implications of the Emerging Paradigm of Evolution*. New York: Pergamon Press, 1980.

Kardashev, N. S. "Transmission of Information by Extraterrestrial Civilizations." In *Extraterrestrial Civilizations*, ed. G. M. Tovmasyan, trans. Z. Lerman, 19–29. Jerusalem: Israel Program for Scientific Translations, 1967.

Katz, Michael, William P. Marsh, and Gail Gordon Thompson, eds. *Earth's Answer: Explorations of Planetary Culture at the Lindisfarne Conferences*. New York: Harper & Row, 1977.

Kauffman, Louis H. "Self-Reference and Recursive Forms." *Journal of Social and Biological Structure* 10 (1987): 53–72.

Kiang, Nancy Y., Shawn Domagal-Goldman, Mary N. Parenteau, David C. Catling, Yuka Fujii, Victoria S. Meadows, Edward W. Schwieterman, and Sara I. Walker.

"Exoplanet Biosignatures: At the Dawn of a New Era of Planetary Observations." *Astrobiology* 18, no. 6 (2018): 619–29.

Kingsland, Sharon E. *The Evolution of American Ecology, 1890–2000*. Baltimore: John Hopkins University Press, 2005.

Kirk, Andrew G. *Counterculture Green: The "Whole Earth Catalog" and American Environmentalism*. Lawrence: University Press of Kansas, 2007.

Kline, Ronald. "Where Are the Cyborgs in Cybernetics?" *Social Studies of Science* 39, no. 3 (June 2009): 331–62.

Kuhn, Thomas. *The Structure of Scientific Revolutions*. 2nd ed. Chicago: University of Chicago Press, 1970.

Lane, Nick. *Power, Sex, Suicide: Mitochondria and the Meaning of Life*. 2nd ed. Oxford: Oxford University Press, 2018.

Latour, Bruno. *Aramis, or The Love of Technology*. Trans. Catherine Porter. Cambridge: Harvard University Press, 1996.

Latour, Bruno. *Facing Gaia: Eight Lectures on the New Climatic Regime*. Trans. Catherine Porter. Medford, Mass.: Polity, 2017.

Latour, Bruno. *Facing Gaia: Six Lectures on the Political Theology of Nature*. Gifford Lectures on Natural Religion, version 10-3-2013. https://eportfolios.macaulay .cuny.edu/wakefield15/files/2015/01/LATOUR-GIFFORD-SIX-LECTURES_1.pdf.

Latour, Bruno. *Pandora's Hope: Essays on the Reality of Science Studies*. Cambridge: Harvard University Press, 1999.

Latour, Bruno. *Politics of Nature: How to Bring the Sciences into Democracy*. Trans. Catherine Porter. Cambridge: Harvard University Press, 2004.

Latour, Bruno. *Reassembling the Social: An Introduction to Actor-Network Theory*. Oxford: Oxford University Press, 2007.

Latour, Bruno. *Reset Modernity! The Field Book*. Karlsruhe: ZKM | Center for Art and Media, 2016.

Latour, Bruno. "Why Gaia Is Not a God of Totality." *Theory, Culture & Society* 34, nos. 2–3 (2017): 61–81.

Latour, Bruno, Pablo Jensen, Tommaso Venturini, Sebastian Grauwin, and Dominique Boullier. "The Whole Is Always Smaller than Its Parts—A Digital Test of Gabriel Tarde's Monads." *British Journal of Sociology* 63, no. 4 (2012): 590–615.

Latour, Bruno, and Timothy M. Lenton. "Extending the Domain of Freedom, or Why Gaia Is So Hard to Understand." *Critical Inquiry* 45, no. 3 (Spring 2019): 659–80.

Latour, Bruno, and Steve Woolgar. *Laboratory Life: The Construction of Scientific Facts*. Beverly Hills, Calif.: Sage Publications, 1979.

Lazier, Benjamin. "Earthrise; or, The Globalization of the World Picture." *American Historical Review* 116, no. 3 (June 2011): 602–30.

Lem, Stanisław. *The Cyberiad: Fables for the Cybernetic Age*. Trans. Michael Kandel. 1965. New York: Harvest, 1985.

Lemmens, Pieter. "Re-orienting the Noösphere: Imagining a New Role for Digital Media in the Era of the Anthropocene." *Glimpse: Publication of the Society for Phenomenology and Media* 19 (2018): 8–17.

Lemmens, Pieter, and Yuk Hui. "Reframing the Technosphere: Peter Sloterdijk and Bernard Stiegler's Anthropotechnological Diagnoses of the Anthropocene." *Krisis* 2 (2017): 25–40.

Lenton, Timothy M., Stuart J. Daines, James G. Dyke, Arwen E. Nicholson, David M. Wilkinson, and Hywel T. P. Williams. "Selection for Gaia across Multiple Scales." *Trends in Ecology & Evolution* 33, no. 8 (August 2018): 633–45.

Lenton, Timothy M., and Bruno Latour. "Gaia 2.0: Could Humans Add Some Level of Self-Awareness to Earth's Self-Regulation?" *Science* 361, no. 6407 (September 14, 2018): 1066–68.

Lenton, Timothy M., and Andrew Watson. *Revolutions That Made the Earth.* Oxford: Oxford University Press, 2011.

"Lindisfarne." *CoEvolution Quarterly,* no. 2 (Summer 1974): 130.

Litfin, Karen. "Principles of Gaian Governance: A Rough Sketch." In Crist and Rinker, *Gaia in Turmoil,* 196–219.

Litfin, Karen. "Thinking Like a Planet: Gaian Politics and the Transformation of the World Food System." In *Handbook of Global Environmental Politics,* 2nd ed., ed. Peter Dauvergne, 419–30. Cheltenham, UK: Elgar, 2012.

Lotka, Alfred J. *Elements of Physical Biology.* Baltimore: Williams and Wilkins, 1925.

Lovelock, James E. *The Ages of Gaia: A Biography of Our Living Earth.* New York: Norton, 1988.

Lovelock, James E. "Daisyworld: A Cybernetic Proof of the Gaia Hypothesis." *CoEvolution Quarterly,* no. 38 (Summer 1983): 66–72.

Lovelock, James E. "Gaia: A Model for Planetary and Cellular Dynamics." In W. I. Thompson, *Gaia,* 89–90.

Lovelock, James E. *Gaia: A New Look at Life on Earth.* 1979. New York: Oxford University Press, 1989.

Lovelock, James E. "Gaia: A Planetary Emergent Phenomenon." In W. I. Thompson, *Gaia 2,* 30–49.

Lovelock, James E. "Gaia as Seen through the Atmosphere." *Atmospheric Environment* 6 (1972): 579–80.

Lovelock, James E. "Gaia as Seen through the Atmosphere" (1983). In Westbroek and De Jong, *Biomineralization,* 15–25.

Lovelock, James E. *Gaia: The Practical Science of Planetary Medicine.* Oxford: Oxford University Press, 2000.

Lovelock, James E. *Homage to Gaia: The Life of an Independent Scientist.* London: Oxford University Press, 2000.

Lovelock, James E. "The Independent Practice of Science." *CoEvolution Quarterly,* no. 25 (Spring 1980): 22–30, reprinted from *New Scientist,* September 6, 1979.

Lovelock, James E. "James Lovelock Responds." *CoEvolution Quarterly,* no. 29 (Spring 1981): 62–63.

Lovelock, James E. "More on Gaia and the End of Gaia." *CoEvolution Quarterly,* no. 31 (Fall 1981): 36–37.

Lovelock, James E. *Novacene: The Coming Age of Hyperintelligence.* With Bryan Appleyard. Cambridge: MIT Press, 2019.

Lovelock, James E. "Our Sustainable Retreat." In Crist and Rinker, *Gaia in Turmoil,* 21–24.

Lovelock, James E. Papers. Science Museum, London.

Lovelock, James E. "A Physical Basis for Life-Detection Experiments." *Nature* 207 (August 7, 1965): 568–70.

Lovelock, James E. *The Revenge of Gaia: Earth's Climate in Crisis and the Fate of Humanity.* New York: Basic Books, 2006.

Lovelock, James E., and C. E. Giffin. "Planetary Atmospheres: Compositional and other Changes Associated with the Presence of Life." *Advances in the Astronautical Sciences* 25 (1969): 179–93.

Lovelock, James E., and James P. Lodge Jr. "Oxygen in the Contemporary Atmosphere." *Atmospheric Environment* 6 (1972): 575–78.

Lovelock, James E., R. J. Maggs, and R. A. Rasmussen. "Atmospheric Dimethyl Sulphide and the Natural Sulphur Cycle." *Nature* 237 (1972): 452–53.

Lovelock, James E., and Lynn Margulis. "Atmospheric Homeostasis by and for the Biosphere: The Gaia Hypothesis." *Tellus* 26 (1974): 2–10.

Lovelock, James E., and Lynn Margulis. "Homeostatic Tendencies of the Earth's Atmosphere," *Origins of Life* 5 (1974): 93–103.

Lovelock, James E., and Michael Whitfield. "Life Span of the Biosphere." *CoEvolution Quarterly*, no. 31 (Fall 1981): 37–38.

Luhmann, Niklas. "The Autopoiesis of Social Systems." In *Essays on Self-Reference*, 1–20. New York: Columbia University Press, 1990.

Luhmann, Niklas. "The Cognitive Program of Constructivism and a Reality that Remains Unknown." In Luhmann, *Theories of Distinction*, 128–52.

Luhmann, Niklas. "Introduction: Paradigm Change in Systems Theory." In Luhmann, *Social Systems*, 1–11.

Luhmann, Niklas. "The Modern Sciences and Phenomenology." In Luhmann, *Theories of Distinction*, 33–60.

Luhmann, Niklas. "Self-Organization and Autopoiesis." In Clarke and Hansen, *Emergence and Embodiment*, 143–56.

Luhmann, Niklas. *Social Systems*. Trans. John Bednarz Jr. with Dirk Baecker. Stanford: Stanford University Press, 1995.

Luhmann, Niklas. *Theories of Distinction: Redescribing the Descriptions of Modernity*. Ed. William Rasch. Stanford: Stanford University Press, 2002.

Luhmann, Niklas. *Theory of Society, Volume 1*. Trans. Rhodes Barrett. Stanford: Stanford University Press, 2012.

Luisi, Pier Luigi. "Autopoiesis—The Invariant Property." In *The Emergence of Life: From Chemical Origins to Synthetic Biology*, 2nd ed., 119–56. Cambridge: Cambridge University Press, 2016.

Luisi, Pier Luigi. "Autopoiesis: The Logic of Cellular Life." In *The Emergence of Life: From Chemical Origins to Synthetic Biology*, 155–81. Cambridge: Cambridge University Press, 2006.

"Lynn Margulis." *The Telegraph*, December 13, 2011.

Lytkin, Vladimir, Ben Finney, and Liudmila Alepko. "Tsiolkovsky, Russian Cosmism and Extraterrestrial Intelligence." *Quarterly Journal of the Royal Astronomical Society* 36, no. 4 (1995): 369–76.

MacArthur, Robert H., and Edward O. Wilson. *The Theory of Island Biogeography*. Princeton: Princeton University Press, 1967.

Maher, Neil. *Apollo in the Age of Aquarius*. Cambridge: Harvard University Press, 2017.

Maher, Neil. "Shooting the Moon." *Environmental History* 9, no. 3 (2004): 526–31.

Maniaque-Benton, Caroline, with Meredith Gaglio, eds. *Whole Earth Field Guide*. Cambridge: MIT Press, 2016.

Mann, Charles. "Lynn Margulis: Science's Unruly Earth Mother." *Science* 252 (April 19, 1991): 378–81.

Margalef, Ramón. "Perspectives in Ecological Theory." *CoEvolution Quarterly*, no. 6 (1975): 49–66.

Margulis, Lynn. "Big Trouble in Biology: Physiological Autopoiesis versus Mechanistic Neo-Darwinism." In Brockman, *Doing Science*, 211–35; reprinted in Margulis and Sagan, *Slanted Truths*, 265–82.

Margulis, Lynn. "Early Life: The Microbes Have Priority." In W. I. Thompson, *Gaia*, 98–109.

Margulis, Lynn. Family Papers. Ashland, Oregon.

Margulis, Lynn. "Gaia Is a Tough Bitch." In Brockman, *The Third Culture,* 129–46.

Margulis, Lynn. "Kingdom Animalia: The Zoological Malaise from a Microbial Perspective." *American Zoologist* 30, no. 4 (1990): 861–75; reprinted in Margulis and Sagan, *Slanted Truths,* 91–111.

Margulis, Lynn. "Lynn Margulis Comments on James Lovelock." *CoEvolution Quarterly,* no. 25 (Spring 1980): 24–25.

Margulis, Lynn. "Lynn Margulis Responds." *CoEvolution Quarterly,* no. 29 (1981): 63–65.

Margulis, Lynn. "On the Evolutionary Origin and Possible Mechanism of Colchicine-Sensitive Mitotic Movements." *BioSystems* 6 (1974): 16–36.

Margulis, Lynn. *Origin of Eukaryotic Cells: Evidence and Research Implications for a Theory of the Origin and Evolution of Microbial, Plant, and Animal Cells on the Precambrian Earth.* New Haven: Yale University Press, 1970.

Margulis, Lynn. Papers. Library of Congress, Washington, D.C.

Margulis, Lynn, ed. *Proceedings of the Second Conference on Origins of Life: Cosmic Evolution, Abundance, and Distribution of Biologically Important Elements.* New York: Gordon and Breach, 1971.

Margulis, Lynn. Review of *Theory and Experiment in Exobiology,* vol. 2, ed. A. W. Schwartz. *Quarterly Review of Biology* 49 (1974): 55–56.

Margulis, Lynn. *Symbiosis in Cell Evolution: Life and Its Environment on the Early Earth.* San Francisco: Freeman, 1981.

Margulis, Lynn. *Symbiosis in Cell Evolution: Microbial Communities in the Archean and Proterozoic Eons.* 2nd ed. New York: Freeman, 1993.

Margulis, Lynn. *Symbiotic Planet: A New Look at Evolution.* New York: Basic Books, 1998.

Margulis, Lynn, and Gregory Hinkle. "The Biota and Gaia: One Hundred and Fifty Years of Support for Environmental Sciences." In Margulis and Sagan, *Slanted Truths,* 207–20.

Margulis, Lynn, and James E. Lovelock. "The Atmosphere as Circulatory System of the Biosphere: The Gaia Hypothesis." *CoEvolution Quarterly,* no. 6 (Summer 1975): 31–40.

Margulis, Lynn, and James E. Lovelock. "Biological Modulation of the Earth's Atmosphere." *Icarus* 21 (1974): 471–89.

Margulis, Lynn, and James E. Lovelock. "Gaia and Geognosy." In *Global Ecology: Towards a Science of the Biosphere,* ed. M. B. Rambler, L. Margulis, and R. Fester, 1–30. San Diego: Academic Press, 1989.

Margulis, Lynn, and James E. Lovelock. "Is Mars a Spaceship, Too?" *Natural History* 85 (1976): 86–90.

Margulis, Lynn, Clifford Matthews, and Aaron Haselton, eds. *Environmental Evolution: Effects of the Origin and Evolution of Life on Planet Earth.* 2nd ed. Cambridge: MIT Press, 2000.

Margulis, Lynn, and Dorion Sagan. *Dazzle Gradually: Reflections on the Nature of Nature.* White River Junction, Vt.: Chelsea Green, 2007.

Margulis, Lynn, and Dorion Sagan. *Microcosmos: Four Billion Years of Evolution from Our Microbial Ancestors.* New York: Summit Books, 1986.

Margulis, Lynn, and Dorion Sagan. *Microcosmos: Four Billion Years of Microbial Evolution.* 2nd ed. Berkeley: University of California Press, 1997.

Margulis, Lynn, and Dorion Sagan. *Origins of Sex: Three Billion Years of Genetic Recombination.* New Haven: Yale University Press, 1986.

Margulis, Lynn, and Dorion Sagan. *Slanted Truths: Essays on Gaia, Symbiosis, and Evolution*. New York: Copernicus, 1997.

Margulis, Lynn, and Dorion Sagan. *What Is Life?* Berkeley: University of California Press, 2000.

Margulis, Lynn, and John F. Stolz. "Microbial Systematics and a Gaian View of the Sediments." In Westbroek and De Jong, *Biomineralization*, 27–54.

Maturana, Humberto. "Everything Is Said by an Observer." In W. I. Thompson, *Gaia*, 75.

Maturana, Humberto. "Neurophysiology of Cognition." In *Cognition: A Multiple View*, ed. Paul L. Garvin, 3–23. New York: Spartan Books, 1970.

Maturana, Humberto, and Francisco J. Varela. *Autopoiesis and Cognition: The Realization of the Living*. Boston: Riedel, 1980.

Maturana, Humberto, and Francisco J. Varela. *Autopoietic Systems: A Characterization of the Living Organization*. Biological Computer Laboratory, Report No. 9.4, September 1, 1975.

Maturana, Humberto, and Francisco J. Varela. *The Tree of Knowledge: The Biological Roots of Human Understanding*. Rev. ed. Boston: Shambhala, 1998.

Matyssek, Rainer, and Ulrich Lüttge. "Gaia: The Planet Holobiont." *Nova Acta Leopoldina* NF 114, no. 391 (2013): 325–44.

McClure, Michael. Review of *Origin of Eukaryotic Cells*, by Lynn Margulis. *CoEvolution Quarterly*, no. 6 (1975): 41.

McConville, David. "On the Evolution of the Heavenly Spheres: An Enactive Approach to Cosmography." Doctoral thesis, Planetary Collegium, Faculty of Arts, Plymouth University, 2014.

McCray, W. Patrick. *The Visioneers: How a Group of Elite Scientists Pursued Space Colonies, Nanotechnology, and a Limitless Future*. Princeton: Princeton University Press, 2012.

McDougall, Walter A. *The Heavens and the Earth: A Political History of the Space Age*. Baltimore: Johns Hopkins University Press, 1985.

McFall-Ngai, Margaret. "Truth Straight On: Reflections on the Vision and Spirit of Lynn Margulis." *Biological Bulletin* 223 (August 2012): 1.

McFall-Ngai, Margaret, et al. "Animals in a Bacterial World: A New Imperative for the Life Sciences." *PNAS* 110, no. 9 (February 26, 2013): 3229–36.

Midgley, Mary, ed. *Earthy Realism: The Meaning of Gaia*. Charlottesville, Va.: Societas, 2007.

"Mind/Body Dualism Conference: Position Papers." *CoEvolution Quarterly*, no. 11 (Fall 1976): 56.

Mitchell, W. J. T., and Mark B. N. Hansen, eds. *Critical Terms for Media Studies*. Chicago: University of Chicago Press, 2010.

Morin, Edgar. *La méthode, 1: La nature de la nature*. Paris: Éditions du Seuil, 1977.

Morin, Edgar, and Anne Brigitte Kern. *Homeland Earth: A Manifesto for the New Millennium*. New York: Hampton Press, 1999.

Morin, Marie-Eve. "Cohabitating in the Globalized World: Peter Sloterdijk's Global Foams and Bruno Latour's Cosmopolitics." *Environment and Planning D: Society and Space* 27 (2009): 58–72.

Morowitz, Harold J. *Beginnings of Cellular Life: Metabolism Recapitulates Biogenesis*. New Haven: Yale University Press, 1992.

Morowitz, Harold J. *Energy Flow in Biology: Biological Organization as a Problem in Thermal Physics*. New York: Academic Press, 1968.

Morowitz, Harold J., John P. Allen, Mark Nelson, and Abigail Alling. "Closure as a Scientific Concept and Its Application to Ecosystem Ecology and the Science of the Biosphere." *Advances in Space Research* 36 (2005): 1305–11.

Morton, Timothy. "Imperial Measures: *Dune*, Ecology and Romantic Consumerism." *Romanticism on the Net* 21 (February 2001). id.erudit.org/iderudit/005966ar.

Munns, David P. D., and Kärin Nickelsen. "To Live among the Stars: Artificial Environments in the Early Space Age." *History and Technology* 33, no. 3 (2017): 272–99.

Neyrat, Frédéric. "The Birth of Immunopolitics." Trans. Arne de Boever. *Parrhesia* 10 (2010): 31–38.

Neyrat, Frédéric. *The Unconstructable Earth: An Ecology of Separation.* Trans. Drew S. Burk. New York: Fordham University Press, 2018.

Nicholson, Arwen E., David M. Wilkinson, Hywel T. P. Williams, and Timothy M. Lenton. "Gaian Bottlenecks and Planetary Habitability Maintained by Evolving Model Biospheres: The ExoGaia Model." *Monthly Notices of the Royal Astronomical Society* 477, no. 1 (June 11, 2018): 727–40.

Ntyintyane, Lucas. "Earth Is One Big Family." *Business Day*, July 28, 2011.

O'Neill, Gerard K. "The High Frontier." In Brand, *Space Colonies*, 8–11.

O'Neill, Gerard K., with Stewart Brand. "Is the Surface of a Planet Really the Right Place for an Expanding Technological Civilization?" *CoEvolution Quarterly*, no. 6 (Fall 1975): 20–28.

Onori, Luciano, and Guido Visconti. "The GAIA Theory: From Lovelock to Margulis; From a Homeostatic to a Cognitive Autopoietic Worldview." *Rendiconti Lincei: Scienze Fisiche e Naturali* 23 (2012): 375–86.

Pace, Norman R., Jan Sapp, and Nigel Goldenfeld. "Phylogeny and Beyond: Scientific, Historical, and Conceptual Significance of the First Tree of Life," *PNAS* 109, no. 4 (January 24, 2012): 1011–18.

Partridge, Christopher. "Mother Earth, Goddess Gaia." In *The Re-Enchantment of the West, Volume II: Alternative Spiritualities, Sacralization, Popular Culture, and Occulture*, 61–65. New York: T&T Clark, 2005.

Poole, Robert. *Earthrise: How Man First Saw the Earth.* New Haven: Yale University Press, 2008.

Potter, Christopher. *The Earth Gazers: On Seeing Ourselves.* New York: Pegasus Books, 2018.

Prigogine, Ilya, and Isabelle Stengers. *Order Out of Chaos: Man's New Dialogue with Nature.* New York: Bantam, 1984.

Reider, Rebecca. *Dreaming the Biosphere: The Theater of All Possibilities.* Albuquerque: University of New Mexico Press, 2009.

Richter, Hannah. "Beyond the 'Other' as Constitutive Outside: The Politics of Immunity in Roberto Esposito and Niklas Luhmann." *European Journal of Political Theory* (July 2016). https://doi.org/10.1177/1474885116658391.

Riegler, Alexander, Karl H. Müller, and Stuart A. Umpleby, eds. *New Horizons for Second-Order Cybernetics.* Hackensack, N.J.: World Scientific, 2018.

Robinson, Kim Stanley. *Aurora.* New York: Orbit, 2015.

Roughgarden, Joan, Scott F. Gilbert, Eugene Rosenberg, et al. "Holobionts as Units of Selection and a Model of Their Population Dynamics and Evolution." *Biological Theory* 13, no. 1 (March 2018): 44–65.

Ruse, Michael. *The Gaia Hypothesis: Science on a Pagan Planet.* Chicago: University of Chicago Press, 2013.

Sagan, Carl. "Three from Space." *CoEvolution Quarterly*, no. 6 (1975): 28–29.

Sagan, Dorion. *Biospheres: Metamorphoses of Planet Earth*. New York: McGraw-Hill, 1990.

Sagan, Dorion. "What Narcissus Saw: The Oceanic 'I'/Eye." In Brockman, *Speculations*, 247–66.

Sagan, Dorion, and Lynn Margulis. "Gaia and Philosophy" (1984). In Margulis and Sagan, *Dazzle Gradually*, 172–84.

Sagan, Dorion, and Lynn Margulis. "Gaia and the Evolution of Machines." *Whole Earth Review* 55 (Summer 1987): 15–21.

Sagan, Dorion, and Lynn Margulis. "The Uncut Self." In Margulis and Sagan, *Dazzle Gradually*, 16–26.

Scharmen, Fred. "Jeff Bezos Dreams of a 1970s Future." May 13, 2019. https://www .citylab.com/perspective/2019/05/space-colony-design-jeff-bezos-blue-origin -oneill-colonies/589294.

Scharmen, Fred. *Space Settlements*. New York: Columbia Books on Architecture and the City, 2019.

Schiltz, Michael. "Space Is the Place: The *Laws of Form* and Social Systems." In Clarke and Hansen, *Emergence and Embodiment*, 157–78.

Schneider, Stephen H., and Penelope J. Boston, eds. *Scientists on Gaia*. Cambridge: MIT Press, 1993.

Schneider, Stephen H., James R. Miller, Eileen Crist, and Penelope J. Boston, eds. *Scientists Debate Gaia: The Next Century*. Cambridge: MIT Press, 2004.

Schweickart, Russell. "Who's Earth?" *CoEvolution Quarterly*, no. 6 (1975): 42–45.

Serres, Michel. *Biogea*. Trans. Randolph Burks. Minneapolis: Univocal, 2012.

Serres, Michel. *Hermes: Literature, Science, Philosophy*. Ed. Josué V. Harari and David F. Bell. Baltimore: Johns Hopkins University Press, 1982.

Serres, Michel. *The Natural Contract*. Trans. Elizabeth MacArthur and William Paulson. Ann Arbor: University of Michigan Press, 1995.

Shapiro, James A. *Evolution: A View from the 21st Century*. Upper Saddle River, N.J.: FT Press Science, 2011.

Sherwood, Martin. "Case for 'Counterfoil Research.'" *New Scientist* 61, no. 883 (January 31, 1974): 285.

Sleep, Norman H., Dennis K. Bird, and Emily Pope. "Paleontology of Earth's Mantle." *Annual Review of Earth and Planetary Sciences* 40 (2012): 277–300.

Sloterdijk, Peter. *In the World Interior of Capital: For a Philosophical Theory of Globalization*. Trans. Wieland Hoban. Malden, Mass.: Polity, 2013.

Sloterdijk, Peter. *Spheres, Volume 1: Bubbles—Microspherology*. Trans. Wieland Hoban. Los Angeles: Semiotext(e), 2011.

Sloterdijk, Peter. *Spheres, Volume 3: Foams—Plural Spherology*. Trans. Wieland Hoban. Los Angeles: Semiotext(e), 2016.

Sonea, Sorin, and Leo G. Mathieu. *Prokaryotology: A Coherent View*. Montreal: University of Montréal Press, 2000.

Sonea, Sorin, and Maurice Panisset. *A New Bacteriology*. Boston: Jones and Bartlett, 1983.

Space Science in the Twenty-First Century: Imperatives for the Decades 1995 to 2015: Life Sciences. Washington, D.C.: National Academy Press, 1988.

Spencer-Brown, George. *Laws of Form*. London: Allen and Unwin, 1969.

Stamets, Paul. *Mycelium Running*. Berkeley, Calif.: Ten Speed Press, 2005.

Stengers, Isabelle. "Accepting the Reality of Gaia: A Fundamental Shift?" In Hamilton, Bonneuil, and Gemenne, *The Anthropocene*, 134–44.

Stengers, Isabelle. *Cosmopolitics I.* Trans. Robert Bononno. Minneapolis: University of Minnesota Press, 2010.

Stengers, Isabelle. *Cosmopolitics II.* Trans. Robert Bononno. Minneapolis: University of Minnesota Press, 2011.

Stengers, Isabelle. *In Catastrophic Times: Resisting the Coming Barbarism.* Trans. Andrew Goffey. Ann Arbor, Mich.: Open Humanities Press, 2015.

Stengers, Isabelle. *Thinking with Whitehead: A Free and Wild Creation of Concepts.* Trans. Michael Chase. Cambridge: Harvard University Press, 2011.

Stiegler, Bernard. *The Neganthropocene.* Ed., trans., and intro. Daniel Ross. London: Open Humanities Press, 2018.

Stolz, John F. "Climate Change and the Gaia Hypothesis." In *The Urgency of Climate Change: Pivotal Perspectives,* ed. Gerard Magill and Kiarash Aramesh, 50–72. Newcastle upon Tyne: Cambridge Scholars, 2017.

Stolz, John F. "Gaia and Her Microbiome." *FEMS Microbiology Ecology* 93, no. 2 (2017): 1–13.

Strick, James. "Creating a Cosmic Discipline: The Crystallization and Consolidation of Exobiology, 1957–1973." *Journal of the History of Biology* 37, no. 1 (March 2004): 131–80.

Suzuki, David. "Journey into New Worlds: Lovelock's Gaia." *The Sacred Balance.* Toronto, 2002. VHS.

Swyngedouw, Erik, and Henrik Ernstson. "Interrupting the Anthropo-obScene: Immuno-biopolitics and Depoliticizing Ontologies in the Anthropocene." *Theory, Culture & Society* 35, no. 6 (2018): 3–30.

Syvitski, James P. M. "Anthropocene: An Epoch of Our Making." *Global Change* 78 (2012): 12–15.

Szerszynski, Bronislaw. "Viewing the Technosphere in an Interplanetary Light." *Anthropocene Review* 4, no. 2 (2017): 92–102.

Tauber, Alfred I. "The Cognitivist Paradigm 20 Years Later: Commentary on Nelson Vaz." *Constructivist Foundations* 6, no. 3 (2011): 342–44.

Tauber, Alfred I. "The Immune System and Its Ecology." *Philosophy of Science* 75 (April 2008): 224–45.

Tauber, Alfred I. "Immunology's Theories of Cognition." *History and Philosophy of the Life Sciences* 35 (2013): 239–64.

Thomas, Lewis. *The Lives of a Cell: Notes of a Biology Watcher.* New York: Viking, 1974.

Thompson, Evan. "Life and Mind: From Autopoiesis to Neurophenomenology; A Tribute to Francisco Varela." *Phenomenology and the Cognitive Sciences* 3, no. 4 (2004): 381–98.

Thompson, Evan. "Life and Mind: From Autopoiesis to Neurophenomenology." In Clarke and Hansen, *Emergence and Embodiment,* 77–93.

Thompson, Evan, and Mog Stapleton. "Making Sense of Sense-Making: Reflections on Enactive and Extended Mind Theories." *Topoi* 28 (2009): 23–30.

Thompson, William Irwin. *At the Edge of History.* New York: Harper & Row, 1971.

Thompson, William Irwin, ed. "From Biology to Cognitive Science: General Symposium on the Cultural Implications of the Idea of Emergence in the Fields of Biology, Cognitive Science, and Philosophy." In W. I. Thompson, *Gaia 2,* 210–48.

Thompson, William Irwin. "Gaia and the Politics of Life: A Program for the Nineties?" In W. I. Thompson, *Gaia,* 167–214.

Thompson, William Irwin, ed. *Gaia: A Way of Knowing—Political Implications of the New Biology.* Great Barrington, Mass.: Lindisfarne Press, 1987.

Thompson, William Irwin. "A Gaian Politics." *Whole Earth Review* 53 (Winter 1986): 4–15.

Thompson, William Irwin, ed. *Gaia 2: Emergence—The New Science of Becoming*. Hudson, N.Y.: Lindisfarne Press, 1991.

Thompson, William Irwin. *Imaginary Landscape: Making Worlds of Myth and Science*. New York: St. Martin's Press, 1989.

Thompson, William Irwin. "Introduction: The Cultural Implications of the New Biology." In W. I. Thompson, *Gaia*, 11–34.

Thompson, William Irwin. "Introduction: The Imagination of a New Science and the Emergence of a Planetary Culture." In W. I. Thompson, *Gaia 2*, 11–29.

Thompson, William Irwin. *Passages about Earth: An Exploration of the New Planetary Culture*. New York: Harper & Row, 1974.

Thompson, William Irwin. Preface. In W. I. Thompson, *Gaia*, 7–10.

Thompson, William Irwin. *Thinking Together at the Edge of History: A Memoir of the Lindisfarne Association*. Traverse City, Mich.: Lorian Press, 2016.

Thompson, William Irwin. *The Time Falling Bodies Take to Light: Mythology, Sexuality, and the Origins of Culture*. New York: St. Martin's Press, 1981.

Todd, John. Correspondence. In Brand, *Space Colonies*, 48–49.

Tresch, John. "We Have Never Known Mother Earth." Review of *Facing Gaia: Eight Lectures on the New Climate Regime*, by Bruno Latour. *Public Books*, December 1, 2017. http://www.publicbooks.org/we-have-never-known-mother-earth/.

Tsing, Anna. *The Mushroom at the Edge of the World*. Princeton: Princeton University Press, 2017.

Turner, Fred. *From Counterculture to Cyberculture: Stewart Brand, the Whole Earth Network, and the Rise of Digital Utopianism*. Chicago: University of Chicago Press, 2006.

Tyrrell, Toby. *On Gaia: A Critical Investigation of the Relationship between Life and Earth*. Princeton: Princeton University Press, 2013.

Varela, Francisco J. "A Calculus for Self-Reference." *International Journal of General Systems* 2 (1975): 5–24.

Varela, Francisco J. "Describing the Logic of the Living: The Adequacy and Limitations of the Idea of Autopoiesis." In *Autopoiesis: A Theory of the Living Organization*, ed. Milan Zeleny, 36–48. New York: Elsevier North-Holland, 1981.

Varela, Francisco J. "The Early Days of Autopoiesis." In Clarke and Hansen, *Emergence and Embodiment*, 62–76.

Varela, Francisco J. "The Emergent Self." In Brockman, *The Third Culture*, 209–22.

Varela, Francisco J. "Excursus into Dialectics." In Varela, *Principles of Biological Autonomy*, 99–102.

Varela, Francisco J. "Laying Down a Path in Walking." In W. I. Thompson, *Gaia*, 48–64.

Varela, Francisco J. "Not One, Not Two." *CoEvolution Quarterly*, no. 11 (Fall 1976): 62–67.

Varela, Francisco J. "Patterns of Life: Intertwining Identity and Cognition." *Brain and Cognition* 34 (1997): 72–87.

Varela, Francisco J. *Principles of Biological Autonomy*. New York: Elsevier North Holland, 1979.

Varela, Francisco J., and Mark Anspach. "Immu-knowledge: The Process of Somatic Individuation." In W. I. Thompson, *Gaia 2*, 68–85.

Varela, Francisco J., Antonio Coutinho, Bruno Dupire, and Nelson M. Vaz. "Cognitive Networks: Immune, Neural, and Otherwise." In *Theoretical Immunology, Part Two*, ed. A. S. Perelson, 159–75. Boston: Addison-Wesley, 1988.

Varela, Francisco J., with Donna Johnson. "On Observing Natural Systems."
CoEvolution Quarterly, no. 10 (Summer 1976): 26–31.

Varela, Francisco J., Humberto M. Maturana, and Ricardo Uribe. "Autopoiesis: The
Organization of Living Systems, Its Characterization and a Model." *BioSystems* 5
(1974): 187–96.

Varela, Francisco J., Evan Thompson, and Eleanor Rosch. *The Embodied Mind: Cognitive Science and Human Experience.* Cambridge: MIT Press, 1991.

Vaz, Nelson. "The Enactive Paradigm 33 Years Later: Response to Alfred Tauber."
Constructivist Foundations 6, no. 3 (2011): 345–51.

Vaz, Nelson. "The Specificity of Immunologic Observations." *Constructivist Foundations* 6, no. 3 (2011): 334–42.

Vaz, Nelson, and Francisco Varela. "Self and Non-Sense: An Organism-Centered Approach to Immunology." *Medical Hypothesis* 4 (1978): 231–67.

Vernadsky, Vladimir. *The Biosphere.* Foreword by Lynn Margulis et al. Intro.
Jacques Grinevald. Trans. David B. Langmuir. Rev. and annotated by Mark A. S.
McMenamin. New York: Copernicus, 1998.

Volk, Tyler. *Gaia's Body: Toward a Physiology of Earth.* Cambridge: MIT Press, 2003.

von Foerster, Heinz. *The Beginning of Heaven and Earth Has No Name: Seven Days with
Second-Order Cybernetics.* Ed. Albert Müller and Karl H. Müller. Trans. Elinor
Rooks and Michael Kasenbacher. New York: Fordham University Press, 2015.

von Foerster, Heinz. "Cybernetics." In *Encyclopedia of Artificial Intelligence,* ed. S. C.
Shapiro, 1:225–26. New York: Wiley, 1990.

von Foerster, Heinz, ed. *Cybernetics of Cybernetics: Or, The Control of Control and the
Communication of Communication.* Biological Computer Laboratory, University of
Illinois at Urbana–Champaign, 1974. Reprint, Minneapolis: Future Systems, 1995.

von Foerster, Heinz. "Gaia's Cybernetics Badly Expressed." *CoEvolution Quarterly,*
no. 7 (Fall 1975): 51.

von Foerster, Heinz. "Laws of Form." Review of *Laws of Form,* by George Spencer-
Brown. *Whole Earth Catalog,* Spring 1970, 14.

von Foerster, Heinz. "Objects: Tokens for (Eigen)-Behaviors" (1974). In von Foerster,
Understanding Understanding, 261–71.

von Foerster, Heinz. *Observing Systems: Selected Papers of Heinz von Foerster.* Intro.
Francisco J. Varela. Salinas, Calif.: Intersystems Publications, 1981.

von Foerster, Heinz. "On Self-Organizing Systems and Their Environments" (1960). In
von Foerster, *Understanding Understanding,* 1–19.

von Foerster, Heinz. Papers. University of Illinois at Urbana–Champaign.

von Foerster, Heinz. *Understanding Understanding: Essays on Cybernetics and Cognition.* New York: Springer, 2003.

Wark, McKenzie. "Bruno Latour: Occupy Earth." Blog post, October 10, 2017, available
at https://www.versobooks.com/blogs/3425-bruno-latour-occupy-earth.

Wark, McKenzie. *Molecular Red: Theory for the Anthropocene.* New York: Verso, 2016.

Watson, Andrew J., and James E. Lovelock. "Biological Homeostasis of the Global
Environment: The Parable of Daisyworld." *Tellus* 35B (1983): 284–89.

Watson, Matthew C. "Derrida, Stengers, Latour, and Subalternist Cosmopolitics."
Theory, Culture & Society 31, no. 1 (2013): 75–98.

Westbroek, Peter. *Life as a Geological Force: Dynamics of the Earth.* New York: Norton.
1991.

Westbroek, Peter, and Elisabeth W. De Jong, eds. *Biomineralization and Biological
Metal Accumulation: Biological and Geological Perspectives.* Dordrecht: D. Reidel,
1983.

Wiener, Norbert. *Cybernetics: Or, Control and Communication in the Animal and the Machine.* Cambridge: MIT Press, 1948.

Wiener, Norbert. *The Human Use of Human Beings: Cybernetics and Society.* Boston: Houghton Mifflin, 1950.

Williams, Mark, Jan Zalasiewicz, P. K. Haff, Christian Schwägerl, Anthony D. Barnosky, and Erle C. Ellis. "The Anthropocene Biosphere." *Anthropocene Review* 2, no. 3 (2015): 196–219.

Wilson, Alexander. "Biosphere, Noosphere, Infosphere: Epistemo-Aesthetics and the Age of Big Data." *Parallax* 23, no. 2 (2017): 202–19.

Wolfe, Cary. "(Auto)Immunity, Social Theory, and the 'Political.'" *Parallax* 23, no. 1 (2017): 108–22.

Wolfe, Cary. *Before the Law: Humans and Other Animals in a Biopolitical Frame.* Chicago: University of Chicago Press, 2013.

Wolfe, Cary. *What Is Posthumanism?* Minneapolis: University of Minnesota Press, 2010.

Zalasiewicz, Jan, et al. "Scale and Diversity of the Physical Technosphere: A Geological Perspective." *Anthropocene Review* 4, no. 1 (2017): 9–22.

Zilber-Rosenberg, Ilana, and Eugene Rosenberg. "Role of Microorganisms in the Evolution of Animals and Plants: The Hologenome Theory of Evolution." *FEMS Microbiology Review* 32 (2008): 723–35.

INDEX

Abelson, Philip H., 32
Abraham, Ralph, 148
actor-network theory (ANT), 64, 68–69
adaptationism, 145, 150, 244
albedo, 135, 151
AI (artificial intelligence), 270; as literary characters, 203, 252–53. *See also* cyborgs; intelligence: artificial
Aldrin, Edwin ("Buzz"), 206–7; *Return to Earth,* 207
animals, 11, 33, 38, 51–52, 74, 169–76, 191, 196, 226, 235, 248, 271, 290n30, 294n44, 296n32; as holobiont hosts, 213, 236–38; immune system of, 233, 238; nervous system of, 233; origin of, 173, 236; and the technosphere, 258–59
animism, 89, 152–53
Anthropocene, the, 9, 13, 24, 47, 169, 243–48, 251–54, 263–64; alternative names for, 244; and the biosphere, 256–64; combustion in, 261; defined, 243; end of, 12, 254, 256, 269; and the Gaian system, 12, 18, 46, 146, 207, 256–65; mature, 265. *See also* geology; Grinspoon, David; Latour, Bruno; Stengers, Isabelle
Anthropocene technosphere, 19, 169, 246, 257–64; absorptive propensity in, 258; biomorphic description of, 259–60, 263; boundaries of, 258; purported autonomy of, 258–61; recycling deficits of, 259–61; role of humans in, 257–58, 260–62; solar energy and, 263. *See also Aurora*; organicism; technosphere; Vernadsky, Vladimir
archaea, 10, 235

Archean eon, 44, 93, 130, 171, 173, 185, 240, 270
Armstrong, Neil, 207
Ashby, H. Ross, 106
astrobiology, 9, 18–19, 204, 208, 245–56; the astrobiological gaze, 205–6; Gaia and, 26, 245–46, 255–56, 264–67; planetary evolution and, 262–64; SETI and, 247, 263–64; as speculative practice, 247–48; and technological civilizations, 247, 250, 263–64; and the technosphere, 246, 256, 262–65. *See also* exobiology; NASA
Atlan, Henri, 11, 73, 141, 145, 289n17
atmosphere, 11, 23–27, 33, 36, 45, 62, 72, 94, 111, 123, 149, 201, 221, 260, 262; biogenic, 10, 27–28, 32, 146, 230–31, 281n9; chemical disequilibrium of, 25, 134, 218; extraterrestrial, 25, 184, 262, 265; greenhouse gasses in, 54; life detection by, 25–26, 114, 122; regulation (homeostasis) of, 2, 8, 26–29, 48, 116, 146, 166
"Atmosphere as Circulatory System of the Biosphere, The" (Margulis and Lovelock), 114–18, 122, 133
Atmospheric Environment, 26–32, 36
"Atmospheric Homeostasis by and for the Biosphere" (Lovelock and Margulis), 33, 122, 124, 133
Aurora (Kim Stanley Robinson), 19, 249–56; as Anthropocene allegory, 251–52, 256; as Gaian critique, 249, 254–55; generation ship in, 19, 250, 254; "islanding," 251; metabolic rifts, 251–52, 256. *See also* humanity
autology, 6, 122
autonomy, 77–78, 143, 147, 241, 273;

and Gaia, 55, 58, 79, 152–55, 220; in neocybernetic systems theory (NST), 17, 38, 53–54, 65, 83, 86–87, 92; and technology, 252–53, 258–61, 268–70; in Francisco Varela, 17–18, 84, 90, 127, 141, 144, 151–55, 203, 233
autopoiesis, 7–8, 11, 15, 84–94, 132, 148, 206, 257, 285n9, 285n20, 295n32; biotic, 4, 44; component, 159, 164–67; contingency of, 164; as a criterion of life, 51, 76, 124, 166; and the cyborg, 51–52; Donna Haraway and, 50–53, 64; and the holobiont, 236; and immunity, 227, 231, 233–34, 241; impatience of, 55; Bruno Latour and, 65, 69, 283n27; Lynn Margulis and, 4, 17, 71, 76–78, 140, 157–80, 270, 273, 281n23, 285n10; Maturana and Varela on, 5, 51, 54, 84–87, 144, 148, 163, 172, 206, 285n9; metabiotic, 5, 16, 42–44, 54, 69, 91–93, 194; mortality and, 173; Francisco Varela and, 125, 142, 151–56, 233. *See also* Gaia: autopoietic; self-production
Autopoiesis and Cognition (Maturana and Varela), 84, 145, 285n9
"Autopoiesis: The Organization of Living Systems" (Varela, Maturana, and Uribe), 85–86, 166
Avatar (dir. James Cameron), 57

bacteria, 4, 10, 11, 61, 158, 160, 173, 179, 228, 245; and evolution, 51, 77, 160, 164–65, 235, 240, 270, 294n38, 294n44; and Gaia, 10, 29, 74, 76, 144, 146, 149, 167, 171, 218, 230, 237; and the Great Oxidation Event, 264; and the holobiont, 236–38. *See also* microbes
Baecker, Dirk, 42, 93
Ballard, J. G, 196
Barlow, Connie, 219–21, 249, 259, 293n18
Bateson, Gregory, 16, 243; Stewart Brand on, 109–11, 125; cybernetics of, 102, 109–11, 125, 142–43, 188, 191; and the Lindisfarne Association, 17, 139–45; *Mind and Nature*, 141; and the Mind-Body Dualism Conference, 127, 288n52; *Steps to an Ecology of Mind*, 188, 191–94; and the systems counter-

culture, 101–2, 109–11, 131. *See also* epistemology
Bernal, J. D., 219
Berry, Wendell, 111, 118
Bezos, Jeff, 119, 280n38, 287n29, 288n36
Big Bang, the, 208
biocentrism, 11, 89, 158
Biological Computer Laboratory, 106–7, 194, 285n9
"Biological Modulation of the Earth's Atmosphere" (Margulis and Lovelock), 33
biology, 1, 17, 24, 29, 65, 109, 119, 201, 244, 247–48, 266; in *Aurora* (Kim Stanley Robinson), 251–55, 273; of cognition, 124; and Gaia, 30–33, 115–16, 124, 133–35, 144, 166, 217–19, 235, 246, 257, 269; of immunity, 226–33, 237–38; Lynn Margulis and, 157, 171–80, 189, 245; mechanistic, 166; microbiology, 10, 24, 177, 218; molecular, 163; physical, 96–97, 105; and sociology, 70; and symbiosis, 235–38; systems, 76, 144, 147; and the technosphere, 259–60. *See also* neo-Darwinism
biomineralization, 133
biopolitics, 18, 46; Gaian, 62, 74, 225, 230, 239–41; and immunitary discourse, 224–30, 238–41; in Bruno Latour, 62; in Lindisfarne discourse, 144–47, 229–30, 239–40; thanatopolitics, 226–27
biosignature, 25–26, 114, 246
biosphere, 8, 10, 14, 24, 95, 115, 119, 133, 157, 179–80, 225, 231–32, 237, 286n28; artificial, 251; as autopoietic system, 5, 94; bacterial, 10, 146; boundaries of, 213–20; as cybernetic system, 26, 29, 36–37, 41, 48, 88, 117; as ecosystem, 10; and the geosphere, 97; immunity and, 239–41; modeled (in Daisyworld), 136; symbiosis and, 239; and the technosphere, 12–13, 19, 168–69, 256–62, 273
Biosphere 2, 202, 292n22
biota, 4, 6, 24, 36, 40, 74, 134, 162, 239–40, 249; Gaian perception of, 13, 35, 42, 44, 55, 67–68, 88–89, 116, 147, 150, 157, 172, 229, 257; microbiota, 18, 253; modeled (in Daisyworld), 132, 135–36

Bolin, Bert, 33
boundaries, 12, 42, 96, 105, 123, 168, 193–94, 208, 248; cellular, 77, 84, 161; between disciplines, 33; formal (virtual), 45; of Gaia, 18, 48–53, 61, 123, 214–24, 241, 272; operational (autopoietic), 39, 61, 77–80, 97–98, 129, 172, 176, 257–58, 270, 292n18. *See also* closure
Brand, Stewart, 16, 17, 132; and Gregory Bateson, 102, 109–11, 127; and *CoEvolution Quarterly,* 111, 115; on cybernetics, 109–10; and Daisyworld, 137; and the Gaia hypothesis, 111–14; and Lindisfarne Association, 140; and Lynn Margulis, 112; and space colonies, 118, 120–21, 197–98; and William I. Thompson, 139–40; *II Cybernetic Frontiers,* 108–11, 157; and Francisco Varela, 125–26; and Heinz von Foerster, 106, 108, 122, 287n9; and the *Whole Earth Catalog,* 102, 106, 108–9, 111, 117, 140
Brockman, John, 170, 290n8
Byurakan Conference on Extraterrestrial Civilizations and Interstellar Communication, 263

Canguilhem, Georges, 89, 285n16, 285n18
Cannon, Walter, 41
capitalism, 55–58, 261
cells, living, 1, 3, 6, 65, 83, 115, 125, 139, 146, 153, 220, 238, 241, 259, 271; autopoiesis of, 4, 7, 68, 76, 85–86, 90, 94, 155, 158; cognition of, 84, 233; eukaryotic, 51, 77–78, 158, 164–65, 167, 171, 235–36, 294n44; evolution of, 2, 25; and Gaia, 4, 45, 214, 234; lymphocytes, 231–33, 239; Lynn Margulis on, 51, 76–78, 144–45, 157, 160–80, 285n10, 294n39; mitochondria in, 77–78, 164, 240, 284n41; prokaryotic, 10, 33, 51, 78, 158, 160, 165, 167, 171, 236. *See also* bacteria; membrane; organism; symbiogenesis; symbiosis
chaos theory, 1, 47, 132, 148, 151
chemistry, 1, 9, 27, 35–37, 45, 150, 160, 172–74, 218, 230, 271; atmospheric, 10, 144, 218

circularity, 80, 83, 139; circular causality, 38, 84, 88; in first-order cybernetics, 6, 8, 29, 83–84, 110, 153; in Gaia, 116, 220, 225, 231; logic of, 6, 83–84, 133–34, 137–38; in neocybernetic systems theory, 83–91, 95, 131, 155, 161–62, 176
civilization, 214, 264; dominant, 178; extraterrestrial, 247, 256, 263–64; industrial, 24, 143; materialistic, 178; mechanistic, 180; sustainable, 264; technological, 120, 178, 180, 247, 250, 254, 263, 272; urban, 192
climate, 134, 189; change, 47, 54, 56, 60, 241; Gaian effects on, 35–36, 62, 72, 80, 150, 269; virtual (in Daisyworld), 135–36
closure, 8, 13, 38–39, 44, 81, 83–86, 107, 110, 116, 131, 205, 219, 223–24, 261, 290n38, 292n18; environmental, 7, 16–19, 118–21, 196, 200–203, 219–22, 248–55, 259, 266, 295n24; immunitary, 216–17, 228–34; systemic (operational), 6, 12, 24, 29, 40–41, 62, 65, 72, 77, 80, 90–98, 124–28, 145, 152–54, 158, 163, 167–69, 175, 193, 236, 241, 244–45, 256, 260, 286n36, 293n32. *See also* boundaries; living systems
CoEvolution Quarterly, 8, 16, 102, 111–38, 165, 207; contributors, 111–12; Daisyworld in, 132–38; the Gaia hypothesis in, 112–18, 122–24, 287n19; and the Lindisfarne Association, 139–42; and *Neuromancer*'s high-orbital habitats, 197–202; on space colonies, 17–18, 118–22; space-colony debates, 201–2; and the systems counterculture, 114–18, 192; Francisco Varela in, 125–32, 152; Heinz von Foerster in, 122–24
cognition, 62, 89, 90, 98, 175, 184, 204, 206, 210, 280n36; autopoiesis and, 7, 11, 84–85, 94, 147, 179, 206, 233; and the immune system, 230, 232–33, 294n33; in neocybernetic systems theory, 83, 84, 107–8, 122, 124–32, 143–43, 148; planetary (Gaian), 12, 14, 43–46, 53, 62–63, 80, 93, 162, 232
Cold War, 48, 50, 246, 292n18

colonization: bacterial, 237; planetary, 121, 248–49

communication, 143, 169, 183–84, 194, 246, 259; bacterial, 160, 218; in cybernetics, 110, 188, 191, 292n18; extraterrestrial, 263; and Gaia, 42–46, 49, 62, 93; in social systems, 5, 7, 37–38, 91–92, 168, 283n27

community: alternative, 188; biospheric, 240; ecology, 165–66, 220, 229; human, 109, 141, 200; and immunity, 226, 240; microbial, 51, 149, 164, 171, 237, 240; symbiotic, 87, 273

complexity theory, 1, 71, 148

constructivism. *See* epistemology

control, 38, 44, 57, 85, 87, 94, 101, 161, 187–90, 203, 238, 284n37; environmental, 248–52; feedback, 83–84; Gaian, 28, 36–37, 70, 81, 88, 116–17, 121, 134, 211, 239, 266; social, 287n16; systems, 34, 40, 68, 71, 75, 105, 132, 134–36, 249, 268, 292n18; and the technosphere, 260–61; theory, 19, 38, 76, 109–110, 122, 134, 290n30. *See also* cybernetics

Copernicus, 207, 211; Copernican revolution, 209

correspondence: between James Lovelock and Lynn Margulis, 15, 28–34, 112–14, 165, 266, 270–71, 281n9, 288n45

cosmic microwave background, 208–9

cosmic spheres, 209–10, 216

cosmism, Russian, 248–251, 254–55

cosmopolitics, 15–16

counterculture: American, 15, 105, 109, 117–18, 188; autopoiesis and, 89; holism and, 95. *See also* Bateson, Gregory; *CoEvolution Quarterly*; Lindisfarne Association; Lovelock, James; systems counterculture

counterfoil research, 170, 291n19

Coutinho, Antonio, 230

critical zone, 10, 18, 62, 82, 214–17, 222, 245

culture: academic, 174; cyber-, 102, 165, 195; food, 202; global, 183; human, 141, 191, 216; masculinist, 174; modern, 174, 177; mono-, 102; planetary, 17, 142–45, 148, 150, 178, 180, 187;

popular, 50, 101; pure, 173; spacefaring, 65, 183; technical, 50, 196; Western. *See also* counterculture; systems counterculture

cybernetics: biological, 6, 9, 16, 24, 28, 40, 51, 103, 145, 193–94; first-order, 6, 8, 33, 40–41, 53, 64, 69, 76, 83–88, 188; informatic, 142, 195; organic, 102, 111, 129, 157; second-order, 4, 6, 9, 15–16, 42, 53, 83–87, 106, 122–30, 194, 206, 287n11, 292n18. *See also* autopoiesis; circularity; control; machines; neocybernetic systems theory (NST)

cyberspace, 34, 195–96, 203

cyborgs, 252, 268, 292n18; autopoiesis and, 51–52; exobiology and, 268; Donna Haraway and, 15, 47–53, 64–65, 76, 184; James Lovelock and, 47, 76, 268–73

"Cyborgs and Space" (Clynes and Kline), 49, 269.

cycles: atmospheric, 146; biogeochemical (Gaian), 10, 12, 32, 80, 116, 186, 203, 217, 222–24; ecological, 115, 120, 221, 256; natural, 63; systemic, 110, 131; in Vladimir Vernadsky, 257

Daisyworld, 16, 41, 47, 88, 111, 132–38, 151, 157, 282n46, 290n30; Francisco Varela's critique of, 151–54. *See also* scale

Dalby, Simon, 261, 263

Darwin, Charles, 11, 150

Darwinism, 71, 160, 235; social, 224. *See also* neo-Darwinism

Dawkins, Richard, 168. *See also* selfish gene

Derrida, Jacques, 225, 286n36

dialectics: classical, 128, 145; cybernetic, 129; immunitary, 226–28, 238

Dick, Philip K., 196

differentiation, 97, 129, 145, 220, 294n46; cell, 154; functional, 283n27; operational, 92, 167, 272. *See also* system differentiation

Digital Universe Atlas (NASA visualization), 208–11

distinction, 68, 79, 89–90, 106, 152, 220–21, 225, 266; conceptual, 52, 61, 69; events of, 45, 92; epistemological, 126; in *Laws of Form,* 107–8, 126, 204; "leaky," 52; observation as, 205; systemic (operational), 12, 16, 30, 39, 45, 52–54, 57, 62, 69, 75–77, 92–98, 125, 131, 153–54, 176, 193, 219, 227, 239, 257, 281n23, 286n28; theological, 70; unity of the, 129, 210

Doolittle, W. Ford, 134, 287n19

double positivity, 129

Dune (Frank Herbert), 17, 186–93; "The Ecology of Dune," 189, 192–93; and Gaia, 193. *See also* ecology; ecosystem; living systems; psychedelics; society

Dutreuil, Sébastien, 23, 287n67, 298n63

dynamical systems theory. *See* chaos theory

Earth, return to, 206–11, 253, 256; in Peter Sloterdijk, 206–7

"Earth as a Hybrid Planet: The Anthropocene in an Evolutionary Astrobiological Context" (Frank, Kleidon, and Alberti), 262, 264

Earthrise, 102–6, 117, 185–86, 207, 246, 279n3, 286n4, 286n5; cybernetics of, 102–5

Earth seen from space, 112, 114, 183–85, 207, 216, 286n4

Earth system, 9, 15, 96, 184, 220, 243–47, 257, 259, 269; in Bruno Latour, 18, 66; science, 9, 18, 23, 72, 186, 206, 244, 246; in Isabelle Stengers, 54–58

"Ecological Considerations for Space Colonies" (Margulis et al), 118–22, 201

ecology: and the Anthropocene, 19, 260; and astrobiology, 263, 266; in *Aurora* (Kim Stanley Robinson), 249–56; closed, 18, 118–21, 200–203, 220, 248–49, 252; and cybernetics, 17, 291n9, 292n18; deep, 68; in *Dune* (Frank Herbert), 186–94; ecosystem ecology, 10, 119, 188–89, 192–93; and Gaia discourse, 10, 16, 37, 46, 55, 58–59, 115, 164, 169, 218, 225, 235; and holism, 95, 97, 101, 214; and the holobiont, 213, 235, 237; of ideas, 190–93; and immunity, 227, 229, 232, 294n46; and the Lindisfarne Association, 141–48; and Lynn Margulis, 9,

29, 149, 157, 165–67, 180, 266; of mind (Gregory Bateson), 109–10, 125, 188, 193–94, 243; and neocybernetic systems theory, 42–44, 125–27, 293n32; in *Neuromancer* (William Gibson), 194–204; population, 290n30; in the systems counterculture, 101–3, 114, 139, 204. *See also* ecosystem

ecomodernism, 244

ecosystem, 10, 44, 104, 109, 119, 130, 165; artificial, 119–20, 196, 198–203, 249, 266; in *Aurora* (Kim Stanley Robinson), 251; and cybernetics, 16, 114, 188; in *Dune* (Frank Herbert), 188–93; and Gaia discourse, 59, 75, 96, 105, 163, 166, 219–21, 273; in immunitary discourse, 231–40; in *Neuromancer* (William Gibson), 196–203. *See also* holobiont; symbiont

Ehrlich, Paul and Anne, 111, 118, 201; *The Population Bomb,* 201

Ellul, Jacques, 258

emergence, 33, 37, 42, 92, 98, 139, 147–49, 183, 262; domain of, 72, 149, 155–56; in Gaia, 3, 10, 12, 68, 79–80, 116, 162, 221, 224, 230; in living systems, 5, 171, 236, 241, 264; in James Lovelock, 39–40, 72–74, 88, 132, 150–51, 154–55; in Lynn Margulis, 86–87; systemic, 55, 70, 93, 129–31, 260; in Francisco Varela, 126, 153–56, 232–34. *See also* Thompson, William Irwin

energy, 65, 91, 176, 183, 189, 219; disequilibrium, 146; flow, 105, 202, 257, 286n6; free, 76, 88, 96–97, 122; industry, 48; and information, 267, 284n36; and matter, 25, 97, 161, 219–20, 272; metabolic, 24, 92; solar, 3–4, 162, 217, 262, 267; and the technosphere, 257–64

entropy, 188, 196, 241, 252: atmospheric, 25; and Gaia, 123–24, 218–19; informatic, 96, 123, 272; in living systems, 95–96; reduction, 25, 122–23; thermodynamic, 25, 55, 122–23, 272

environment, 27, 29, 46, 59, 79, 128, 188, 193, 236, 254; body as (for the immune system), 231–34, 238–39; closed, 84, 118–19, 200–203, 220, 226, 248–49, 259, 295n24; consciousness of, 103; control of, 28, 88, 116, 192, 238, 248; crisis of, 9, 56, 102, 175, 191, 264; evolution of, 157–58; extraterrestrial, 49, 161, 168, 263; as Gaia, 30, 45, 72, 126, 149, 162–63, 167, 245; of Gaia, 10–11, 13, 18, 30, 48, 53–55, 58, 61, 67, 70, 73, 80–81, 88, 105–6, 151, 213, 220–22, 225, 240; human, 50, 192, 244; of life, 8, 10–11, 33, 55, 75–76, 88, 92–93, 97, 122, 150, 162, 179, 188, 217, 228, 241, 246, 266, 270, 280n36, 284n37; living, 78; orbital, 119–21, 196–202; of systems, 4–7, 16, 39–40, 43–45, 54, 67, 69, 72, 77–78, 84–85, 91–98, 105–7, 123, 129–32, 145, 179, 229, 240, 260, 273, 286n36, 293n32; technological, 213, 249, 252, 257, 268, 270–71; virtual, 195, 203. *See also* closure: environmental

environmental: management, 190, 261–63; movement, 105, 118, 214

epistemology, 7, 144, 179, 208; in Gregory Bateson, 188–93; challenges to, 101; cybernetic, 188; ecology as, 193; limits to, 98, 108, 186–87, 202, 210–11; neocybernetic (constructivist), 107–8, 124–27, 131, 145, 175; in Francisco Varela, 126–27. *See also* cognition; *Laws of Form*

Esposito, Roberto, 225–30, 239, 293n32; *Bíos: Biopolitics and Philosophy*, 226–27; *Immunitas: The Protection and Negation of Life*, 225–27, 230, 239

ethics, 94, 109, 127, 180, 193–94

event, 51, 76, 93, 131–32, 183, 231, 265; cellular, 7, 51; Gaian, 79, 81, 134; of meaning, 5, 45, 92, 108. *See also* Great Oxidation Event

evolution: in Gregory Bateson, 191–94; biological, 33, 86–88, 95, 97, 119, 130, 166, 169, 173, 251, 270; cellular, 26, 164; cosmic, 19, 26, 185, 204–5, 246, 250, 262, 267–68, 292n23; Darwinian, 11; microbial, 9–10, 27, 61, 147, 240. *See also* Margulis, Lynn; symbiogenesis

exobiology, 10, 137, 205, 245, 268; and Gaia, 24, 265–68. *See also* astrobiology

exoplanet, 26, 246, 250, 295n18

"Extending the Domain of Freedom" (Latour and Lenton), 23–24, 267

feedback, 38–39, 83–84, 125, 155; in Gregory Bateson, 110; in Daisyworld, 88, 135, 153–54; in Gaian systems, 29, 40–41, 48, 81, 93, 115–16, 146, 169, 234, 245; Bruno Latour on, 81, 216–17. *See also* control

Fleck, Ludwik, 177–78; *Genesis and Development of a Scientific Fact*, 177

fossil fuels, 24, 259–63

"From Biology to Cognitive Science" (ed. W. I. Thompson), 150–56

Fuller, Richard Buckminster, 101, 106, 109, 118, 131–32, 196

fungi, 11, 61, 74, 171, 228, 235–37, 250; mycorrhizae, 235

Gaia: and the Anthropocene, 12, 18, 46, 146, 207, 256–65; astrobiological, 265–67; autopoietic, 5, 7, 17, 34, 43, 53–54, 69, 71, 80, 89, 101, 111, 126, 139, 141, 152, 157–79, 240–41, 257, 272; boundaries of, 213–24; evolution of, 3, 6, 8, 12, 17, 44–45, 66, 72–73, 150–51, 169, 179, 240, 245; as goddess, 36, 46, 50, 60, 65, 105, 115, 140; and issues of totality, 214–16; Bruno Latour and, 7, 14–16, 18, 23–24, 46, 53–54, 58–82, 130, 207, 210, 227, 260–61, 267, 284n37, 295n32, 296n38; mediations of, 37, 65, 183–85; metabiotic, 5, 7, 15, 18, 36–37, 42–45, 54–55, 71–73, 77, 80–81, 89, 93–94, 152–53, 158, 163, 166, 220, 222, 232, 273; microbiome of, 218; name of, 1–3, 8–9, 12, 15, 26, 30, 34–37, 48, 117, 282n3; not an organism, 3, 8, 89, 162–63, 221–22; observed by Star cybernetics, 130; as organism, 1, 34–36, 48, 59, 67, 117–18, 214–15; reception of, 30, 60, 65, 115, 132–34, 137; as responsive, 11–14, 53–55, 58, 80, 179, 232 (*see also* cognition: planetary); as a self-referential system, 93–94; and spaceborne Earth imagery, 183–84; Isabelle Stengers and, 46, 53–58; as a system, 3–4, 7, 15, 29, 42–44, 69, 71, 81, 89–94; uniqueness of, 267. *See also* biology; communication; critical zone; Gaia discourse; Gaia hypothesis; Gaia theory; immunity; living systems; Lovelock, James; Margulis, Lynn; ontology; space colonies, system differentiation; technosphere; Thompson, William Irwin "Gaia and the Evolution of Machines" (D. Sagan and Margulis), 13–14, 76, 159, 165–69, 248, 280n31

Gaia discourse, 1, 5, 11, 15–18, 23, 36, 46, 74, 97, 142, 156, 225; in Donna Haraway, 50, 53, 65; in Bruno Latour, 65, 68–71, 227, 261; in James Lovelock, 3, 23, 87, 233; in Lynn Margulis, 86, 144, 220; in W. I. Thompson, 140, 143–45, 147–49, 179; in the *Whole Earth* network, 103, 117, 132, 165

Gaia: Goddess of the Earth (*Nova* documentary), 105, 229–30

Gaia hypothesis, 1–2, 8–11, 15–16, 18, 23, 26–37, 41, 81, 87–88, 101, 103, 130–31, 139, 146, 150, 196, 238, 246, 265, 283n28; in *CoEvolution Quarterly*, 111–26, 132–37; Donna Haraway on, 48–51, 183; Bruno Latour on, 58–60, 64–67; Lynn Margulis on, 162, 172, 180, 294n39; Isabelle Stengers on, 53; W. I. Thompson on, 139–41, 144; Francisco Varela on, 225, 231. *See also* Lovelock, James

Gaian being, 46, 62, 205–6, 211

Gaian matrixes, 11, 43, 45, 49, 88, 166

Gaian thought, 12–15, 34–35; and astrobiology, 266–67; in *Aurora* (Kim Stanley Robinson), 255; Bruno Latour and, 64; at the Lindisfarne Association, 11, 139–56, 280n31; in Margulis and Sagan, 12–13, 273, 280n32; new geocentrism as, 18, 207–10. *See also* perception

Gaia theory, 3, 15, 28, 34, 47, 53, 68, 78, 88–89, 133, 140, 207, 264, 286n28, 289n63, 289n69; and biopolitics, 18, 31, 224–25, 230, 234, 241; divergence between Lovelock and Margulis on, 75–78, 142, 178, 273; Earth system science and, 18, 207; Bruno Latour on, 66–67, 72, 207, 215–17, 283n29, 295n32; at the Lindisfarne Association, 140–56; James Lovelock on, 150, 158; Lynn Margulis on, 3, 157–60, 165, 174, 273; as systems theory, 4, 9–10, 17, 37–42, 54, 69, 94, 98, 292n18

Galileo, 207; Galilean paradigm, 210

gender, 12

genome, 8, 78, 164–65, 234–35; human, 175; metagenome, 87

geocentrism, 207; new, 18, 208–11

geology, 28, 37, 42–45, 48, 50, 52–53, 116, 137, 213, 225, 238–40, 281n9; and the Anthropocene, 18, 146, 169, 243–46, 256–63, 282n1, 282n12; and biology, 31, 33, 93, 158, 229, 257, 292n1; "geological force," 12, 24, 246, 251, 256; "geological time," 13, 25, 41, 116, 243, 245, 251, 271; "geological turn," 18, 244, 261; global scale of, 56

geosphere: and the Anthropocene, 257, 264; and the biosphere, 97, 136, 168, 217, 225; postbiotic, 133. *See also* International Geosphere–Biosphere Program

globalization, 31, 56–58, 184, 187, 206, 214–16

global warming, 43, 47, 102, 254

globe, 160, 183, 186–87, 214–16; globality, 214

Golding, William, 2, 26, 27, 36

governor, Watt's, 154–55

gravity, 121, 196–200, 224, 248

Great Oxidation Event, 240, 262, 264

great silence, the, 255–56

Grinspoon, David, 246, 264, 295n26; and the Anthropocene, 246, 248, 256, 265; on geocentrism, 207–8

Günther, Gotthard, 107, 287n13

Habermas, Jürgen, 143

habitability, 28, 119, 121, 134, 151, 160, 208, 213, 246–47, 255, 269, 291n26

Hadean eon, 173

Haff, Peter, 258–63

Hagen, Joel, 10, 189

Haraway, Donna, 15, 47–53, 64–65, 76, 183–84, 244, 282n3, 283n12; and autopoiesis, 15, 50–52, 64; on cybernetics, 48–50; "Cyborg Manifesto," 52, 268; "Cyborgs and Symbionts," 15, 47, 50–52, 64, 76, 183; on Gaia, 48–53; and Lynn Margulis, 51–52; *Staying with the Trouble*, 52; *When Species Meet*, 52

Harvey, William, 115

Heisenberg, Werner, 148

heliocentrism, 207

Henkel, Anna, 42, 54, 286n31

holarchy: Gaia as, 98, 221
holism, 16, 77, 160–61; and Gaia, 35–37, 116, 213–14, 221, 224, 296n38; James Lovelock on, 39–41, 150–51; and neocybernetic systems theory, 42, 71, 93–97, 116–17; and the technosphere, 257–61; in Francisco Varela, 127–32, 156; in the *Whole Earth* network, 95, 101, 108, 132. *See also* Latour, Bruno
holobiont, 18, 87, 213, 235–40; as ecosystem, 237–238; planetary, 239–40
Holocaust, the, 177
Holocene, the, 243
homeostasis, 8, 36, 38–41, 67–70, 83, 88, 106, 134, 11; atmospheric, 8, 33, 115–16, 122, 124, 133; in Daisyworld, 135, 151
humanity, 24, 32, 35, 54, 74, 144, 146, 150, 167, 178, 187, 192; Anthropocene, 46, 243–46, 251, 261, 264; in *Aurora*, 250–54; cosmic horizon of, 208–9; Earth as cradle of, 254; Gaia's transcendence of, 55–58, 257; geological force of, 9, 12; in *Neuromancer*, 200–203; uniqueness of, 247, 267

Icarus (scientific journal), 33, 85, 112
idealism: holistic, 108; philosophical, 128, 206, 286n36
Illich, Ivan, 112, 170, 291n19
imaginary, the: anthropotechnic, 216; classical, 31; cyborg, 273; fractal, 148; global, 186–87; physical, 80; planetary (Gaian), 17–18, 46, 49, 57, 79, 118, 121, 140, 183–211; technological, 23; transhumanist, 273; Whole Earth, 184
"Immu-knowledge: The Process of Somatic Individuation" (Varela and Anspach), 230–32
immune system, 18, 149, 153, 156, 213, 227–41, 253; as a cognitive system, 232–33; as an ecosystem, 231–32, 234, 238–39; compared to Gaia, 231–34, 240; lymphocytes, 231–33, 239; and the microbiome, 238; compared to the nervous system, 232–34
immunity, 18, 260; autoimmunity, 225, 228; and autopoiesis, 227, 231, 233, 293n32; in biopolitics, 224–29; and

community, 226; and the holobiont, 238; immune self, 229, 233, 237–39; planetary (Gaian), 18, 31, 61, 149, 227, 229–34, 239–41; in Peter Sloterdijk, 31, 46, 206, 216, 283n31; symbiosis and, 234, 237–41. *See also* Esposito, Roberto; Varela, Francisco; Wolfe, Cary
individuality: biological, 235–38, 248; distributed, 239
information theory, 1, 23, 40, 76, 122–24, 142, 161, 272–73, 284n36, 288n43
intelligence, 19, 74, 256, 264, 268, 292n23, 295n19; artificial, 76, 87, 177, 188, 203, 253, 269, 273; extraterrestrial, 255–56; hyperintelligence, 268–69; technological, 246
International Geosphere–Biosphere Program (IGBP), 244
islands, 214; island biogeography, 251
"Is Mars a Spaceship, Too?" (Margulis and Lovelock), 114

Jackson, Wes, 147–48
Jantsch, Erich, 141; on autopoiesis and Gaia, 158, 160–61; *The Self-Organizing Universe*, 141, 158, 160
Jerne, Niels, 230
Jet Propulsion Laboratory (JPL), 24, 26, 103, 245, 265
Johnson, Donna, 125

Kant, Immanuel, 206, 285n16, 286n36
Kardashev, Nikolai S., 263; Kardashev scale, 256, 263–64
Kesey, Ken, 118
Known Universe, The (planetarium program), 208, 210
Kuhn, Thomas, 31

Laboratory Life: The Construction of Scientific Facts (Latour and Woolgar), 177
Latour, Bruno, 46, 134, 177, 187, 227, 259; and actor-network theory (ANT), 64, 69, 73; and the Anthropocene, 63–64; on critical zones, 18, 82, 216–17, 222; contra cybernetics and systems theory, 64–75, 81, 130, 283n27; *Facing Gaia: Eight Lectures,*

60–67, 70–74, 79–81, 283n32; *Facing Gaia: Six Lectures* (Gifford Lectures), 60–61; on Gaia, 7, 15–16, 47, 58–82, 207–8, 215, 217; on geocentrism, 207; "God of Totality," 60, 215; contra holism, 59, 67–71, 79–80, 215, 227, 259, 261, 296n38; *Inside* (with Frédérique Aït-Touati), 215; and James Lovelock, 23–24, 62–81, 207, 210, 283n29, 284n37; and Lynn Margulis, 62–63, 74–80, 284n37; on the new climatic regime, 14; political ecology and, 54–55, 59, 62; 284n32; *Politics of Nature*, 58–59; *Reassembling the Social*, 64–65; *Reset Modernity!*, 217; on Peter Sloterdijk, 46, 80, 216, 293n6; and Isabelle Stengers, 53–54, 63, 282n3. *See also* Gaia; Gaia discourse; Gaia hypothesis; Gaia theory; Lovelock, James

Laws of Form (George Spencer-Brown): as neocybernetic epistemology, 108, 125–26, 204–5; on self-reference, 106–8, 131, 204, 287n11, 287n14; in the *Whole Earth Catalog*, 106–8, 122. *See also* distinction; observation; Varela, Francisco; von Foerster, Heinz

Lecon sur la statique chimique des êtres organisés (Dumas and Boussingault), 32–33

Lederberg, Joshua, 266

Lem, Stanisław, 268, 272; *The Cyberiad,* 268–69, 272

Lenton, Timothy M., 23–24, 136, 260, 267

Levi, Primo, 177

L5 (Lagrange libration point), 121, 197

lichen: Gaia compared to, 222; as a holobiont, 236

life detection, 114, 246, 265; biological, 24–25; by entropy reduction, 25, 122–23. *See also* astrobiology; biosignature; exobiology

Lilly, John, 106

Lindisfarne Association, 11, 16–18, 139–56, 229–30; James Lovelock and, 16–17, 73, 138, 141–55, 290n30; Lynn Margulis and, 138, 141–49, 170, 178; the systems counterculture and, 140–44, 170, 229–30; William I. Thompson and, 139–55, 187; Francisco Varela and, 89, 141–49, 229–34

lithosphere, 213, 218, 245, 262

living systems (biotic, autopoietic), 4–6, 51–52, 65, 69, 76, 84–96, 124, 155, 159, 161, 172, 179, 241, 280n36; and the Anthropocene, 258; in Henri Atlan, 73; and cognition (sentience), 7; in *Dune*, 189; entropy and, 123; and Gaia, 3–5, 30, 40, 44, 115–16, 129–31, 166–67, 203, 219–21, 292n18; in Erich Jantsch, 161; in Bruno Latour, 79; and machines, 270; operational closure of, 77, 119, 158, 169; as self-immunizing, 241; thermodynamics of, 219, 286n6; in W. I. Thompson, 145. *See also* biota; nonliving systems

Lotka, Alfred J., 96–98, 189; *Elements of Physical Biology,* 96, 189

Lovelock, James: *Ages of Gaia,* 43, 158–59, 188, 271–72; on autopoiesis, 50, 158; collaboration with Lynn Margulis, 2–3, 9–10, 15, 24, 26–34, 36–37, 88, 96, 112, 133, 137; and cybernetic systems, 6, 8, 23–24, 29, 33–34, 40–42, 75–76, 79, 83, 87–88, 137, 176–77; on cyborgs, 47, 268–72; and Daisyworld, 47, 132–37; "Daisyworld: A Cybernetic Proof of the Gaia Hypothesis," 111, 132, 137; and the electron-capture detector, 23–24; and esobiology, 266–67; and exo/astrobiology, 18, 24–25, 245–46, 265–66, 268; on the fate of Gaia, 269–72; "Gaia: A Model for Planetary and Cellular Dynamics," 146; *Gaia: A New Look,* 41, 81, 87; "Gaia: A Planetary Emergent Phenomenon," 150; on Gaia as a coupled system, 11, 97–98, 158, 163; "Gaia as Seen through the Atmosphere," 26–32, 133; and Gaian regulation, 8, 36, 41, 81, 88, 105–6, 146; on Gaia's boundaries, 218–19; *Gaia: The Practical Science of Planetary Medicine,* 74; Donna Haraway on, 48–50; on holism, 39–40, 150–51; "The Independent Practice of Science," 23; as an independent scientist, 2, 23, 137, 170; on information in the cosmos, 267, 272–73; and Bruno Latour, 59–63, 65–68, 70–76, 207, 210; on living within Gaia, 14; on the name of Gaia, 36–37; and NASA, 23–24, 48, 245–46, 265–66, 271; *Novacene,* 12, 19, 177, 244, 265–73; "A Physical Basis for Life-Detection Experiments," 25; and the systems counterculture, 137; on the thermodynamics of Gaia, 18, 45, 122–23, 218–19; Francisco Varela on, 151–56, 225, 230–31. *See also* correspondence; cyborgs; emergence; holism; Latour, Bruno; Lindisfarne Association; Margulis, Lynn; recursion

Luhmann, Niklas: on autopoiesis, 42, 91–92, 227; in Bruno Latour, 64, 283n27; neocybernetics and, 125, 206, 287n11; and social systems theory, 5, 91, 94–96, 293n32; on system differentiation, 72, 93, 131

Luisi, Pier Luigi, 90, 285n20

machines, 69, 178, 183, 189, 203, 272; cybernetics of, 6, 38, 41, 50–52, 75–76, 81, 105, 110–11, 157, 161, 268; evolution of, 13–14, 157, 159, 165–69, 248, 269; machinic phylum, 168; as metabiotic systems, 166–69, 270–72. *See also* reproduction

Macy conferences on cybernetics, 122, 125, 284n1

Margalef, Ramón, 114, 288n52

Margulis, Lynn: and the Anthropocene, 256; and autopoiesis, 4–5, 16–17, 51–55, 76–77, 85–87, 94, 101, 140–41, 157–80, 220, 270–71, 273, 285n10, 291n21; "Big Trouble in Biology," 157, 170, 174–80, 245; and *CoEvolution Quarterly,* 112–26; collaboration with James Lovelock, 2, 6, 8–10, 24–34, 40, 62–63, 67, 88; collaboration with Dorion Sagan, 159; and cybernetics, 157; and Daisyworld, 133, 138, 157; "Early Life: The Microbes Have Priority," 146–47; and evolution, 2–3, 8, 17, 51, 77, 146, 157, 160, 171, 235–36, 285n10; and exo/astrobiology, 18, 245–46, 266; "Gaia and the Evolution of Machines" (with D. Sagan), 159, 165–69; on Gaian thought, 12–13, 35, 50; on Gaia theory, 3–4, 7, 12, 15, 89, 217, 229–230, 256; on the holobiont,

18, 235–38; "Kingdom Animalia," 169–74; and the Lindisfarne Association, 11, 16–17, 141–49, 152, 156; on James Lovelock, 162, 165; on machines, 165–69, 270–71; *Microcosmos* (with D. Sagan), 93, 159–63; on the name of Gaia, 2–3; and NASA, 245, 266; and neo-Darwinism, 87, 169–80, 289n63; *Origin of Eukaryotic Cells,* 2, 77, 114; *Origins of Sex* (with D. Sagan), 51, 159, 163–65; on space colonies, 118–22; on symbiosis and symbiogenesis, 10, 51, 77–78, 234, 239; *Symbiosis in Cell Evolution,* 157, 235–38; *Symbiotic Planet,* 2–3, 10–12, 74–75, 162, 189, 281n13; on technology and the technosphere, 12–13, 165–69, 257, 270–71, 273; on tests of Gaia, 229–30, 240; and William I. Thompson, 170, 180; *What Is Life?* (with D. Sagan), 2, 5, 7, 51–52. *See also* autopoiesis; biology; cells, living; correspondence; ecology; emergence; evolution; Gaia discourse; Gaia hypothesis; Gaia theory; Haraway, Donna; Latour, Bruno; Lovelock, James; membrane; microbes; neo-Darwinism

Mars, 24–26, 114, 121, 218, 252, 265–66; Viking landers on, 24–25, 114, 168–69, 246, 265, 271

materialism: autopoietic, 172–73, 176; cultural, 178; scientific, 148

materiality, 5, 39, 42, 45, 52–53, 66, 76, 131, 184, 203, 249, 261, 268, 272, 283n12; of the environment, 150–51, 162; of Gaia, 57–63, 79, 116, 158, 213, 224, 267, 286n31; of living systems, 85, 90, 92, 222, 241, 255–56; material closure, 12, 18–19, 44, 119–21, 200, 202–3, 220–21, 248–49, 259; of the technosphere, 256–60, 270; transformativity of, 43

Maturana, Humberto, 4–5, 11, 101, 124–25, 141, 206; "Everything Is Said by an Observer," 145–46; "Neurophysiology of Cognition," 89–90, 124. *See also* autopoiesis; *Autopoiesis and Cognition*

Maxwell's Demon, 272

McClure, Michael, 114

McConville, David, 208–10
McCulloch, Warren, 106
Mead, Margaret, 125
meaning, 31, 42–46, 54, 108, 110, 226, 273;
    corporeal, 46, 286n31; medium of, 43,
    91, 93; systems, 43, 91–93, 108
mechanism: biological (process), 90, 155,
    163, 228, 234, 237; cybernetic, 83, 84,
    88, 153, 155; Gaian, 8, 29, 40–42, 48,
    115, 121, 134–36, 202, 266; philo-
    sophical (theory), 76, 84, 86, 161, 166,
    170–71, 174–80; technological, 38, 261
membrane, 218, 221–22, 241; atmosphere
    as, 94; cellular, 45, 77, 86, 90, 164, 217,
    219, 232, 241; critical zones as, 82,
    217; Gaia as, 149, 217, 222, 224, 245,
    283n31; mitochondrial, 77–78
metabolism, 5–7, 24, 55, 78, 86, 115, 133,
    176, 178, 218, 222, 249, 259, 262, 270,
    272; and autopoiesis, 90, 155, 163,
    217; Gaian, 214, 245; metabolic rift,
    251–52, 256, 261; microbial, 29, 33, 63
metadynamics, 154, 156, 233–34
meteoric impacts, 230, 240
microbes, 10, 24, 61, 112, 164, 237, 240,
    247–48, 250; and immunity, 238;
    Lynn Margulis and, 27–28, 74, 146,
    149, 167; planetary (Gaian) role of, 10,
    27, 63, 146, 218, 245, 256, 264
microcosm, 10, 87, 144, 149, 160–62,
    232–41, 256, 266; Archean, 44;
    human, 49; planetary (Gaian), 44, 61,
    74, 146, 160, 264
mitochondrion, 77–78, 240, 284n41
modernity, 35, 243; scientific, 62; in Peter
    Sloterdijk, 206; Western, 60, 256
Moon, the, 103–5, 121, 186, 197, 207, 279n3
Morin, Edgar, 187
Morowitz, Harold, 42, 105
mutation: genetic, 87, 166, 169; techno-
    logical, 166, 168
myth, 35, 61, 66, 85, 115, 177–78; of Gaia,
    37, 65; and science, 15, 46. *See also*
    Thompson, William Irwin

NASA: Apollo 13 mission, 254; and
    astrobiology, 19, 26, 204–6, 245–49,
    266, 268; ATS-III weather satellite,
    183–85; CELSS program (controlled
    ecological life-support system),

248–49, 252; and Gaia, 15, 49, 182,
    207, 245–49, 266, 271; imagery, 17, 49,
    103, 105, 118, 183–85, 197–99, 204–5.
    See also astrobiology; exobiology;
    Lovelock, James; Margulis, Lynn
Native Americans, 178–79
Natural History (magazine), 112–14
nature, 14, 57–60, 63, 148, 175, 179, 200,
    202, 208, 292n30, 296n32: laws of,
    57; mind and, 143; (non)domination
    of, 13, 35, 56; protection of, 13–14; as
    resource, 13; and society, 58, 69; in
    Baruch Spinoza, 226
neocybernetic systems theory (NST),
    4–7, 16–18, 34, 38, 53, 83–98, 112, 124,
    129–32, 157, 159, 161, 168, 174, 206; and
    actor-network theory (ANT), 64, 69;
    and cyborgs, 52; and Gaia, 7, 40–45,
    62, 69, 83–98, 132, 180, 292n18; and
    holism, 71; and machines, 76; and
    the systems counterculture, 101, 106;
    in Francisco Varela, 126, 129, 143,
    152–56. *See also* autonomy; auto-
    poiesis; *Laws of Form*; self-reference
neo-Darwinism, 34, 68, 168, 235, 284n37,
    289n63; defined, 169; Lynn Margulis's
    critique of, 87, 169–80
nervous system, 90, 107, 145, 232–34,
    237, 271
Neuromancer (William Gibson), 18, 165,
    194–204, 253, 291–92n17, 292n18; arti-
    ficial intelligences in, 203; ecological
    closure in, 199–203; as cyberpunk,
    198; Earthbound settings, 195–96;
    and Gaia, 195, 202–3; high-orbital
    settings, 196–203; repurposing
    of O'Neill's high-frontier designs,
    196–201. See also *CoEvolution
    Quarterly*; cyberspace, ecology;
    ecosystem; humanity; society; space
    colonies
New Age, the, 1, 3
new climatic regime, 14–15, 243
Neyrat, Frédéric, 225, 284n32, 293n32
nonliving systems: autopoietic (meta-
    biotic), 5, 42, 45, 69, 91; in distinction
    from living systems, 30, 45, 51–52, 76,
    176, 290n3; nonautopoietic (abiotic,
    physical), 42, 45. *See also* machines
noosphere, the, 258, 296n34

Novacene epoch, the, 268–69; fate of
    Gaia in, 269, 272

objects, 42, 71, 108, 126, 142, 149, 175, 185,
    244; astrobiological, 246–47; cosmic,
    44; physical, 255; quasi-, 64; techno-
    logical, 38, 68, 258–59
observation, 39, 49, 61, 77, 85, 95, 103, 127,
    131, 148, 177, 221, 247–48; finitude
    of, 94; in *Laws of Form*, 108, 126, 204;
    in neocybernetic systems theory, 7,
    15, 81, 86, 88, 101, 107–8, 122, 126–28,
    205–10, 258; non-observation, 108,
    233, 292n18; observer, the, 11, 49,
    64, 84, 93, 96, 108, 110, 112, 124–27,
    130–32, 145–47, 175, 178–79, 183, 186,
    194, 205, 273; second-order, 205; self-
    observation, 187; unobservability, 43,
    210, 214. *See also* paradox; systems:
    observing
O'Neill, Gerard K., 16, 118–22, 197–201;
    "The High Frontier," 118, 121; NASA
    support for, 118. See also *Neuro-
    mancer*; space colonies
ontology, 131, 175, 283n32; cybernetic,
    287n13; and Gaia, 48, 54, 60, 68–70;
    and the immune self, 229; and the
    "Whole Earth," 98
organicism, 68–69, 162, 285n16; in the
    description of the technosphere, 259
organism, 70, 90, 95, 160, 222, 224, 235:
    and autopoiesis, 161, 164, 192; and
    environment, 75, 97, 151, 218, 280n36;
    eukaryotic, 10, 33, 44, 158, 235–36;
    and machine, 6, 48, 268; as open
    system, 90
Origins of Life (scientific journal), 112
Oyama, Susan, 148

Paleolithic age, 243
paradox, 110, 208, 215; in autopoietic
    Gaia, 43, 81, 94; in the concept of
    nature, 14; in Daisyworld, 41; in the
    name of Gaia, 60; of observation,
    206, 210; self-referential, 108, 122,
    126, 286n36; "Vernadsky's," 219–20
Pask, Gordon, 106, 288n52
perception, 41, 103, 149, 210, 215; in Gaian
    thought, 11–14, 35, 179–80
phylogeny, 10, 146, 172, 236–37, 294n38

physics, 1, 37, 45, 80, 150, 174, 208, 220,
    255; "physics envy," 174, 179. *See also*
    energy; entropy; thermodynamics
physiology, 9, 168, 226; and autopoi-
    esis, 171, 176–80; of cognition, 90;
    planetary (Gaian), 10, 162–63; and
    homeostasis, 68
planetization, 143, 187–88
planets, 168, 203, 208, 247; evolutionary
    state of, 262; habitable, 255; lifeless,
    115, 255; living, 7, 27, 42, 61, 179, 186,
    218, 246, 248, 255. *See also* exoplanet
plants, 11, 74, 171–72, 190–91, 202, 248,
    252; as holobiont hosts, 213, 235–36;
    plant–animal distinction, 171; and
    the technosphere, 258
Pleistocene, the, 243
pollution, 24, 259
posthuman, the, 183, 256, 271, 295n31;
    posthumanism, 62, 207, 225,
    295n32
Prigogine, Ilya, 158, 160
proprioception, 11–12, 41
protists, 51, 164, 294n44
psychedelics, 105, 118; in *Dune* (Frank
    Herbert), 189

Reality Club, The, 170, 174
recursion, 47, 83, 93, 107, 205, 287n11;
    in Gaia, 12, 94, 147, 211; in living sys-
    tems, 4–5, 130, 179; James Lovelock
    on, 134, 137–38; in systems theory, 6,
    80, 83–91, 95, 101, 124, 131, 154, 161
reentry, 45, 83, 205, 287n13, 287n14
reductionism, 39–41, 66, 95, 174, 180
reduction to practice, 23
remediation, 46; environmental, 252;
    mycelial, 237
representationalism, 145
reproduction, 73, 91; biological, 86–87,
    119, 159, 163–67, 284n37; and Gaia, 3,
    8, 45, 63; of machines, 166–68

sacred, the, 109
Sagan, Carl, 24, 26, 33, 112, 114, 118, 247
Sagan, Dorion: collaboration with Lynn
    Margulis, 2, 17, 76, 157, 159, 280n32,
    290n10, 293n5; on Gaia's offspring,
    248. *See also* Margulis, Lynn
Salk, Jonas, 141

scale, 4, 56, 76, 120, 160, 176, 185, 202, 260; Daisyworld and, 136; Gaia and, 7, 58, 79–80, 167

Schumacher, E. F., 141

Schweickart, Rusty, 114, 118

science. *See* biology; chaos theory; chemistry; complexity theory; cybernetics; ecology; evolution; Gaia theory: Earth system science; geology; information theory; physics; physiology; serial endosymbiosis theory; thermodynamics

science fiction, 2, 23, 121, 183, 253, 269, 273; and astrobiology, 247. See also *Aurora*; *Dune*; Lem, Stanisław: *The Cyberiad*; *Neuromancer*

selfish gene, 117, 168, 287n27

self-organization, 6, 55, 84, 123, 144, 188, 259, 289n17; Gaia and, 88, 219, 290n5. *See also* Jantsch, Erich

self-production (autopoiesis), 5, 7, 55, 76, 78, 85–86, 96, 159, 163, 166, 175–76, 217, 260, 270; Gaian, 18, 165, 220, 236; in meaning systems, 92

self-recognition, 183

self-reference: biotic, 7; in neocybernetic systems theory, 6, 83–85, 91, 97, 122, 153, 161, 179, 194, 206, 283n27, 286n36, 287n11,13; 293n32; planetary (Gaian), 14, 63, 80, 93–96, 211, 220; in Francisco Varela, 125–31, 142, 152, 231–33. *See also* closure: systemic (operational); *Laws of Form*

self-regulation, 48, 50, 249, 268; cybernetic, 33, 38–41, 84; in Daisyworld, 132–34, 151, 290n30; Gaian, 8, 25–29, 41, 72, 88, 105, 111, 116, 118, 150–51, 238, 294n39. *See also* homeostasis

serial endosymbiosis theory (SET), 10, 158, 235. *See also* Margulis, Lynn

Serres, Michel, 73, 187, 284n36, 298n17

SETI (search for extraterrestrial intelligence), 247, 256, 263

Shannon, Claude, 272

Shock of the Anthropocene, The (Bonneuil and Fressoz), 244

Sloterdijk, Peter, 187, 222; on immunitary enclosures, 46, 216, 224; *In the World Interior of Capital*, 31, 206–7; Bruno Latour on, 46, 80, 216, 293n6;

on modernity, 206–7, 292n30; on the Ptolemaic spheres, 31; *Spheres, Volume 1: Bubbles*, 216, 224; *Spheres, Volume 3: Foams*, 283n31

Snyder, Gary, 112, 118

society, 6, 58, 69, 94, 170, 178, 280n31; American, 109; in the Anthropocene, 248, 257–58; in *Aurora*, 253; in *Dune*, 190; in *Neuromancer*, 200; patriarchal, 174; technological, 168, 258

sociology of scientific knowledge (SSK), 177

Soleri, Paolo, 118, 141, 195–96

space colonies, 16, 18, 118–22, 282n38, 288n29; as miniature Gaias, 119–20, 248–49, 254; in *Neuromancer*, 196–201. *See also* "Ecological Considerations for Space Colonies"

space program: Apollo, 103, 105, 118, 246, 254; international, 246

space sciences, 246

Spaceship Earth, 118

space travel, 248–50, 263, 268

Spacewar, 110

Spencer-Brown, George, 16, 101, 106–8, 125–27, 131, 204, 287n13. See also *Laws of Form*

Spinoza, Baruch, 226

Star cybernetics, 143, 145, 148. *See also* Varela, Francisco

Stengers, Isabelle, 46–47, 83, 232, 282n3; and the Anthropocene, 56; on capitalism, 54–58; contra cybernetics, 53–54; on Gaia, 53–58; on Gaia the Intruder, 56–58; *In Catastrophic Times*, 34, 55–58; and Bruno Latour, 15–16, 53–54, 63–64; *Thinking with Whitehead*, 55. *See also* transcendence

Steps to an Ecology of Mind (Gregory Bateson): and cybernetics, 191, 194; and ecology, 191–94; "Ecology and Flexibility in Urban Civilization," 192–93; and the "ecology of ideas," 191–93; relation to *Dune*, 191–93

stewardship, 14, 248, 280n33

Stolz, John F., 218

sublime, the, 31, 173; astronautical, 49; cosmic, 204; ethical, 194; technological, 250, 253–54

superorganism, 70; bacterial, 160; Gaia
    as, 45, 93, 158, 162
Suzuki, David, 2, 36
symbiogenesis, 10, 51, 75–78, 87, 171, 235,
    283n12. *See also* serial endosymbiosis
    theory
symbiont, 75, 78, 87, 213, 236–38; endo-
    symbiont, 77–78, 235
symbiosis: and autopoiesis, 87, 157–58,
    236; in cell evolution, 2, 18, 51, 74–75,
    294n39; and ecology, 237; endosym-
    biosis, 10, 51, 77–78, 164; and Gaia,
    10, 80, 94, 130, 145, 167, 218, 234–35,
    239–41; and the holobiont, 213,
    235–39; non-obligatory, 236; obliga-
    tory, 78, 236. *See also* immunity;
    Margulis, Lynn
"Symbiotic View of Life: We Have Never
    Been Individuals, A" (Gilbert, Sapp,
    and Tauber), 235–38
system differentiation, 43–44, 96; in the
    Gaian instance, 44, 54, 68, 71, 131
systems: abiotic, 43–44; biotic, 5, 43–45,
    93, 259, 273; metabiotic, 5, 45, 54, 152;
    meaning, 43, 91–93, 108; observing
    (cognitive), 43, 84–85, 107, 125, 179;
    open, 95–96, 110, 219–20, 284n36; psy-
    chic, 5, 42–45, 69, 91–94, 190, 270; self-
    referential, 6, 83–84, 91, 93, 96–97,
    107–8, 125–26, 206, 231; social, 5, 38,
    42–45, 69, 90–94, 190, 270; systems
    thinking, 9, 38, 68, 106, 109, 137, 184,
    211; technological, 6, 13, 38, 51, 84,
    105, 166, 169, 233, 260; whole, 37–39,
    43, 84, 96–97, 103–9, 125–27, 132,
    137–38, 184, 188, 202, 211, 214. *See also*
    autopoiesis; cybernetics; environ-
    ment; living systems; machines; neo-
    cybernetic systems theory; nonliving
    systems; self-reference; systems
    counterculture; systems theory
systems counterculture, 9, 11, 16–17,
    101–3, 131, 140, 188, 204, 285n20; and
    Stewart Brand, 109, 111; and cyber-
    space, 195; ecological sensibility of,
    204; and the Lindisfarne Association,
    144, 229–30; and James Lovelock, 137;
    and Lynn Margulis, 170, 180, 234; and
    W. I. Thompson, 144; and Heinz von
    Foerster, 111, 122–31, 141

systems theory, 4–8, 15–16, 37–42, 50–55,
    76, 293n32; Bruno Latour and, 64–71;
    semantics of, 38–40; social, 5, 42–45,
    64, 90–94, 206, 279n12. *See also* auto-
    poiesis; differentiation; neocyber-
    netic systems theory

technology, 13, 19, 23, 43, 48–53, 95, 102,
    158, 169, 178, 195, 244, 247–48, 251,
    256–57, 259, 264–65, 270, 292n23;
    astronautic, 184, 263; autonomy of,
    260; communications, 184; compu-
    tational, 184; of fire, 261; molecular-
    sequencing, 294n38; nano-, 245;
    narrative, 49; natural, 266; observa-
    tional, 107; physical, 49; semiotic, 49;
    space, 24, 248; speculative, 118
technosphere, 5, 13, 15, 214, 270; and the
    Anthropocene, 19, 246, 256–61; and
    Gaia, 12–13, 19, 165–69, 256–64, 273.
    See also Anthropocene technosphere
teleology: and Daisyworld, 134–35; and
    Gaia, 42, 72, 88, 134
Tellus (scientific journal), 33, 85, 112,
    124, 133
thermodynamics, 1, 25, 40, 55, 95, 122,
    218–19, 262
Thom, René, 148
Thomas, Lewis, 1, 17, 77, 141, 279n2
Thompson, Evan, 90, 139, 142, 147
Thompson, William Irwin, 11, 17, 111,
    118, 139–51, 154, 187, 229, 239; *At the
    Edge of History,* 140; on autopoiesis,
    147–48, 154; and counterfoil research,
    170; on Daisyworld, 154; on Gaia, 140,
    143–45, 147–49, 179; "Gaia and the
    Politics of Life," 147; *Gaia: A Way of
    Knowing,* 17, 141–47; *Gaia 2: Emer-
    gence,* 17, 139, 141–42, 147–51, 233, 239;
    Imaginary Landscape, 141, 143; on
    myth and science, 140, 143, 147–48;
    and the systems counterculture,
    144–45; *The Time Falling Bodies Take
    to Light,* 178. *See also* Lindisfarne As-
    sociation; Margulis, Lynn
thought collectives, 140, 177–78
Todd, John, 141, 148, 202
totality, 37, 59–60, 68–73, 80, 98, 108, 227,
    283n32; Gaian, 48, 116–18; holistic,
    35, 67, 94, 221; totalization, 43, 68,

95–98, 130–31, 213–16; universe as, 107, 175, 211
transcendence: in *Aurora*, 250; capitalist, 57–58; Gaian, 55, 57; liberation from, 46. *See also* Stengers, Isabelle
Tsiolkovsky, Konstantin, 248–50, 254–55, 295n19, 295n26
Turner, Fred, 102

"Understanding Whole Systems": in *CoEvolution Quarterly,* 111, 114; in the *Whole Earth Catalog,* 103–5

Varela, Francisco, 16, 101, 111, 206, 287n11; and autopoiesis, 4–5, 7, 11, 51, 54, 84–86, 89–91, 155, 233; on the development of neocybernetics, 83–84, 88; "Immu-knowledge: The Process of Somatic Individuation" (with Mark Anspach), 230–33; and immunitary discourse, 18, 227, 229–34, 239, 294n33; and *Laws of Form,* 125–26; "Laying Down a Path in Walking," 145; and the Lindisfarne Association, 17, 141–49; on Lovelock's Gaia theory, 151–56, 225; and Lynn Margulis, 126, 157; "Not One, Not Two," 127–32, 142; "On Observing Natural Systems," 125–27; *Principles of Biological Autonomy,* 127, 141, 144, 287n11; Star cybernetics, 127–31; and the *Whole Earth* network, 125–32; on whole systems, 125–26, 128. *See also* autonomy; autopoiesis; Daisyworld; emergence; holism; immune system; neocybernetic systems theory; self-reference
Vaz, Nelson, 230, 294n33
Venus, 24–25, 218, 265
Vernadsky, Vladimir, 257–58, 295n26; "Vernadsky paradox," 219
view from nowhere, 208–11
Viking missions, 24–25, 114, 168–69
Viola, Bill, 148

vitalism, 3, 66, 72, 96; "geometric" (Sloterdijk), 224; neovitalism, 227
Volk, Tyler, 11, 249, 259, 293n18; *Gaia's Body,* 221, 223; on Gaia's boundaries, 218–24; on Gaia's two surfaces, 222–24
von Bertalanffy, Ludwig, 95, 106, 219
von Foerster, Heinz, 11, 16, 101, 194; on the development of neocybernetics, 6, 84, 122; on Gaia's neocybernetics, 123–24, 158; on *Laws of Form,* 106–8, 122, 204 ; and the systems counterculture, 111, 122–31, 141
von Humboldt, Alexander, 214, 292n30
von Lewenheimb, Sachs, 114

Wark, McKenzie, 251, 261, 283n32
Westbroek, Peter, 187
Wittgenstein, Ludwig, 125
Whole Earth Catalog, 9, 16, 102–9, 111, 122, 126; Earthrise and, 102–6, 185–86; R. Buckminster Fuller in, 131–32; *Laws of Form* in, 106–8, 204; *Next Whole Earth Catalog,* 117–18. *See also* Brand, Stewart; *Laws of Form*; "Understanding Whole Systems"
Whole Earth image, 183–86
Whole Earth Jamboree, 140
Whole Earth network, 16, 89, 95, 101–2, 117, 165, 230
Whole Earth Review, 102, 111, 139, 165
Whole University Catalog, 106, 287n9, 293n32
Wiener, Norbert, 38, 106
Woese, Carl, 235, 294n38
Wolfe, Cary, 225–27, 286n36; *Before the Law,* 225–27

Youngblood, Gene, 148

Zajonc, Arthur, 147
Zalasiewicz, Jan, 257–59
zoocentrism, 171, 236

(continued from page ii)

46 *Biology in the Grid: Graphic Design and the Envisioning of Life*
PHILLIP THURTLE

45 *Neurotechnology and the End of Finitude*
MICHAEL HAWORTH

44 *Life: A Modern Invention*
DAVIDE TARIZZO

43 *Bioaesthetics: Making Sense of Life in Science and the Arts*
CARSTEN STRATHAUSEN

42 *Creaturely Love: How Desire Makes Us More and Less Than Human*
DOMINIC PETTMAN

41 *Matters of Care: Speculative Ethics in More Than Human Worlds*
MARIA PUIG DE LA BELLACASA

40 *Of Sheep, Oranges, and Yeast: A Multispecies Impression*
JULIAN YATES

39 *Fuel: A Speculative Dictionary*
KAREN PINKUS

38 *What Would Animals Say If We Asked the Right Questions?*
VINCIANE DESPRET

37 *Manifestly Haraway*
DONNA J. HARAWAY

36 *Neofinalism*
RAYMOND RUYER

35 *Inanimation: Theories of Inorganic Life*
DAVID WILLS

34 *All Thoughts Are Equal: Laruelle and Nonhuman Philosophy*
JOHN Ó MAOILEARCA

33 *Necromedia*
MARCEL O'GORMAN

32 *The Intellective Space: Thinking beyond Cognition*
LAURENT DUBREUIL

31 *Laruelle: Against the Digital*
ALEXANDER R. GALLOWAY

30 *The Universe of Things: On Speculative Realism*
STEVEN SHAVIRO

29 *Neocybernetics and Narrative*
BRUCE CLARKE

28 *Cinders*
JACQUES DERRIDA

27 *Hyperobjects: Philosophy and Ecology after the End of the World*
TIMOTHY MORTON

26 *Humanesis: Sound and Technological Posthumanism*
DAVID CECCHETTO

25 *Artist Animal*
STEVE BAKER

24 *Without Offending Humans: A Critique of Animal Rights*
ÉLISABETH DE FONTENAY

23 *Vampyroteuthis Infernalis: A Treatise, with a Report by the Institut Scientifique de Recherche Paranaturaliste*
VILÉM FLUSSER AND LOUIS BEC

22 *Body Drift: Butler, Hayles, Haraway*
ARTHUR KROKER

21 *HumAnimal: Race, Law, Language*
KALPANA RAHITA SESHADRI

20 *Alien Phenomenology, or What It's Like to Be a Thing*
IAN BOGOST

19 *CIFERAE: A Bestiary in Five Fingers*
TOM TYLER

18 *Improper Life: Technology and Biopolitics from Heidegger to Agamben*
TIMOTHY C. CAMPBELL

17 *Surface Encounters: Thinking with Animals and Art*
RON BROGLIO

16 *Against Ecological Sovereignty: Ethics, Biopolitics, and Saving the Natural World*
MICK SMITH

15 *Animal Stories: Narrating across Species Lines*
SUSAN MCHUGH

14 *Human Error: Species-Being and Media Machines*
DOMINIC PETTMAN

13 *Junkware*
THIERRY BARDINI

12 *A Foray into the Worlds of Animals and Humans*, with *A Theory of Meaning*
JAKOB VON UEXKÜLL

11 *Insect Media: An Archaeology of Animals and Technology*
JUSSI PARIKKA

10 *Cosmopolitics II*
    ISABELLE STENGERS

9   *Cosmopolitics I*
    ISABELLE STENGERS

8   *What Is Posthumanism?*
    CARY WOLFE

7   *Political Affect: Connecting the Social and the Somatic*
    JOHN PROTEVI

6   *Animal Capital: Rendering Life in Biopolitical Times*
    NICOLE SHUKIN

5   *Dorsality: Thinking Back through Technology and Politics*
    DAVID WILLS

4   *Bíos: Biopolitics and Philosophy*
    ROBERTO ESPOSITO

3   *When Species Meet*
    DONNA J. HARAWAY

2   *The Poetics of DNA*
    JUDITH ROOF

1   *The Parasite*
    MICHEL SERRES